Instabilities, Chaos and Turbulence

2nd Edition

ICP FLUID MECHANICS

Vol. 1   Instabilities, Chaos and Turbulence (2nd Edition):
         by Paul Manneville

ICP Fluid Mechanics Vol. I

Instabilities, Chaos and Turbulence

2nd Edition

Paul Manneville
Ecole Polytechnique, France

Imperial College Press

Published by

Imperial College Press
57 Shelton Street
Covent Garden
London WC2H 9HE

Distributed by

World Scientific Publishing Co. Pte. Ltd.
5 Toh Tuck Link, Singapore 596224
USA office: 27 Warren Street, Suite 401-402, Hackensack, NJ 07601
UK office: 57 Shelton Street, Covent Garden, London WC2H 9HE

British Library Cataloguing-in-Publication Data
A catalogue record for this book is available from the British Library.

ICP Fluid Mechanics — Vol. 1
INSTABILITIES, CHAOS AND TURBULENCE (2nd Edition)
Copyright © 2010 by Imperial College Press

All rights reserved. This book, or parts thereof, may not be reproduced in any form or by any means, electronic or mechanical, including photocopying, recording or any information storage and retrieval system now known or to be invented, without written permission from the Publisher.

For photocopying of material in this volume, please pay a copying fee through the Copyright Clearance Center, Inc., 222 Rosewood Drive, Danvers, MA 01923, USA. In this case permission to photocopy is not required from the publisher.

ISBN-13 978-1-84816-392-8
ISBN-10 1-84816-392-4
ISBN-13 978-1-84816-393-5 (pbk)
ISBN-10 1-84816-393-2 (pbk)

Printed in Singapore.

To Jacqueline,

Jean-Baptiste, Sébastien, Alexis, and Claire.

Preface

During the last quarter of the Twentieth Century, the study of nonlinear and complex systems has experienced unprecedented development. In the following pages, we present a first approach to standard results and recent advances in this field, applied to phenomena at our scale in our surrounding world. The pretext chosen is mainly instabilities in out-of-equilibrium systems, and more specifically the transition to turbulence in flowing fluids. It should however be clear that the methods used are of fully general use and can be applied as soon as a mathematical model of the problem at hand has been constructed, entering the dynamical framework that we shall develop.

The book is based on lecture notes for a short optional course given to second-year students in an engineering school, École Nationale Supérieure de Techniques Avancées, in Paris. This school is devoted to the training of high level engineers in fields including applied mathematics, mechanics and hydrodynamics, electronics,..., oceanography, and management. At the time of the course, students have not yet chosen their specialisation, so that the course has to be sufficiently general and without too specific requisites. Accordingly, the book should be of interest to nearly any science-oriented undergraduate student and, potentially, to everybody wanting to learn about recent advances in the field of applied nonlinear dynamics. Technicalities are not completely avoided but they are explained as simply as possible using heuristic arguments and specific worked examples, while openings on different topics can be gained by solving exercises at the end of each chapter using the same methods as those explained in the text.

At first, the problem of *chaos* that one has to face very early in this field may seem abstract and difficult. Even if the treatment of examples is not complete, the reader should get a concrete and operational mastery of concepts and techniques to be used from them. As far as the difficulty is concerned, our aim is to transmit knowledge rather informally and without full mathematical rigour. With respect to mathematics and physics, only basic understanding is required, at the level of what is currently known after one or two years of undergraduate training. In mathematics, this does not go further than elementary algebraic calculus, basic notions of linear algebra and ordinary differential calculus. For physics, it should suffice to follow one's intuition and to admit a few fundamental equations without discussion. Adaptation of the approach to any other field of interest should thus be envisaged without excessive anxiety.

A first brief chapter situates the context of the study: evolutionary problems involving a specific independent variable called *time*, distinguishing *discrete* systems governed by finite sets of ordinary differential equations and *continuous* media described by partial differential equations. It serves as an introduction to the rest of the book, explaining in particular that continuous media driven out of equilibrium may experience instabilities inducing structures that further break down, leading to turbulence. It is concluded by a few remarks on the modelling of processes in more general systems, from chemistry to mathematical ecology, and beyond.

The second chapter is devoted to a preliminary study of dynamical systems *with a small number of degrees of freedom*. The archetype of such systems is the oscillator which serves to introduce the first manifestations of nonlinear effects, e.g. the occurrence of self-oscillations or the relation between amplitude and frequency.

The way to complex behavior is then apparently left at a too early stage, before the occurrence of chaos. In Chapter 3 we turn to a specific but particularly simple and intuitive physical problem, the stability of a fluid layer heated from below and entering a convection regime. The first part of the chapter is devoted to the elementary study of this paradigm of self-organising systems: the instability mechanism leading to the formation of convection cells out of thermal fluctuations, and an approximate determination of the threshold. In the second part, a description is given of the "death" of the so-formed *dissipative structure* and of the steps toward turbulence.

After this detour, we come back to the mathematics of the transition within the dynamical systems framework. A preliminary step has to be performed, namely a *mode reduction* resting on the distinction between *driving* and *enslaved* modes, and on the elimination of the latter. The emergence of complexity is then analyzed as a result of the increase of the number of driving modes. This is done in two steps, related to the role of *confinement effects*, most easily understood in the context of convection: depending on its size, the convection box can contain a few cells or, on the contrary, many cells. In the first case, lateral boundaries freeze the spatial structure of the thermo-hydrodynamic fields and one is left with a few interacting modes characterised by their amplitudes as a function of time. This leads to the concept of *temporal chaos*, the occurrence of which is examined in Chapter 4 where we introduce classical *scenarios*. Tools used to identify chaos and measure its intensity are then presented in Chapter 5 where we also discuss empirical aspects and direct applications to chaotic mixing in fluids and chaos control.

The results and methods described up to that point will work provided that the underlying dynamics actually involves a small number of modes, which will only be recognised afterwards. In convection, when lateral walls are far enough and confinement is accordingly weak, many cells are present with variable sizes and positions. *Spatio-temporal chaos* becomes the relevant concept. Chapter 6 introduces the tools appropriate to describe *extended* systems, especially universal envelope equations accounting for modulations and *patterns* in the general case.

In common language, the third element of the title, *turbulence*, refers to the irregular, highly fluctuating, behavior of most of the flows surrounding us (the opposite situation of so-called *laminar* flows is rather exceptional). This problem is tackled in two steps. Instability and transition in *open flows* is examined in Chapter 7. In contrast to systems considered in Chapter 3, where the fluid remained confined to an enclosure, it now circulates from upstream to downstream, the consequences of which will be sketchily discussed. In Chapter 8, we consider *developed turbulence*, again along two paths. First we analyze the different scales, from the largest where energy is injected into eddies by instability mechanisms to the smallest where it is consumed by viscous dissipation. In a second instance, we turn to the statistical problem of predicting the average properties of a given turbulence flow, of utmost practical interest for an engineer.

Chapter 9 recapitulates the results and opens the perspective toward a complex dynamical system of contemporary interest, the climate of the Earth, and the problem of understanding/modeling its past and present trends.

A first appendix is devoted to a summary of linear algebra results that are useful throughout the book. As far as the understanding of nonlinear phenomena is concerned, the recourse to computers has been of considerable help at different levels. This is the reason why a second appendix, introducing hands-on computer sessions, is devoted to elementary methods that can be developed, with sufficient common sense and no superfluous specialisation, to extract useful information from numerical simulations of simple, even simplistic, but well-designed generic models of nonlinear dynamics and pattern formation. The course was completed by laboratory sessions on topics, the theory of which is considered in some of the exercises.

Palaiseau, April 2010.

Contents

Preface   vii

1. Introduction and Overview   1
    1.1 Dynamical Systems as a Context 2
    1.2 Continuous Media as a Subject 5
    1.3 From Simple to Complex 9
        1.3.1 Thermal convection: the instability mechanism . . 10
        1.3.2 Nonlinear convection and dynamical systems . . . 12
        1.3.3 Complexity in dynamical systems 14
        1.3.4 Stability and instability of open flows 19
        1.3.5 Beyond the transition: fully developed turbulence 21
    1.4 More on Complexity and its Modelling 24
        1.4.1 Reacting systems, from chemistry to ecology . . . 25
        1.4.2 Modelling distributed systems 27
        1.4.3 Phase transitions and critical phenomena 32
        1.4.4 Concluding remarks 35
    1.5 Exercises 35

2. First Steps in Nonlinear Dynamics   39
    2.1 From Oscillators to Dynamical Systems 39
        2.1.1 First definitions 39
        2.1.2 Formalism of analytical mechanics 45
        2.1.3 Gradient systems 47
    2.2 Stability and Linear Dynamics 49
        2.2.1 Formulation of the linear stability problem 49
        2.2.2 Two-dimensional linear systems 51

		2.2.3	Stability of a time-independent regime	55
	2.3	Two-dimensional Nonlinear Systems		57
		2.3.1	Two examples of oscillators	57
		2.3.2	Amplitude and phase of nonlinear oscillators	63
	2.4	What Next?		73
	2.5	Exercises		73

3. Life and Death of Dissipative Structures — 83

    3.1 Emergence of Dissipative Structures 83
        3.1.1 Qualitative analysis of the instability mechanism . 83
        3.1.2 Simplified model 85
        3.1.3 Normal mode analysis, general perspective 86
        3.1.4 Back to the model 89
        3.1.5 Vicinity of the threshold: linear stage 93
        3.1.6 Classification of unstable modes 95
    3.2 Disintegration of Dissipative Structures 97
        3.2.1 Simplified model of nonlinear convection 97
        3.2.2 Transition to turbulence of convection cells 100
        3.2.3 Transition toward chaos in confined systems . . . 104
        3.2.4 Dynamics of textures in extended systems 107
        3.2.5 Turbulent convection 111
    3.3 Exercises 113

4. Nonlinear Dynamics: from Simple to Complex — 125

    4.1 Reduction of the Number of Degrees of Freedom 126
        4.1.1 Role of aspect ratios 126
        4.1.2 Low-dimensional effective dynamics 128
        4.1.3 Centre manifolds and normal forms 131
    4.2 Transition to Chaos 135
        4.2.1 Time-independent and periodic regimes 135
        4.2.2 Quasi-periodicity and resonance 140
        4.2.3 Quasi-periodicity and locking 146
    4.3 Beyond Quasi-Periodicity 154
        4.3.1 Dynamics on stable and unstable manifolds 156
        4.3.2 Nature of chaotic regimes 157
        4.3.3 Two classical scenarios of transition to chaos . . . 159
    4.4 Exercises 163

5. Characterising and Using Chaos 171
   5.1 Quantifying Chaos 171
       5.1.1 Instability of trajectories and Lyapunov exponents 171
       5.1.2 Fractal features 174
   5.2 Empirical Approach to Chaotic Systems 177
       5.2.1 Standard analysis using Fourier transforms 179
       5.2.2 Reconstruction using delays 181
       5.2.3 Sampling frequency and embedding dimension .. 184
       5.2.4 Application 188
   5.3 Mixing by Stirring 191
       5.3.1 Mixing: diffusion or/and advection? 191
       5.3.2 Time-periodic two-dimensional configurations .. 195
       5.3.3 Time-dependent two-dimensional open flows ... 201
       5.3.4 Time-independent three-dimensional open flows . 204
   5.4 Control of Chaos 205
       5.4.1 Synchronisation of dynamical systems 205
       5.4.2 Delay systems and chaos control using delays ... 211
       5.4.3 The OGY method 215
   5.5 Conclusions 220
   5.6 Exercises 220

6. Nonlinear Dynamics of Patterns 227
   6.1 Quasi-one-dimensional Cellular Structures 228
       6.1.1 Steady states 228
       6.1.2 Amplitude equations 230
   6.2 Dissipative Crystals 232
   6.3 Short Term Selection of Patterns 235
   6.4 Modulations and Envelope Equations 236
       6.4.1 Quasi-one-dimensional cellular patterns 237
       6.4.2 Two-dimensional modulations of quasi-1D patterns 238
       6.4.3 Quasi-two-dimensional cellular patterns 240
       6.4.4 Oscillatory patterns and dissipative waves 242
       6.4.5 Universal long-wavelength instabilities 244
   6.5 What Lies Beyond? 249
   6.6 Exercises 250

7. Open Flows: Instability and Transition 255
   7.1 Base Flow Profiles 257

		7.1.1 Strictly one-dimensional flows	257
		7.1.2 More general velocity profiles	259
		7.1.3 Extension to arbitrary profiles	262
	7.2	Linear Stability	262
		7.2.1 General framework	262
		7.2.2 Inviscid flows	266
		7.2.3 Viscous flows	273
		7.2.4 Instability and downstream transport	276
	7.3	Transition to Turbulence	282
		7.3.1 Nonlinear development of instabilities	282
		7.3.2 Inviscidly unstable flows	286
		7.3.3 Inviscidly stable flows	291
		7.3.4 Turbulent spots and intermittency	295
	7.4	Exercises	307
8.	Developed Turbulence		313
	8.1	Scales in Developed Turbulence	314
		8.1.1 Production scale	314
		8.1.2 Inertial scales and the Kolmogorov spectrum . . .	316
		8.1.3 Dissipation scales	318
		8.1.4 Remarks	320
	8.2	Mean Flow and Fluctuations	321
		8.2.1 Statistical approach	321
		8.2.2 Reynolds averaged Navier–Stokes equation	322
		8.2.3 Energy exchanges in a turbulent flow	324
	8.3	Mean Flow and Effective Diffusion	325
		8.3.1 Mixing length and eddy viscosity	325
		8.3.2 Determination of the mean flow	327
	8.4	Beyond the Elementary Approach	332
		8.4.1 The structure of turbulent flows	332
		8.4.2 Turbulence modelling	336
		8.4.3 Large eddy simulations	339
	8.5	Exercises	343
9.	Summary and Perspectives		347
	9.1	Dynamics, Stability and Chaos	348
	9.2	Continuous Media, Instabilities and Turbulence	352
	9.3	Approach to a Complex System: the Earth's Climate . . .	356

9.4　Exercise: Ice ages as catastrophes 371

Appendix A　Linear Algebra　375

A.1　Vector Spaces, Bases, and Linear Operators 375
A.2　Structure of a Linear Operator 378
　A.2.1　Jordan normal form 379
　A.2.2　Exponential of a matrix 382
　A.2.3　Perturbation of a linear problem 383
A.3　Metric Properties of Linear Operators 387
　A.3.1　Scalar products, adjoints, non-normal operators . 387
　A.3.2　Fredholm Alternative 389
　A.3.3　Boundary value problems and adjoint operators . 390
A.4　Application to linear control in state space 391
A.5　Exercises 395

Appendix B　Numerical Approach　399

B.1　Treatment of the Time Dependence 400
B.2　Treatment of Space Dependence in PDEs 406
　B.2.1　Finite difference methods 406
　B.2.2　Spectral methods 411
B.3　Exercises 417
B.4　Case studies 418
　B.4.1　ODEs 1: Forced pendulum 419
　B.4.2　ODEs 2: Lorenz model 420
　B.4.3　ODEs 3: Rössler and Chua models 424
　B.4.4　PDEs 1: SH model, finite differences 425
　B.4.5　PDEs 2: SH model, pseudo-spectral method ... 428

Bibliography　431

Index　435

Chapter 1
Introduction and Overview

In a linear world, the effects being always proportional to their causes, everything would be *simple* since the tools that would allow us to represent it as a *superposition* of elementary states are rather well mastered. Unfortunately the world is *complex*. We indeed often observe effects that saturate despite an increase of their causes, or which go in different and somehow unexpected ways. All this is ascribed to nonlinearity. Of course, if we succeed in determining the state of a nonlinear system, we immediately try to go back to a problem that we know how to handle by linearising the dynamics around it and treating small departures from it by perturbation, and next to reproduce this scheme as far as possible in order to reach other fully nonlinear states, eventually "far from" the initial one.

Most recently, our knowledge of intrinsically nonlinear phenomena has made much progress (even though it rests in part on the use of linear tools) and investing it in applications is of primary interest. The aim of these lecture notes is therefore to introduce the reader to this breakthrough by taking the problem of macroscopic instabilities as a pretext. In thermodynamic systems close to equilibrium the response to excitations is essentially linear, i.e. proportional to the (sufficiently small) amplitude of the applied stress with a proportionality factor called a *susceptibility*. As a consequence of linearity, the regime that develops is *unique*. When the applied stress increases, the system is driven farther from equilibrium and nonlinearity can no longer be neglected. This opens the possibility of *bifurcation* towards different regimes that can coexist and compete. As a result of such catastrophes, losing reference to the initial state, the dynamics becomes increasingly complicated, typically from *laminar* to *turbulent* flow.

1.1 Dynamical Systems as a Context

In order to study the causes of the complexity induced by nonlinearity, we shall concentrate our attention on problems defined in terms of the evolution of a set of *state variables* functions of a single independent variable called *time*. Mechanical systems in the ordinary sense obviously belong to this class. Their archetype is the oscillator, and in its simplest linear expression, the *harmonic oscillator*:

$$m\frac{\mathrm{d}^2}{\mathrm{d}t^2}X = F = -kX\,, \tag{1.1}$$

where X measures the departure from the equilibrium length of the spring, m is the mass attached to its end, and F is the restoring force, here taken proportional to X with (constant) stiffness coefficient k (Figure 1.1, left). As a result, the oscillation period is independent of the amplitude: $T = T_0 = 2\pi/\omega_0$, with $\omega_0^2 = k/m$.

At this stage, nonlinearity can be introduced in two ways. First it is easy to imagine that the stiffness coefficient is a function of X itself. Assuming $k = k_0(1 + \alpha X^2)$ leads to what is known as the *Duffing oscillator*. Another possibility comes from the existence of mechanical constraints. The ideal rigid *pendulum*, a mass m at the end of a weightless rod of length l and revolving around a horizontal axis in the field of gravity, is a good example (Figure 1.1, right). Denoting its position by the angle θ the rod forms with the vertical axis, one obtains:

$$J\frac{\mathrm{d}^2}{\mathrm{d}t^2}\theta + mgl\sin\theta = 0\,, \tag{1.2}$$

where $J = ml^2$ is the moment of inertia and $-mgl\sin\theta$ the torque exerted by the weight of the mass. The nonlinearity built in the sine function is a consequence of the geometrical constraint keeping the mass rigidly at a

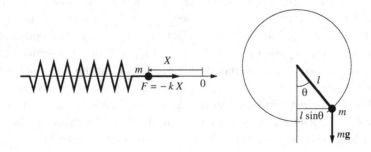

Fig. 1.1 Left: Linear spring. Right: Rigid pendulum.

constant distance from the rotation axis and plays a role in case of large deviations. As an outcome, several equilibrium positions exist. The 'down' position is a stable equilibrium point, with small amplitude oscillations around it governed by the same equation as the harmonic oscillator. The 'up' position is unstable and the pendulum departs from it upon perturbations of any amplitude, even infinitesimal. Furthermore, low energy states oscillate around the 'down' position with a period depending on the energy, while at larger energies, the system rotates always in the same direction, accelerating and slowing down periodically as it passes through the 'down' and 'up' positions, respectively. The behaviour of nonlinear oscillators will be further considered in Chapter 2.

Such mechanical examples can serve us to introduce a geometrical description of the dynamics in a space called the *state space*, here with coordinates the positions X or θ and the momenta $m\frac{d}{dt}X$ or $J\frac{d}{dt}\theta$, and to account for the dynamics in terms of trajectories in that space. In practice, any system involving a finite set of dependent variables serving to characterise its state, its *state variables*, governed by differential equations, belongs to the class we consider here. Examples are intensities and potential differences in an electrical circuit (§2.3.1.2), concentrations of reactants in chemistry, population densities in ecology,..., see §1.4. In analytical mechanics, a pair 'position + conjugate momentum' is called a *degree of freedom*. The state space is the product of the configuration space and the momentum space and is usually called the *phase space*. In a more general context, what is called a degree of freedom is simply a state variable.

In general, it is advantageous to write down the dynamical equations as a system of first order differential equations

$$\tfrac{d}{dt}X_j = \mathcal{F}_j\left(\{X_i(t); i=1,\ldots,d\},t\right), \qquad j=1,\ldots,d,$$

where the integer d is called the *dimension* of the system. Using the notations $\mathbf{X} \equiv \{X_j, j=1,\ldots,d\}$ for a point in the state space $\mathbb{X}$, and $\mathcal{F} \equiv \{\mathcal{F}_j, j=1,\ldots,d\}$ for the set of functions governing the dynamics, we have:

$$\tfrac{d}{dt}\mathbf{X} = \mathcal{F}(\mathbf{X},t). \tag{1.3}$$

For a system such as (1.3) time is a *continuous* variable. In contrast, a *discrete-time* system is defined as an iteration:

$$\mathbf{X}_{k+1} = \mathbf{\Phi}\left(\mathbf{X}_k\right), \tag{1.4}$$

or, in expanded form,

$$X_{j,k+1} = \Phi_j\left(\{X_{i,k}; i=1,\ldots,d\},k\right). \qquad j=1,\ldots,d.$$

While appearing as a topic in itself in mathematics, the study of discrete-time dynamical systems turns out to be essential in physics and engineering owing to their occurrence as a result of the *stroboscopic analysis* of periodically forced continuous-time dynamical systems, or of a *Poincaré section* in self-oscillating systems. They can also introduce themselves as the outcome of modelling effort of specific phenomena, e.g. seasonal counts in population dynamics (§1.4.1). All this will be considered in detail in Chapters 4 and 5.

Systems considered here all carry out the intuitive concept of *determinism* which imply the accurate prediction of a state at time t or k beyond some time t_0 or k_0 at which some initial condition is specified. This *initial value problem* presents itself formally or explicitly as a computational problem, directly if it is defined as an iteration (1.4) or indirectly through some numerical approximation for a continuous-time system (1.3), e.g. Euler extrapolation,

$$\mathbf{X}(t_{k+1}) = \mathbf{X}(t_k) + \Delta t\, \mathcal{F}\left(\mathbf{X}(t_k), t_k\right), \qquad (1.5)$$

since analytical integration is rarely possible, see §B.1 for an introduction. In this perspective, the dimension d is just the number of initial conditions necessary to start the evolution. In mechanics, it is thus twice the number of degrees of freedom since positions and velocities have both to be specified, see §2.1.2.

Returning to the geometrical perspective introduced above, we are now interested in the properties of *trajectories* followed by the system in its state space. This study rapidly points to the key role played by the concept of *stability* taken in the broad sense of *resistance to perturbations*. In practice, this vague definition refers to two different viewpoints.

The first one, rather quantitative, applies to specific trajectories: the solutions found, which depend on initial conditions, have to withstand small perturbations imposed either at the start or during the subsequent evolution, owing to unavoidable disturbances, either intrinsic (thermodynamic fluctuations) or extrinsic (noise). From this first viewpoint, only *stable* solutions (that resist) are physically observable. Further, a sufficiently large set of experimental conditions should make them attainable.

The second viewpoint is more qualitative. It refers to the very definition of the system itself: when compared to the real world, any abstract implementation is intrinsically blurred with numerous approximations and its control parameters are not determined with infinite precision. Accordingly, in order to be of help, the analytical model must be *robust*, i.e. its predictions must not be too sensitive to these different sources of inaccuracy. This property indeed fails at *bifurcation points* where the system experiences qualitative changes of behaviour. At such points one says that it is *structurally unstable*: the nature of the state depends sensitively on the perturbation.

It can be easily understood that these two facets of the term 'stability' are equally important for the applications. They will be at the heart of the most abstract part of this book, first in Chapter 2 and next in Chapter 4 and 5 where we shall give a more precise meaning to the word 'prediction' when the considered system evolves chaotically, that is to say in a way *unpredictable* in the long term in spite of short term *determinism*, due to an instability of trajectories that is the essence of *chaos*.

1.2 Continuous Media as a Subject

Dynamical systems considered up to now were endowed with a supposedly small number of dependent variables, thus living in spaces of low enough dimensions. Once the microscopic structure of matter has been recognised, in principle one should turn to the study of systems made of a large number of components at the molecular scale, each with its own degrees of freedom in the mechanical sense. When considering the motion of a simple solid, with fixed relative positions of its elementary constituents, the effective number of degrees of freedom naturally decreases to just 6: 3 coordinates for the position of its centre of mass and 3 angles for its orientation. In more general cases (soft materials, liquids, gases,...) a refined description of the microscopic configurations is however generally useless due to chaos at this scale. Our ignorance of dynamical details can be circumvented by adopting a statistical point of view from which the less probable state (specific initial condition) evolves into the most probable state (equilibrium compatible with conservation laws), founding the *thermodynamic* approach. In this framework, microscopic variables are replaced by average quantities such as energy, entropy, temperature or pressure (see also §1.4.3).

In practice, *global* thermodynamic equilibrium is not of much interest since it accounts for a world eaten by the worm of the equiprobability of microscopic states and, in some sense, completely dead. On the contrary, an inhomogeneous world traversed by various fluxes making it alive (which by the way allows us to study it!) is much richer and more interesting. We are indeed confronted with a large class of transport processes in media that are out of equilibrium on a macroscopic scale and there is a wide range of time and space scales for which the concept of *continuous medium* is appropriate, that is to say the description of a system according to which its state variables are functions of the position in physical (three-dimensional) space, $\mathbf{X}(\mathbf{x},t)$, governed by partial differential equations:

$$\partial_t \mathbf{X} = \boldsymbol{\mathcal{G}}(\mathbf{X}, \boldsymbol{\nabla}\mathbf{X}, t). \tag{1.6}$$

From a mathematical perspective, such systems are infinite-dimensional since we need to specify the value of these variables at every point in space. On the other hand the validity of the description relies on the definition of a *mesoscopic* scale in-between the microscopic (molecular) level and the (macroscopic) size of the system taken as a whole. The concept of *material point* and the assumption of *local equilibrium* make sense precisely on this intermediate scale, sufficiently small to be considered as infinitesimal but large enough to contain as many molecules as necessary for the laws of thermodynamics to be relevant.

Equations such as (1.6) are in general derived from the macroscopic balance of extensive thermodynamic variables, say Z, that in differential form read

$$\partial_t \rho_Z + \boldsymbol{\nabla} \cdot \mathbf{J}_Z = \Sigma_Z, \tag{1.7}$$

where ρ_Z is the density of Z, $\mathbf{J}_Z$ its flux, and Σ_Z a source term that cancels when Z is a conserved quantity (momentum, energy). The simplest example is the heat equation accounting for thermal diffusion:

$$C \partial_t T = \chi \nabla^2 T \tag{1.8}$$

which governs the temperature field in a solid submitted to moderate gradients. It derives from the *Fourier law*:

$$\mathbf{J}_Q = -\chi \boldsymbol{\nabla} T, \tag{1.9}$$

a phenomenological relation linking the heat flux to the temperature gradient and defining the thermal conductivity χ. Quantity C in (1.8) is the specific heat defined from the thermodynamic relation $\delta U = C \delta T$ which relates the change of internal energy δU to the temperature variation δT. Inserting this into the balance equation for the internal energy density[1] $\partial_t \rho_U + \boldsymbol{\nabla} \cdot \mathbf{J}_Q = 0$, immediately yields (1.8).

Quantity $\kappa = \chi/C$ is the thermal diffusivity. Dimensionally, this coefficient is homogeneous to $[L]^2/[T]$, like the diffusivity D appearing in

---

[1] The energy density ρ_U at a point $\mathbf{x}$ in physical space is defined as the limit of the ratio $\delta U/\delta V$ of the energy δU contained in a small volume δV around $\mathbf{x}$ as $\delta V \to 0$.

the *Fick law* governing *molecular diffusion*. In this context, an important quantity is the time scale τ for diffusive relaxation over a typical space scale ℓ, directly obtained from the dimensional relation above, $\tau = \ell^2/\kappa$ in the thermal case. The relaxation of T is examined in the most elementary case in Exercise 1.5.1 which further illustrates a rare situation where one can find an analytic transform allowing the exact linearisation of a parent nonlinear problem.

Equation (1.8) is indeed linear provided that the specific heat C and the thermal conductivity χ are constant, which is reasonable not too far from equilibrium. The Fourier law (1.9) is the prototype of a phenomenological relations linking fluxes (effects) to gradients (causes) at a linear level. In practice, the validity range of such a linear law is generally rather wide because the constraints we are able to apply to continuous media are usually weak when compared to molecular interactions. As counter-examples one could cite electronic systems containing active elements with threshold effects (diodes), or chemically reacting media (§1.4.1), and *a fortiori* biological systems. However, in macroscopic media strong nonlinearity can in fact arise from global considerations. This is particularly the case in fluid systems where macroscopic flow deeply affects the transport properties, rendering a locally linear medium effectively nonlinear.

So, in hydrodynamics the concept of the material point transforms itself into that of *fluid particle*, the position of which becomes a function of time:

$$\mathbf{M}(t) = (x_\mathrm{M}(t), y_\mathrm{M}(t), z_\mathrm{M}(t)) \ .$$

It is most often useful to pass from this *Lagrangian* description, to an *Eulerian* approach in terms of the velocity field:

$$\mathbf{v} \equiv (v_x, v_y, v_z) \equiv \left(\tfrac{\mathrm{d}}{\mathrm{d}t} x_\mathrm{M}, \tfrac{\mathrm{d}}{\mathrm{d}t} y_\mathrm{M}, \tfrac{\mathrm{d}}{\mathrm{d}t} z_\mathrm{M}\right) \ .$$

The two descriptions are linked by the definition of the *material derivative* measuring the evolution of any physical quantity Z attached to the fluid particle as it is followed during its motion: $Z = Z(x_\mathrm{M}, y_\mathrm{M}, z_\mathrm{M}, t)$. Mathematically, material differentiation is thus a total differentiation with respect to time, $\tfrac{\mathrm{d}}{\mathrm{d}t} Z$, that is:

$$\tfrac{\mathrm{d}}{\mathrm{d}t} Z = \partial_t Z + \partial_x Z \tfrac{\mathrm{d}}{\mathrm{d}t} x_\mathrm{M} + \partial_y Z \tfrac{\mathrm{d}}{\mathrm{d}t} y_\mathrm{M} + \partial_z Z \tfrac{\mathrm{d}}{\mathrm{d}t} z_\mathrm{M} = \partial_t Z + \mathbf{v} \cdot \boldsymbol{\nabla} Z \ . \quad (1.10)$$

This relation allows a simple expression of the balance equation in differential form provided that the flux of Z is properly split into a *diffusion* part and an *advection* part linked to the macroscopic motion:

$$\mathbf{J}_Z = \mathbf{J}_{Z,\mathrm{diff}} + \rho_Z \mathbf{v} \ .$$

When studying the mechanism of natural convection in fluids originally at rest we shall demonstrate the role of a global nonlinearity played by the advection term in spite of its apparent linearity in ρ_z, but before considering this example, let us recapitulate the equations governing the motion of a single component simple fluid with a constant viscosity. The first one is the continuity equation accounting for the conservation of matter. In full generality it reads:

$$\partial_t \rho + \boldsymbol{\nabla} \cdot \mathbf{J}_\rho = 0\,,$$

(no source term). Noticing that diffusion of matter within itself does not make sense, we get $\mathbf{J}_\rho = \rho \mathbf{v}$ which leads to

$$\partial_t \rho + \mathbf{v} \cdot \boldsymbol{\nabla} \rho + \rho \boldsymbol{\nabla} \cdot \mathbf{v} \equiv \tfrac{\mathrm{d}}{\mathrm{d}t}\rho + \rho \boldsymbol{\nabla} \cdot \mathbf{v} = 0\,. \quad (1.11)$$

In the following we shall consider *incompressible flows* characterised by $\tfrac{\mathrm{d}}{\mathrm{d}t}\rho = 0$, so that the continuity equation more simply reads

$$\boldsymbol{\nabla} \cdot \mathbf{v} = 0\,. \quad (1.12)$$

On the other hand, the compressional viscosity drops out when $\boldsymbol{\nabla} \cdot \mathbf{v} = 0$ and there only remains a shear viscosity which relates the viscous stress tensor to the rate of strain in a linear way:

$$\sigma_{ij} = \mu(\partial_i v_j + \partial_j v_i)\,,$$

where μ is the *dynamic viscosity* (*Newtonian fluid*). Once inserted into the momentum conservation equation, this relation leads to the *Navier–Stokes* equation

$$\rho(\partial_t + \mathbf{v} \cdot \boldsymbol{\nabla})\mathbf{v} = -\boldsymbol{\nabla} p + \mu \boldsymbol{\nabla}^2 \mathbf{v}\,. \quad (1.13)$$

The *kinematic viscosity* $\nu = \mu/\rho$ is the transport coefficient accounting for the diffusive relaxation of velocity gradients (Stokes law). At the inviscid limit ($\nu \to 0$) the Navier–Stokes equation is called the *Euler* equation. The opposite limit where the viscous dissipation dominates is called the *Stokes approximation*. Finally if the heating generated by viscous friction is negligible, energy conservation yields the heat equation (1.8) where $\partial_t T$ is just replaced by $\tfrac{\mathrm{d}}{\mathrm{d}t}T \equiv (\partial_t + \mathbf{v} \cdot \boldsymbol{\nabla})T$, hence:

$$\partial_t T + \mathbf{v} \cdot \boldsymbol{\nabla} T = \kappa \boldsymbol{\nabla}^2 T\,. \quad (1.14)$$

An introduction to concepts with many illustrative applications entering the scope of this book can be found in the DVD [MFM (2007)].

1.3 From Simple to Complex

In a narrow vicinity of thermodynamic equilibrium, the response of a continuous system is simply proportional to the strength of applied constraints and displays the same symmetries, stationary or periodic in time, uniform or periodic in space, for example. However, far from equilibrium, when nonlinearity can no longer be neglected, the system may leave the so-called *thermodynamic branch* and bifurcate toward solutions that breaks some of these symmetries.

The nature of the regime that develops depends on the value of *control parameters* measuring the relative intensity of the different contributions to the dynamics. In fluid mechanics, it will most often be the *Reynolds number* defined as $R = U\ell/\nu$ where U and ℓ represent a typical velocity and a length scale both characteristic of the flow under consideration while ν, the kinematic viscosity introduced above, is a fluid's property. In line with what has been said about the thermal diffusion time, it is enlightening to analyse the Reynolds number as the ratio of the viscous relaxation time over the length ℓ, $\tau_\mathrm{v} = \ell^2/\nu$, to the advection time required to carry velocity fluctuations over the same distance, $\tau_\mathrm{a} = \ell/U$. When the Reynolds number $R = \tau_\mathrm{v}/\tau_\mathrm{a}$ is small, viscosity has time to wipe out inhomogeneity, whereas when it is large, thermodynamic *dissipation* is too slow and the mechanical contribution of *advection* dominates.

The study of the stability of a given regime called the *base state* rests on the definition of perturbations, i.e. departures from this state, and their subsequent evolution. Figure 1.2 illustrates the general situation. Below a value of the control parameter R denoted R_g, 'g' for global, the base state is *unconditionally stable*: whatever the shape and the amplitude of the perturbation, it decays and the system returns to its base state (sufficient condition of stability). Above a second value R_c, 'c' for critical, the system is sensitive to at least one unavoidable perturbation and unconditionally

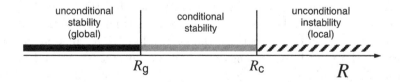

Fig. 1.2 Stability of the base state as a function of some control parameter R, e.g. the Reynolds number.

departs from the base state. The determination of R_c appeals to the so-called *linear stability theory* dealing with the evolution of fluctuations that are *mathematically* infinitesimal. It is indeed sufficient that there exist such a perturbation that is amplified for the base state to be unstable. It may happen that R_g coincide with R_c, in which case $R \geq R_c$ is a necessary and sufficient condition of instability but in general the interval $[R_g, R_c]$ has a finite width, which defines a range of *conditional stability*: the stability of the base state depends on the shape and amplitude of the perturbations to which it is submitted. Most of the time this fundamentally nonlinear problem remains unsolved.

In some favourable cases, one can succeed in computing the state that sets in above the threshold of the *primary* instability by a perturbation method. This so-obtained state is then promoted as a new base state, the stability of which is of interest. This state can in turn become unstable with respect to a *secondary* instability that complicates the dynamics, and so on up to a chaotic state which, for fluid systems, is usually called *turbulent*. In this perspective, turbulence is considered as resulting from various modes at the end of a cascading process. Instead of staying at this formal viewpoint let us rather examine a particularly simple and intuitive concrete mechanism responsible for the instability of a liquid layer heated from below and initially at rest.

1.3.1 *Thermal convection: the instability mechanism*

Let us consider a layer of fluid at rest between two horizontal plates heated from below and thus presenting a density stratification (Figure 1.3). A temperature difference $\Delta T = T_b - T_t > 0$ is applied between the bottom 'b' and the top 't' of the layer. With a few exceptions, the density decreases with increasing temperature and this stratification with heavier fluid on top of lighter fluid is potentially unstable in a vertical gravity field (like the 'up' position of the pendulum).

When ΔT is small, the fluid stays at rest since the gravitational potential energy that would be gained by moving the heavier fluid to the bottom is not sufficient to counterbalance the energy loss by dissipation in that motion. More precisely, as long as the fluid is at rest, the heat transfer is entirely carried out by *conduction*. The temperature profile in the base state $T_0(z)$ is obtained from the Fourier equation (1.14) that simply reduces itself to $d^2 T_0/dz^2 = 0$ since $\mathbf{v} \equiv 0$, hence $T_0(z) = T_b - \beta z$, where $\beta = \Delta T/h$ is the applied temperature gradient, h being the height of the layer.

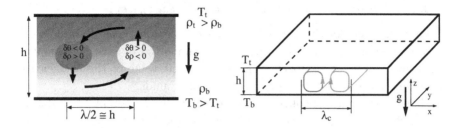

Fig. 1.3 Left: Instability mechanism of thermal convection. Right: Convection cells with wavelength $\lambda_c = 2\pi/k_c$.

Let us suppose a fluctuation localised inside a tiny droplet marked with a slightly higher temperature (perturbation $\theta > 0$). Since this bubble of hotter fluid is surrounded with colder, denser fluid, it experiences an upward differential buoyancy force which makes it moving up, encountering ever colder fluid, which reinforces the motion: this is the *destabilising* part of the mechanism. However, two *stabilising* processes tend to oppose this motion. First, the so-induced velocity tends naturally to decay owing to *viscous friction*. Second, *thermal diffusion* aims at ironing out the horizontal temperature gradient accompanying fluctuation θ. The fluid layer stays at rest as long as dissipation dominates, but convection develops when the destabilisation is sufficient, i.e. when ΔT is larger than some *critical* value ΔT_c called the *instability threshold*.

On the other hand, the stabilising dissipative processes are diffusive in essence, so that their efficiency depends on the horizontal space dependence of the velocity and temperature modulations. Damping is fast on small scales and slow on large scales. It thus depends on the length-scale of the perturbation and it turns out that the onset of the instability corresponds to some optimum scale. For dimensional reasons this scale is of the order of the distance h between the plates, as implied by the sketch in the left part of Figure 1.3. Accordingly, the bifurcated convection state often sets in as a system of periodic rolls, with spatial period $\lambda_c \approx 2h$, as illustrated in its right part. The quantity $k_c = 2\pi/\lambda_c$ which appears when representing the rolls in Fourier series is called the *critical wavevector*. The pattern formed has been called a *dissipative structure* by Prigogine [Glansdorff and Prigogine (1971)].

1.3.2 Nonlinear convection and dynamical systems

Here we account for the dynamics of convection close to the threshold by heuristic arguments that could be entirely supported by a rather complex detailed theoretical approach giving the explicit value of the coefficients that we shall introduce on phenomenological grounds.

The spatiotemporal coherence introduced in the system by the instability mechanism allows us to describe the evolution of the convecting layer using a simple variable $A(t)$ measuring the amplitude of convection. We thus assume that, as far as their horizontal dependence is concerned, the velocity and temperature fields can be taken in the form

$$(v_z, \theta) \propto A(t) \sin(k_c x), \qquad (1.15)$$

where A plays the role of an *effective degree of freedom* and k_c is the critical wavevector introduced earlier. We are now interested in the phenomenological derivation of an evolution equation for this amplitude.

At the infinitesimal stage, the perturbation is governed by a differential equation that should read

$$\tfrac{\mathrm{d}}{\mathrm{d}t} A = \sigma A, \qquad (1.16)$$

where coefficient σ accounts for the growth rate, negative below the threshold (damping) and positive beyond (amplification). Let us define a reduced control parameter $r = (\Delta T - \Delta T_c)/\Delta T_c$ and assume that the behaviour of σ as a function of r is not singular, so that its expression can be restricted to the first term of its Taylor expansion in powers of r:

$$\sigma = r/\tau_0, \qquad (1.17)$$

where τ_0, with the physical dimension of time, characterises the natural evolution of relevant fluctuations.

Equation (1.16) where σ is given by (1.17) accounts for the evolution of coupled velocity-temperature fluctuations as long as the amplitude A is sufficiently small. This does not raise difficulties when $r < 0$ since A decays but when $r > 0$, in the unstable domain, A is exponentially amplified and does not stay small for a long time. In order to model the nonlinear effects, we shall assume that (1.16) remains valid but that σ has to be corrected and becomes a function of A itself. Noticing that the change '$A \mapsto -A$' corresponds to a change of the spinning direction of the rolls (or to a translation of the structure by $\lambda_c/2$) and that the physics of the problem should not be sensitive to such a change, we are led to assume that

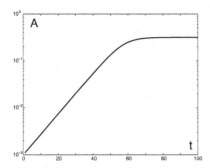 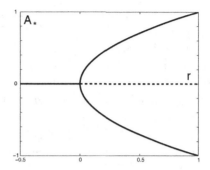

Fig. 1.4 Left: Evolution of the amplitude A as a function of time for model (1.20) with $r = 0.1$ (lin–log plot). Right: Bifurcation diagram giving the amplitude A of the steady state as a function of r. Conventionally, the unstable state $A = 0$ for $r > 0$ is indicated by dashes while a continuous thick line is used for the stable states, $A = 0$ for $r < 0$ and $A = A_* = \pm\sqrt{r}$ for $r > 0$.

the effective growth rate σ_{eff} is an even function of A. At lowest order of an expansion in powers of A, it can be taken as:

$$\sigma_{\text{eff}} = (r - gA^2)/\tau_0, \qquad (1.18)$$

where g is called the *Landau constant*. In the simple case considered here, g is a positive quantity, i.e. $g = 1/\bar{A}^2$, $\bar{A}$ measuring the typical amplitude at which nonlinearity becomes effective. Experiments show (and theory demonstrates) that the convection mechanism is indeed self-limiting, so that the effective growth rate decreases when the amplitude of convection increases. The nonlinear model accounting for convection in the neighbourhood of the threshold then reads:

$$\tau_0 \tfrac{\mathrm{d}}{\mathrm{d}t} A = \mathcal{F}_r(A) = rA - gA^3. \qquad (1.19)$$

Its study is elementary and will be done again later in a more general context. Here it is our first example of an *effective dynamical system*. It governs a simple scalar variable A representing the macroscopic evolution of our system. Upon re-scaling t by τ_0 and A by $\bar{A}$, i.e. making the changes $t \mapsto \tau_0 \tilde{t}$, $A \mapsto \bar{A}\tilde{A}$ and dropping the tildes, equation (1.19) can be recast in a universal form:

$$\tfrac{\mathrm{d}}{\mathrm{d}t} A = rA - A^3. \qquad (1.20)$$

The evolution of A from an initial value $A_0 = 10^{-3}$ for $r = 0.1$ is illustrated in Figure 1.4 (left). As long as $rA \gg A^3$ the growth remains exponential, which appears graphically linear on the lin–log plot in the left part of the figure. When A increases, the slope of the curve[2] σ_{eff} decreases. Ultimately, A saturates to a value $A_* = \sqrt{r}$. Taking a negative initial value would have led to $-A_*$, owing to the symmetry of the equation. The graph giving the nontrivial state as a function of the control parameter, depicted in Figure 1.4 (right), is called a *bifurcation diagram*. On general grounds, the bifurcation diagram is a graph locating the *limit sets* relevant in the $t \to \pm\infty$ limit, as functions of the control parameters.

In the terminology of dynamical systems, A is the *state variable* (or degree of freedom). Accordingly, the real axis is here the *state space* (or phase space) and $\mathcal{F}_r(A)$ is the *vector field* defined on that space. Stationary solutions A_* are called *fixed points*. A stable fixed point is a simple example of *attractor*. The transition 'conduction → convection' is continuous with a stable nontrivial A_* tending to zero as the threshold is approached from above. One then speaks of a *supercritical* bifurcation. Owing to the shape of the graph in Figure 1.4 (right), it is further called a *fork bifurcation*. Equation (1.20) is thus the *normal form* for a supercritical fork bifurcation. In the opposite case, i.e. $g < 0$ in (1.19), the nontrivial solution appears below the threshold and is unstable, the bifurcation is then *sub-critical*, and equation (1.20) has to be further completed to yield a meaningful model. This abstract approach will be carried on in Chapters 2 and 4.

Convection rolls that develop beyond the threshold of the first instability can themselves become unstable with respect to other mechanisms. One thus has to study the flow that results from the superposition of the saturated *primary* mode to the initial base state, using a similar but technically much more complicated approach. *Secondary* modes can then be detected yielding, upon saturation, new base states ready for subsequent destabilisation as the control parameter is further increased.

1.3.3 Complexity in dynamical systems

The convecting layer is a continuous medium described by field variables. The reduction of the dynamics to a single scalar variable A, or a small set of similar variables when other instabilities have taken place, calls for justifications that will be examined in Chapter 4 as a precondition to the use of the concept of *chaos* for interpreting the increasing complexity of the dynamics all along the cascade of instabilities leading to turbulence.

Let us give a few further indications about this step-by-step transition to

---

[2]In a semi-log plot, the slope is $\frac{\mathrm{d}}{\mathrm{d}t} \log A = (1/A) \frac{\mathrm{d}}{\mathrm{d}t} A = r - A^2 = \sigma_{\text{eff}}$.

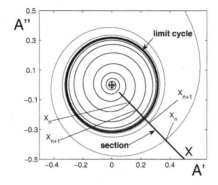

Fig. 1.5 Oscillations governed by Eq. (1.21): when $g_r > 0$ oscillations saturate yielding a limit cycle with radius $\rho = \sqrt{r/g_r}$ for $r > 0$; here $r = 0.1$, $\omega = 1 = g' = g''$.

complexity. Usually after a first bifurcation that has led to the emergence of an asymptotically time-independent nontrivial state, e.g. convection cells, a subsequent instability may introduce some temporal dependence, in the simplest case just periodicity that can be described by an equation analogous to (1.19), namely

$$\tau_0 \tfrac{d}{dt} A = sA - g|A|^2 A, \qquad (1.21)$$

where A now is a complex quantity $A' + iA''$ able to account for phase delays between two state variables A' and A''. The growth rate $s = r - i\omega$ is also complex with a real part $r < 0$ below the instability threshold and > 0 above, and an imaginary part specifying the frequency that sets in. Finally, $g = g' + ig''$ is a complex coupling constant. The sign of g' decides whether the oscillations saturate or not, and the value of g'' measures the nonlinear correction to the linear oscillation frequency ω. Equation (1.21) governs the *Hopf bifurcation*. When oscillations saturate ($g' > 0$) the trajectories come and stick asymptotically in time to a closed curve called a *limit cycle*, see Figure 1.5. The situation is then analogous to what happens under external periodic forcing, except that some internal mechanism here accounts for the occurrence of oscillations. It is conceptually essential to keep a geometrical point of view to analyse the subsequent steps. This is done by performing the analogue of a stroboscopic analysis that consists in taking snapshots of the system at the period of the strobe signal. Here it comes to considering the state of the system when the trajectory cuts some well-chosen *surface of section* and the time series of such intersections, which defines a *return map*

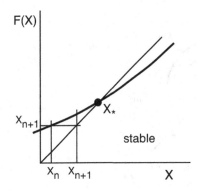 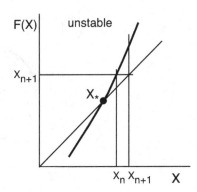

Fig. 1.6 Geometrical analysis of the stability of the limit cycle. Depending on whether the slope of F at X_* is smaller or larger than 1, the limit cycle is stable (left) or unstable (right).

of that surface of section onto itself, as introduced by Poincaré.[3] Taking a line that cuts the limit cycle transversally, here just parameterized by a coordinate X, one can thus define a return map in the form $X_{n+1} = \mathcal{F}(X_n)$. The limit cycle trivially correspond to a fixed point $X_* = \mathcal{F}(X_*)$ and its stability can be decided from the behaviour of iterates starting in its vicinity. When they converge to X_* ($|F'(X_*)| < 1$) the periodic regime is stable, otherwise it is unstable, see Figure 1.6.

Generalising the stroboscopic analysis, the *Poincaré section* technique thus reduces the usual *continuous time* dynamics ($t \in \mathbb{R}$) to a *discrete time* dynamics ($n \in \mathbb{N}$), i.e. the governing differential equations to iterations. At this stage, one can follow the original conjecture of Landau in 1944[4] who proposed that turbulence was the result of an indefinite accumulation of instabilities, each bringing in its own frequency and space dependence so that the dynamics was no longer predictable, see the corresponding cartoon in Figure 1.7 (top). Ruelle and Takens pointed out in 1971[5] that a superposition of motions, in some sense a too linear concept, was not a guarantee of the decay of correlations which seems to be a signature of turbulence. Their scenario, sketched in Figure 1.7 (bottom), rests on another more likely possibility due to *nonlinear* interactions, namely that trajectories in

---

[3] H. Poincaré, Les Méthodes Nouvelles de la Mécanique Céleste (Gauthier-Villars, 1892).
[4] L.D. Landau, "On the problem of turbulence," Akad. Nauk. Doklady **44** (1944) 339, translation in Collected Papers of L.D. Landau, D. ter Haar ed. (Pergamon Press, 1965), pp. 387–391.
[5] D. Ruelle and F. Takens, "On the nature of turbulence," Commun. Math. Phys. **20** (1971) 167–192. Addendum **23** (1971) 343–344.

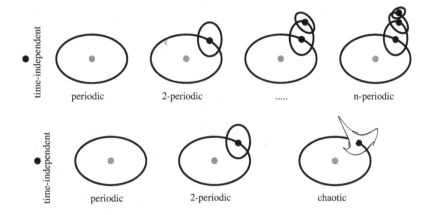

Fig. 1.7 The "nature of turbulence" according to Landau (top) and Ruelle & Takens (bottom). Ellipses attached one to the next in a string feature quasi-periodic behaviour as results from the superposition of periodic motions with incommensurate frequencies. The chaotic end of the Ruelle–Takens cascade is represented by a miniature of the attractor of the Curry–Yorke model that illustrates the disintegration of the ellipse born at the previous step (see Chapter 5, Exercise 5.6.2).

state space diverge due to their *sensitivity to initial conditions*, as a founding property of *deterministic chaos*. This nontrivial property, often called *butterfly effect* after Lorenz, is characteristic of *chaotic behaviour* and is present even in systems with a very small number of active modes, namely *three* state variables in continuous time systems, as shown by Lorenz,[6] only *two* in discrete time systems provided that they are invertible, e.g. the Hénon map (1976),[7] and even *one* variable if this last condition is dropped, which can be done at a heuristic level when modelling the transition to chaos or more plainly taken as a mathematical assumption. In this context, complexity essentially means *lack of (long term) predictability despite determinism*. In the long time limit, chaotic trajectories belong to sets called *strange attractors*. These sets result from a stretch-and-fold process with stretching implied by the instability of trajectories and folding due to overall nonlinearity of the dynamics, bearing some analogy with the kneading of dough (with instability mimicking the dispersion of salt in the bulk of the pastry even if it was initially unevenly distributed). In state space, they display a characteristic foliated structure with a locally self-similar hierar-

---

[6] E.N. Lorenz, "Deterministic nonperiodic flow," J. Atmos. Sci. **20** (1963), 130–141.

[7] M. Hénon, "A two-dimensional mapping with a strange attractor," Commun. Math. Phys. **50** (1976) 69–77.

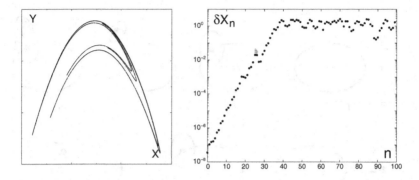

Fig. 1.8 The Hénon map is defined as $X_{n+1} = 1 - aX_n + bY_n$, $Y_{n+1} = X_n$. For $a = 1.4$ and $b = 0.3$, iterates of most initial conditions tend to the attractor depicted in the left panel. Trajectories belonging to this attractor display sensitive dependence on initial conditions as shown in the right panel: the distance between trajectories issued from two close initial conditions first diverge exponentially at a rate ≈ 0.65 per iteration (slope in semilog plot) and saturate at a value corresponding to the diameter of the attractor.

chic organisation called *fractal* by Mandelbrot [Mandelbrot (1982)]. The Hénon attractor displayed in Figure 1.8 illustrates all these characteristics at best. How good is this interpretation of the transition to turbulence in convection will be examined in §3.2.2.3, p. 103.

In practice results obtained using one-dimensional maps are of fundamental interest for the understanding of chaos. Sensitivity to initial conditions is easily visualised using the *dyadic map*, $\mathcal{F}(X) = 2X \pmod{1}$,[8] and scenarios of transition to chaos using other maps such as the *logistic map*, $\mathcal{F}(X) = rX(1 - X)$. Why and how complexity arises in such systems is studied in more detail in Chapter 4, though not too far into mathematical intricacy. Next, Chapter 5 is devoted the characterisation of chaotic trajectories, divergence rates (*Lyapunov exponents*) and occupation of space by strange attractors (*fractal dimensions*), and to a few applications, empirical reconstruction of complex dynamics, mixing by stirring in fluids, and chaos control.

In systems that are closed with respect to the flux of matter (but are maintained in out-of-equilibrium states by other fluxes, e.g. heat), this

---

[8] Just consider two trajectories $\{X_n, n = 0, 1, \ldots\}$ and $\{X'_n, n = 0, 1, \ldots\}$ with initial conditions X_0 and $X'_0 = X_0 + \delta X_0$, where δX_0 is an infinitesimal quantity. Next observe that after one iteration, one gets $\delta X_1 = 2\delta X_0$, and after n iterations, $\delta X_n = 2^n \delta X_0$ which grows exponentially. Taking the modulo keeps the two trajectories within the unit interval but they rapidly become foreign to each other, typically for $n > \log_2(1/\delta X_0)$.

approach is all the more relevant that the instability is constrained to be spatially coherent by *confinement effects*, e.g., by lateral walls at distances of the order of the height h of the convecting layer so that a few cells are expected. When spatial coherence is weak, in general because lateral walls are far apart, modulations are allowed and *patterns* develop. How to deal with this occurrence is introduced in section 1.4.2 below and examined in more depth in Chapter 6. Another kind of difficulties arise from the fact that in many cases we have to deal with *open flows* where matter is transported from up-stream to down-stream the region of interest.

1.3.4 *Stability and instability of open flows*

Natural convection has been the subject of many academic studies since it offers an ideal testing ground for ideas developed in nonlinear dynamics. However, systems of interest in engineering often display a supplementary feature: they involve *open flows* characterised by the existence of a mean current of material properties from upstream to downstream. Whereas the linear stability theory of open flows has a long history, dating back to the end of the XIXth Century, difficulties linked to this specific feature have delayed the nonlinear approach. Rather than a thorough account of results obtained in this field, Chapter 7 should thus be considered as a preliminary presentation aiming at better situating the problem in a nonlinear perspective and making its study easier.

Purely kinetic effects play an important role in open flows and can by themselves already be at the origin of instabilities. Accordingly, a first distinction can be made between flows that are *mechanically unstable* and those that are *mechanically stable*. The former display an inflection point in the base velocity profile which makes them unstable in the absence of viscous dissipation according to the *Rayleigh criterion*. Their prototype is the *mixing layer* illustrated in Figure 1.9 (left). In contrast, mechanically stable flows may become unstable only due to subtle feedback involving a transfer of momentum of viscous origin. A good example is given by the Blasius *boundary layer* flow depicted in Figure 1.9 (right). This classification will affect the whole process of transition to turbulence in an important way (§7.2).

A second distinction has to be made from the very existence of a global flow of matter entering the region of interest and leaving it at the outlet, which makes implicit reference to a frame rigidly attached to the laboratory (obstacle, rigid wall). An important difference indeed appears between

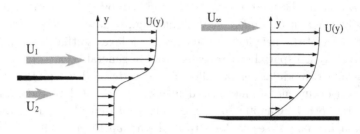

Fig. 1.9 Typical open flows. Left: Mixing layer created at the merging of two fluid veins with different velocities and maintained separated upstream by a splitting plate. Right: Boundary layer developing along a plate at some distance of its leading edge.

instabilities said to be *convective*, those that develop along the flow but are evacuated downstream and do not succeed in going upstream, and instabilities called *absolute* that are sufficiently intense to develop in spite of the global downstream transport. In the first case, the system behaves as a *noise amplifier* and the result downstream essentially depends on the level of background fluctuations (residual turbulence) or of perturbations voluntarily introduced in the system in view of its control. In the second case, one has rather to deal with a genuine *self-sustained oscillator* analogous to those studied in Chapter 2.

The situation is made even more complicated by the fact that the strength of instability mechanisms may vary in space. This is due to the fact that the velocity profile usually evolves downstream as a consequence of viscous dissipation that tends to smooth out the velocity gradients present at the entrance of the flow. Depending on the *local* intensity of the mechanism (linked to the magnitude of the shear) one can observe a change of the character of the mode, usually from absolute to convective, hence the possibility of bifurcated states localised in a given region of space and usually called *global modes*. The wake of a blunt body inserted in an otherwise uniform stream is such an example. A re-circulation takes place close to the obstacle, rendering the overall downstream transport locally sufficiently weak that regularly shed vortices can develop and stay attached to it, forming what is known as the Kármán vortex street, see Figure 1.10.

The understanding of this combined problem "convective/absolute + local/global + linear/nonlinear" has made great progress recently but it requires sophisticated mathematical tools, especially in complex analysis, that would bring us far beyond the limited purpose of the present book.

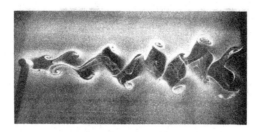

Fig. 1.10 Wake of a cylinder: a vortex array is downstream from the obstacle. (Courtesy P. Le Gal, IRPHE, Marseilles, France.)

The problem will just be evoked at a physically intuitive level in Chapter 7.

In the presentation above the instabilities were tacitly assumed to be linear and supercritical, i.e. gently saturating above threshold. It may however happen that the bifurcation is sub-critical and that several different flow regimes coexist in some range of control parameter, the bifurcated state saturating only "far from" the base state as for the plane channel flow. The base state may also happen to remain stable against infinitesimal fluctuations so that the nontrivial (turbulent) regime cannot be reached by perturbation. Examples are *Poiseuille pipe flow*, the flow in a cylindrical pipe under constant pressure gradient or constant mass flux, and *plane Couette flow*, the flow between two parallel plates moving in opposite directions. Though these flows are known to be linearly stable for all Reynolds numbers, they are expected to become turbulent as the shear rate increases beyond all limit. The transition indeed happens and takes a rather explosive turn: turbulent bursts developing intermittently from localised finite amplitude perturbations show up during the process, coexist with laminar flow and then merge to fill the systems with fully developed turbulence. Figure. 1.11 displays such a turbulent spot in plane Couette flow. It shows that the pocket of turbulence is not structure-less but, on the contrary, present a stream-wise streaky pattern. This type of structure turns out to be omnipresent in turbulent wall flows. Their production and sustainment mechanisms are still the subject of current research (§7.3.4).

1.3.5 *Beyond the transition: fully developed turbulence*

Problems considered up to now all relate to the steps of the cascade leading from a simple and regular base state to a complex and irregular flow still partly ordered but chaotic, that may not yet be called turbulent. It seems

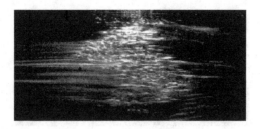

Fig. 1.11 Mature turbulent spot induced by a localised perturbation of finite amplitude in plane Couette flow (courtesy S. Bottin and O. Dauchot, GIT, CE Saclay).

indeed important to understand the steps of the laminar-turbulent transition just sketched in a deterministic framework. The underlying aim is of course to control it as best as we can, delay it if it is harmful, or advance it if we need better mixing. However, it turns out that beyond the transition we have to change our mind and take a fully statistical viewpoint. As a matter of fact, when we pull the considered system (flow) ever farther from equilibrium, more and more degrees of freedom become excited and it no longer makes sense to focus on their individual dynamics, like when passing from few-body systems well described by analytical mechanics to gases for which the thermodynamic approach is more appropriate.

This remark could lead to understanding *developed turbulence* as a new macroscopic state of matter at a scale where the fluid would have transport properties very different from those of ordinary fluids in their laminar state where molecular chaos still control diffusion. Things are unfortunately less simple. The usual distinction between microscopic and macroscopic scales works well for thermodynamics because there is a wide gap between them, the precise reason for which the concept of fluid particle makes sense. In contrast, in turbulence, relevant scales belong to a continuous range from the size of the flow domain to small scales. While the smallest eddies seem to evolve randomly like molecules in a gas, they are still coupled and, in fact, driven by the scales above them so that no de-coupling is truly legitimate.

In Chapter 8, we shall approach the theory of turbulence in a very preliminary way only, without trying to compete with numerous excellent books dealing with it, from conceptual problems to applications. Our limited aim will be to illustrate the idea of an *inertial cascade* (see Figure 1.12) transferring energy from large scales where instability mechanisms generate the eddies, down to the smallest scales where viscosity successfully irons out the fluctuations. It is not difficult to grasp the idea of this transfer by

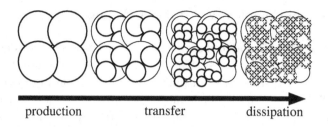

Fig. 1.12 Illustration of the Kolmogorov cascade from large scales to small scales where fluctuating eddies are blurred by viscous dissipation.

the advection term $\mathbf{v} \cdot \boldsymbol{\nabla}\mathbf{v}$ of the Navier–Stokes equations. Considering an eddy motion locally described by a simple trigonometric line $v_x \sim \sin(kx)$, we get $\sin(kx)\partial_x \sin(kx) = k\sin(kx)\cos(kx) = \frac{1}{2}k\sin(2kx)$ that presents itself as a source term in the evolution equation for a mode with a spatial scale half of the initial one. The result of this transfer is a repartition of energy according to the celebrated $k^{-5/3}$ Kolmogorov law that will be derived heuristically.

In the second part, we shall turn to the concrete problem of predicting the lowest order statistics of specific turbulent flows, obtaining the so-called *Reynolds averaged equations* governing the mean flow, which immediately opens the problem of the closure of the statistical description. As a matter of fact, the averaging of primitive equations introduces new higher order statistical quantities called the *Reynolds stresses* that remain to be evaluated in one way or another. We shall then exploit the already mentioned analogy between the kinetic theory of gases and the statistical theory of turbulence to introduce a disputable but heuristically valuable approach to turbulent flows resting on the concept of *turbulent diffusivity* patterned on that of *molecular diffusivity*. This will serve us to make a first evaluation of the average properties of a turbulent flow, taking the turbulent boundary layer as an example, and obtaining the classical Kármán logarithmic law. The main shortcoming of this approach derives from the absence of scale de-coupling mentioned above but we shall not do much more than mentioning it, suggesting further that progress can be obtained in concrete situations by numerical simulations, and observing that the latter always imply the crucial step of sub-grid-scale modelling, i.e. the modelling of smallest scales that cannot be explicitly accounted for in the numerics.

1.4 More on Complexity and its Modelling

Up to now we have introduced some concepts using simple fluids in various situations as working examples. The origin of increasing complexity was to be found in the interplay of transport and diffusion of local physical properties. Diffusion of species in homogenous mixtures is a supplementary source of complication but does not bring radically new features except when *chemical reactions* take place among the species present in the mixture. In the same way *predator-prey* interactions in ecology stray out of the framework of simple diffusive processes. Elements of chemical kinetics and dynamics of populations are given below since this source of complication is important (§1.4.1) but, as long as the system remains well-stirred, molecular concentrations or populations densities are only functions of time and its description stays within the framework of differential dynamical systems studied in Chapters 2, 4, and 5.

When non-uniformity in physical space is allowed, i.e. stirring not efficient enough or confinement effects ineffective to maintain coherence in the system (still considered as a continuous medium), evolution is typically governed by partial differential equations (PDEs) instead of ordinary differential equations (ODEs). *Patterning* is then a manifestation of long-range coherence while interactions are usually still short-range. Elements of the corresponding theory will be introduced in Chapter 6. In spite of its universal contents, useful to treat plain convection, reaction-diffusion or predator-prey, and some more complicated systems, it may still be insufficient. This is particularly the case for what is now understood as *complex systems*, i.e. systems with a large number of *agents* in interaction on *networks*. In such systems, the *global* behaviour may not be straightforwardly inferred from the knowledge of the *local* dynamics (*emerging behaviour* in the absence of external forcing or supervisor). Difficulties arise from the lack of conventional thermodynamic approach but concepts and tools derived from those used in *statistical physics* are of great practical value, especially those involved in the theory of *phase transitions* and *critical phenomena* (§1.4.3).

Different approaches can be developed depending on whether the physical space is continuous (PDEs) or discrete (regular lattices, structure-less networks), whether time is continuous (differential systems) or not (iterations), whether the local state space is continuous (maps) or discrete (cellular automata), whether the coupling is local (i.e. limited to neighbouring agents), or global (everybody interacting with everybody), whether the dynamics is fully deterministic or contains elements of randomness possibly arising from local chaos kept outside of the modelling. A detailed presentation of these topics goes largely beyond our purpose. Accordingly, we shall limit ourselves to the brief presentation in §1.4.2 below and refer to the literature, e.g. [Boccara (2004)], for more detailed approaches. Understanding some of the modelling issues will however be useful when we shall consider the concrete example of Earth as a complex evolving system at the end of Chapter 9.

1.4.1 Reacting systems, from chemistry to ecology

Chemistry has much contributed to the development of applied nonlinear dynamics in offering a way to construct differential systems of practical interest. An elementary step between reactants A_i with concentration A_i can be written as

$$\sum n_i A_i \quad \rightarrow \quad \sum n'_i A_i \tag{1.22}$$

where n_i and n'_i are the stoechiometric coefficients. The corresponding reaction rate is a measure of the number of effective collisions per unit time. A collision is the meeting at a single physical point of all the components on the l.h.s. of (1.22) producing those on the r.h.s. in the proportions given by n_i and n'_i. The probability for reactant A_i to be there is proportional to its concentration and, neglecting the correlation between species, one expects the probability of simultaneous presence of reactants at a point to be the product of individual probabilities, and thus the reaction rate to be $k \prod_i A_i^{n_i}$, involving only the l.h.s. participants and where k is the corresponding rate constant. When such a reaction takes place, the number of molecules of component A_i varies by $n'_i - n_i$, so that one gets:

$$\tfrac{\mathrm{d}}{\mathrm{d}t} A_i = \mathcal{F}_i(\{A_j\}) = (n'_i - n_i) k \prod_i A_i^{n_i}. \tag{1.23}$$

The total variation of the concentration of a given species A_i for a compound reaction is the sum of the variations at each step. Writing equations such as (1.23) for the different species involved in a given reaction leads to a system of ordinary differential equations.

An oscillatory reaction, discovered by Belousov in the 1950s and further studied by Zhabotinsky in the 1960s (BZ reaction) has been the subject of intense laboratory work. It corresponds to the oxidisation of an organic reducer (malonic acid) by BrO_3^- ions with a redox couple (e.g. Ce^{3+}/Ce^{4+}) as a catalyst. It involves about 15 chemical species coupled by an equivalent number of intermediate reaction steps. This system displays chaotic regimes under well chosen experimental conditions. A simplified model of this reaction have been devised with only 5 steps and 3 intermediate variables, called the Oregonator.[9] An even simpler example of oscillating reaction is the subject of Exercise 1.5.2.

Understanding relations between preys and predators as chemical reactions give immediate access to the modelling of population dynamics in ecology. The dynamics of the population X of some species X is usually described in terms of an effective growth rate:

$$\tfrac{\mathrm{d}}{\mathrm{d}t} X = \alpha_{\mathrm{eff}} X \qquad (1.24)$$

where the expression of α_{eff} accounts for the balance between birth and death, which depends on control parameters and the population itself. Assuming simple proportionality to the amount of available food gives unlimited growth at rate α, which may be valid as long as the population is small enough but cannot be sustained for long. So the growth rate has to decrease as the population increases (*limited growth* model). A simple assumption is thus $\alpha_{\mathrm{eff}} = \alpha - \beta X$ where α and β are two positive constants. The corresponding evolution equation is usually called the (continuous-time) *logistic equation*:

$$\tfrac{\mathrm{d}}{\mathrm{d}t} X = \alpha X - \beta X^2 \qquad (1.25)$$

Starting from a low level, the population grows up to an equilibrium value $X_* = \alpha/\beta$. If the initial population is larger than X_*, the growth rate is negative so that X decreases down to X_*. In this simple case an explicit analytic solution is easily found by integrating (1.25).

Now, let us consider that X is a predator and that available food is determined by the population Y of its prey Y, the food is thus not indefinitely available and the term $-\beta X$ is no longer necessary. A simple assumption is that the growth of the population is associated with a food consumption that is proportional to the presence of prey – this is the equivalent of a reactive collision in chemistry – and that the decay of the predator is due to some natural death rate $-\alpha_0$. This leads to:

$$\alpha_{\mathrm{eff}} = \alpha' Y - \alpha_0$$

for α_{eff} in (1.24), α' and α_0 being two positive constants. In order to close the system we now need to write down an equation for the effective growth rate of the prey. By arguments similar to previous ones, one easily gets:

$$\tfrac{\mathrm{d}}{\mathrm{d}t} Y = \gamma Y - \delta X Y . \qquad (1.26)$$

---

[9] R.J. Field and R.M. Noyes, "Oscillations in chemical systems IV. Limit cycle behaviour in a model of a real chemical reaction," J. Chem. Phys. **60** (1974) 1877–1884. For an introduction see: R.F. Field, "Oregonator," Scholarpedia (2007) 2(5):1386.

where γ represents the natural growth rate of the prey and $-\delta X$ the decrease due to predation, parameter δ measuring the hunting talents of the predator. System (1.25,1.26) is the simplest possible example of predator-prey models, generically called *Lotka–Volterra* systems. It can qualitatively explain population oscillation observed in nature, e.g. struggle for life in fox–hare populations detected from fur trade in Canada. This simple model has many weaknesses and can be made more realistic by complicating the expression of the relations between preys and predators. For an exhaustive presentation, consult [Murray (1993)]. Epidemic models basically belong to the same category, the prototype of which is the SIR model (Kermack and McKendrick, 1927), see Exercise 1.5.3.

An obvious limitation is that populations are treated as functions of time only, while the distribution of preys and predators in space must be taken into account since it is the encounter of a prey with its predator at a given place that counts. A more realistic situation is thus obtained when 'geography' is included, that is variables are defined at the nodes of a lattice and interaction becomes a stochastic process involving the state of neighbours. This leads us directly to so-called *agent based* modelling.

1.4.2 *Modelling distributed systems*

Up to now, we have implicitly considered that nonlinear dynamics is developing in a finite dimensional state space, and that the states can be identified using a set of scalar variables, e.g. the amplitude of convection, concentrations of chemical species, populations of predators and preys, etc., but without explicit reference to the physical space where the actual entities evolve. As an example let us consider the case of chemical reactions. As long as the medium remains well-stirred, the concentrations of the reactants can be represented by variables that are independent of the position in space and thus governed by ordinary differential equations as discussed above. On the other hand, in the absence of stirring, the assumption of homogeneity fails and, though the reaction rules still hold locally, molecular diffusion has to be taken into account. Concentrations become functions of space and the ordinary differential equations converted into partial differential equations by the introduction of diffusion terms so that (1.23) is transformed into

$$\partial_t A_i = \mathcal{F}_i(\{A_j\}) + D_i \boldsymbol{\nabla}^2 A_i \quad i = 1, \ldots N, \qquad (1.27)$$

where the coefficients D_i are the molecular diffusivities of the reactants.

A set of partial-differential equations such as (1.27) is called a *reaction-diffusion* system (see Exercise 3.3.4). The extension to population dynamics is straightforward. A more complicated situation takes place in fluid media since, in addition, one has to account for internal macroscopic motions. Keeping the fluid particle concept and the associated velocity field **v** defined earlier, p. 7, one thus has to replace the time derivative ∂_t by the material derivative $\partial_t + \mathbf{v} \cdot \nabla$, which yields a more general *reaction-diffusion-advection* system.

Accounting for spatial modulations of instability modes is much less straightforward in general, since these modes are coherent at the macroscopic scale and non-uniformity in physical space arises from a competition between them. Instability in a *confined* system with lateral extension $\ell \sim h$ (and a few cells of typical width h in it) can be described using a few scalar mode amplitudes X_i with the discrete index i serving to label different solution branches. The threshold corresponds to the most 'dangerous' mode and is usually isolated from the others which become unstable at much larger values of the control parameter. The Landau picture introduced in §1.3.2 holds with just one mode amplitude A governed by (1.19). In contrast, an *extended* system with $\ell \gg h$, is prone to the instability of many modes just differing by their exact number of cells, of the order of $\ell/h \gg 1$. Belonging to the same branch, these modes look alike except for the precise value of their spatial period $\lambda = 2\pi/k$. Modes periodic in the x direction can be labelled by wavevector $A(k)$, i.e. a mode in the form $A(k)\cos(k(x-x_0))$, where x_0 is a constant serving to fix the position of the pattern in an absolute frame of reference ($-kx_0 = \phi$ is the *phase* of that mode). Close to threshold the modes interfere and this interference takes the form of modulations that can be interpreted as beats. As a result, it becomes preferable to describe the beats themselves using the amplitude considered as a variable dependent on both time and space $A(x,t)$, rather than as a scalar just function of time $A(t)$ as in the confined case. Quantity $A(x,t)$ is thus an *envelope* that comes and modulates a background ideal periodic solution $\cos(k_c x)$, where $k_c = 2\pi/\lambda_c$ is the vavevector of the most unstable mode as introduced in Figure 1.3. One is then led to replace (1.19) by a partial differential equation of the reaction-diffusion type:

$$\partial_t A = \mathcal{F}_r(A) + \xi_0^2 \partial_{xx} A, \tag{1.28}$$

where ξ_0 is a *coherence length* measuring the propensity to diffuse modulations brought to the local value of amplitude A. Equation (1.28) is said to account for the spatial unfolding of the bifurcation of dynamical system $\mathcal{F}_r(A)$. Here it is introduced heuristically but it can be derived mathematically through *multiple scale expansion*. The structure of $\mathcal{F}_r$ depends on the nature of the instability, whether it involves spatial periodicity ($k_c \neq 0$) or not ($k_c = 0$), introduces some time dependence ($\omega_c \neq 0$) or not ($\omega_c = 0$). Rayleigh–Bénard convection is an example of *stationary* ($\omega_c = 0$), *cellular* ($k_c \neq 0$) instability and the BZ reaction an example of *homogeneous* ($k_c = 0$), *oscillatory* ($\omega_c \neq 0$) instability. In the case of convection, if the phase of the solution could be clamped (ϕ fixed) one could use a real variable and one would get (1.28) with $\mathcal{F}_r(A) = rA - gA^3$, the strict unfolding of (1.19) but, by definition, extended systems experience little influence of their lateral boundaries and the phase of the pattern cannot be clamped easily. It is then preferable to take the phase into account by considering A as a complex quantity, i.e. change $A(x,t) \in \mathbb{R}$ into $A(x,t) = |A(r,t)|\exp(i\phi(x,t)) \in \mathbb{C}$. After re-scaling time by τ_0/r, length by $\xi_0/\sqrt{r}$, A by $\sqrt{g/r}$, this yields a *universal* equation:

$$\partial_t A = A + \partial_{xx} A - |A|^2 A \qquad (1.29)$$

called the *Real Ginzburg–Landau equation* (RGLE). In the case of a uniform oscillatory instability the starting point is (1.21) and, again after appropriate changes,[10] the spatial unfolding reads:

$$\partial_t A = A + (1 + i\alpha)\partial_{xx} A - (1 - i\beta)|A|^2 A, \qquad (1.30)$$

which is the *Complex Ginzburg–Landau equation* (CGLE). Whereas solutions to the RGLE always behave simply and tend to time-independent states as t goes to infinity, solutions to the CGLE may have complex behaviour. In certain ranges of the parameter space (α, β), the CGLE offers good examples of *spatiotemporal chaos* in the form of *phase turbulence* when $|A|$ remains bounded away from zero, or *amplitude* (or *defect mediated*) *turbulence* when it is no longer the case. Here we have only evoked the anisotropic case where a direction x is singled out by the geometry or the instability mechanism, but the approach also extends to cases of higher effective spatial dimension. Snapshots of solutions to the CGLE in two-dimensions are displayed in figure 1.13.

The derivation of (1.28) is most often not performed and the equation just constructed from general considerations involving the nature of the bifurcation (oscillatory or not), its super- or sub-critical character and the symmetries of the medium, with some coefficients empirically adjusted

---
[10] In addition to re-scaling time, space and amplitude, the imaginary part of the growth rate s in the normal form of the Hopf bifurcation (1.21) has been eliminated through the change to a rotating amplitude $A \mapsto A \exp(i\omega t)$.

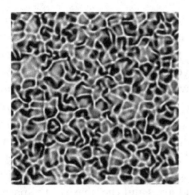 

Fig. 1.13 Complex spatiotemporal dynamics in the two-dimensional complex Ginzburg–Landau equation. Left: $\alpha = 2$ and $\beta = 1$; defects are located at the zeros of the complex amplitude A; in the *defect turbulence regime* illustrated in grey levels of $|A|$ black points correspond to the defects which permanently move erratically, merge, and split. Right: $\alpha = 2$ and $\beta = 0.2$; the *frozen spiral regime* is displayed using the real and imaginary parts of A; the spirals rotate around their cores that stay fixed; it gives a good image of spirals observed in thin layers of the BZ reacting mixture. After H. Chaté & P. Manneville, Physica A **224** (1996) 248–268.

from a few test experiments. The theory is next validated by comparing expectations and findings.

The soundness of the approach rests on a *universality* assumption similar to the one underlying the generality of the routes to chaos. The same belief supports more radical modelling practices in terms of assemblies of sub-systems coupled to each other and placed at the nodes of some network (*agent based modelling*). A simple example, patterned on the search for solutions to partial differential equations in terms of finite differences (§B.2), leads to the consideration of arrays of dynamical systems sitting at the nodes of a regular lattice and just coupled to other systems in their neighbourhood. A one-dimensional *coupled map lattice* (CML) for a single local state variable and nearest-neighbour interaction reads:

$$X_{n+1}^j = \mathcal{F}\left(X_n^j\right) + \left[g_{(+)}\mathcal{F}\left(X_n^{j+1}\right) + g_{(0)}\mathcal{F}\left(X_n^j\right) + g_{(-)}\mathcal{F}\left(X_n^{j-1}\right)\right], \quad (1.31)$$

with $g_{(+)} + g_{(0)} + g_{(-)} = 0$. Pure diffusive coupling corresponds to $g_{(+)} = g_{(-)} = g$ and $g_{(0)} = -2g$ from the discrete approximation of ∂_{xx}. Advection breaking the left-right symmetry implies $g_{(+)} \neq g_{(-)}$ as arising from a term in the form $U\partial_x$ which simply gives $g_{(-)} = -g_{(+)}$. Pattern formation and transition to spatiotemporal chaos in such systems follow the same broad routes as those in systems with continuous time and/or space, see [Kaneko (1993)].

One further step can be accomplished, namely by considering that local states no longer vary continuously but belong to discrete finite sets. *Cellular automata* (CA) are such systems. They evolve according to a rule defined on some neighbourhood $\mathcal{V}^j$ of each lattice node j [Wolfram (1986)]. For example, let us consider k-state automata with $k = 2$ (Boolean automata) on a one-dimensional lattice, $S_{n+1}^j = \mathcal{G}\left(S_n^{j-r}, \ldots, S_n^{j+r}\right)$ with $S = 0$ or 1, where r is the *range* of the coupling rule $\mathcal{G}$, which can be coded according to its action on given neighbourhood configurations. A general coding scheme proposed by Wolfram is illustrated below: The 2^{2r+1} possible configurations C_n^j at j are ordered according to their binary "values" and the resulting state of site j is read as a 2^{2r+1}-digit binary number, further translated in decimal. Here, for the rule formally defined as $S_{n+1}^j = S_n^{j-1} + S_n^j + S_n^{j+1}$ (mod 2),

C_n^j	111	110	101	100	011	010	001	000
S_{n+1}^j	1	0	0	1	0	1	1	0

one gets $\mathcal{R}_\text{g} = 10010110_2 = 150_{10}$, hence rule 150. All the concepts defined in the study of chaos and spatiotemporal chaos in dynamical systems can be adapted to the context of CA. Rules can be classified according to the long-term behaviour of the system from arbitrary initial conditions (concept of attractor) and difference patterns for neighbouring trajectories (concept of stability). Figure 1.14 below displays some typical cases.

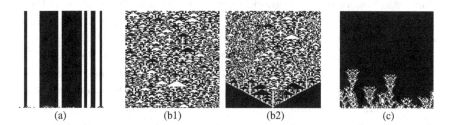

Fig. 1.14 Typical behaviour of 1-dimensional CA with $r = 2$ in Wolfram's classes II, III, and IV: (a) Class II, rule 56 (111000), regular periodic structures with persistent but localised difference patterns, equivalent to limit cycle behaviour. (b) Class III, rule 38 (100110), chaotic aperiodic pattern expanding at constant rate; sensitivity to initial conditions is illustrated by the difference pattern in (b2) between the solution in (b1) and another one issued from an initial condition differing only by the state of the middle site. (c) Class IV, rule 20 (010100), complex localised structures expanding irregularly in time. Periodic boundary conditions, $L = 128$, time n running upwards, $1 \rightarrow$'white', $0 \rightarrow$'black'.

Lattice-gas cellular automata of interest for hydrodynamic applications are special cellular automata describing the molecular dynamics of assemblies of particles with a finite number of states, moving and reacting on a lattice, which makes them popular models of complex media [Rothman and Zaleski (1997)]. Macroscopic (partial differential) evolution equations, e.g. the Navier–Stokes equation, can next be recovered by coarse-grain averaging.

It seems important to stress again that all the concepts and tools that have been sketchily introduced in previous sections and will be studied in more detail in the forthcoming chapters are applicable in the more complex situations described above.

1.4.3 *Phase transitions and critical phenomena*

Up to now, the description framework has remained deterministic. For macroscopic systems, this is acceptable only if we neglect microscopic fluctuations. The law of large numbers, which basically states that the intensity of fluctuations decreases as the inverse-square-root of number of particles, usually supports this assumption since the number of particles involved is large even if, at level of the fluid particle defined earlier (p. 7), it scales as tiny fraction of the Avogadro number. An unavoidable source of *noise* is thus thermal agitation, which brings us back to thermodynamics, and more specifically to phase transitions [Stanley (1988)].

As an example, for magnetic systems made of spins distributed on a regular lattice, short range magnetic interactions favour alignment and thermal agitation tends to destroy it. At high temperature, thermal agitation wins the competition and the system is in the *paramagnetic* state without spontaneous magnetisation. When the temperature T is decreased the ordering effect of local interactions is sufficient to bring the system in a *magnetised* state below some critical temperature T_c called the Curie temperature. The macroscopic magnetisation M is the *order parameter* of the transition. The application of an external magnetic field H gives another control on the magnetisation. Field H is said to be *conjugate* to the order parameter. The competition is expressed in terms of minimisation of a *free energy*, $\mathcal{F}(M, H, T)$, which can be expanded in powers of the departure to equilibrium: $\mathcal{F}(M, H, T) \simeq -\frac{1}{2}rM^2 + \frac{1}{4}gM^4 - MH$, where $r = (T_c - T)/T_c$ is the relative distance to the critical point, g is a coupling constant, and the last term measures the energetic contribution of the interaction with the conjugate field. Thermal fluctuations imply an exploration of the space of spin configurations. The minimisation process amounts to a relaxation of the order parameter toward its equilibrium value, governed by an equation that reads

$$\tau_0 \frac{d}{dt} M = -d\mathcal{F}/dM = rM - gM^3 + H \qquad (1.32)$$

where τ_0^{-1} plays the role of a *transport coefficient*. Let us first consider the case $H = 0$. When $T > T_c$, i.e. $r < 0$, the free energy has a single minimum at $M = 0$, whereas when $T < T_c$ ($r > 0$) there are two nontrivial absolute minima at $M = \pm\sqrt{r/g}$ and a relative maximum at $M = 0$, which accounts for the transition. Next consider the response to the conjugate field H written as $M = \chi H$ where χ is the *susceptibility*. Above the critical point ($r < 0$), the equilibrium response is $M = H/|r|$, so that χ diverges as $1/|r|$ as the critical temperature is approached from above ($r \to 0_-$). If the order parameter is not constant in space, its variations contribute to an increase of the free energy:

$$\mathcal{F}(M, H, T) = -\tfrac{1}{2} r M^2 + \tfrac{1}{4} g M^4 + \tfrac{1}{2} K (\boldsymbol{\nabla} M)^2 - MH \qquad (1.33)$$

where the additional term accounts for the reluctance of the system to accept spatial non-uniformity (K is related to the exchange interaction between spins which favours identical alignment). Quantity $\mathcal{F}$ is the Ginzburg–Landau free energy of the spin problem at hand. Minimisation of the free energy (assuming $H = 0$) yields

$$\tau_0 \partial_t M = rM - gM^3 + K \boldsymbol{\nabla}^2 M \qquad (1.34)$$

similar to (1.29). What has just been sketched is the *Landau theory* of second order transitions with an order parameter.[11] The Landau theory can also deal with 'first order' transitions that are discontinuous, provided that an order parameter can be defined. For the liquid–gas transition, the order parameter is the density difference between the liquid and the gaseous phases and the Van der Waals theory is recovered.

Equation (1.34) is the Ginzburg–Landau equation in its original sense and it is not a matter of chance that the same expression is used for (1.29). Equation (1.30) cannot be obtained from the minimisation of a potential but the extension of the terminology to a different, non-thermodynamic context, is quite natural.

The Landau theory neglects fluctuations: replacing fluctuating quantities by their averages, it is a *mean-field* theory. Close to threshold, the

---
[11] For a review, consult: L.P. Kadanoff *et al.*, "Static phenomena near critical points: theory and experiments," Rev. Mod. Phys. **39** (1967) 395–431.

order parameter varies as $M \propto r^\beta$, with $\beta = 1/2$ for $r > 0$. In the same way, the susceptibility diverges as $\xi \propto |r|^{-\gamma}$ with $\gamma = 1$ for $r < 0$. At the critical point ($r = 0$), the response of the system is given by $M = (H/g)^{1/\delta}$ with $\delta = 3$. From (1.34) one can also derive that typical space variations take place at a scale (coherence length) $\xi \propto r^{-\nu}$ with $\nu = 1/2$.

In the thermodynamic limit of infinitely large systems, at the critical point the free energy is no longer an analytic function and singularities characterised by exponents β, γ, δ, ν (and others) appear in the behaviour of physical observables. However, the situation is more complex and fluctuations cannot be neglected in general. As a result, the power law variations persist, but the exponents shift from their *classical* values given above to nontrivial values. These values depend on the system considered and, above all, on the symmetries of the order parameter[12] and the dimension of physical space. They belong to sets, one for each *universality class*; two systems belonging to the same class, with the same space dimension, have the same set of exponents. Exponents are related by scaling relations such as $\beta = \gamma/(\delta - 1)$, which are trivially fulfilled by the classical values. Other relations involve the dimension of space. Universality stems from the fact that long range coherence builds up in the system, in such a way that global properties depend little on the specificity of local interactions. Furthermore, when these local interactions average over a large number of neighbours, fluctuations may be neglected (again the law of large numbers), which explains that the Landau theory becomes correct above some *upper critical dimension*, d_c: critical exponents then take on their classical, mean-field value. For the simplest magnetic systems $d_c = 4$.[13]

At the microscopic level, thermodynamic systems at equilibrium fulfil a detailed balance principle.[14] Statistical physics of phase transitions also finds application in the more general context of far-from-equilibrium situations typical of instability phenomena, in which systems no longer fulfil detailed balance due to the occurrence of irreversible processes at some stage (dissipation, cut-offs,...). Examples are stochastic cellular automata that

---

[12]To stay with the magnetic example, the macroscopic magnetisation behaves as a vector with 1, 2, or 3 components, according to the existence of an easy direction (Ising model), an easy plane (XY model) or full isotropy (Heisenberg model), respectively.

[13]Physically relevant values are 1, 2, or 3, for linear, planar, or bulk systems, but larger (even infinite), or smaller values (even negative), are considered in theoretical physics.

[14]This principle expresses microscopic reversibility and says that if the transition from one configuration to another is possible with some probability, the reverse transition is also possible with the same probability weighted by an appropriate Boltzmann factor involving the energy difference between the configurations.

are formally close to quantum spin systems (local state can take on a finite number of values, typically two) that can be used to model contamination, percolation. Such systems are defined on lattices at the nodes of which subsystems can be in one of several possible states, and transitions between states are stochastic and depend on the neighbourhood configuration. Epidemics and forest fires can be modelled in this framework, the simplest implementation of which is *directed percolation*,[15] see Exercise 1.5.4.

1.4.4 Concluding remarks

Numerous books develop the topics to be touched upon. A partial list is given in the bibliography. At the risk of repetition, let us emphasise that we have introduced and shall develop in the forthcoming pages a very first approach to current problems, using a wide range of techniques, linear, non-linear, deterministic, statistical, each with its own qualities and limitations, in order to lucidly face complicated situations encountered in our familiar environment, industrial or natural. Accordingly, in the concluding chapter we shall evoke the problem of the Earth's climate as a concentration of the kind of topics examined at one moment or another, with special reference to the predictability problem and modelling issues involved in this nonlinear dynamical system with heterogeneous space-time scales. Appendix A recalls some elementary and not so elementary results of linear algebra including an introduction to control in state space, while Appendix B works in roughly the same direction by giving rudiments of numerical simulation techniques expected to be of help in the understanding of complex processes, provided that we are able to build simplified models with valuable metaphoric value as outlined in the previous sections.

1.5 Exercises

1.5.1 Diffusion equation

The one-dimensional diffusion equation, e.g. the heat equation (1.8), reads

$$\partial_t \vartheta = \kappa \partial_{xx} \vartheta \,. \tag{1.35}$$

Starting from some initial condition $\vartheta_0(x)$

1) determine by substitution the relation between the growth rate s and

---

[15] For a review, consult: H. Hinrichsen, "Non-equilibrium critical phenomena and phase transitions into absorbing states," Advances in Physics **49** (2000) 815–958.

the wavevector k of a periodic fluctuation taken in the form $\vartheta_k(x,t) = \bar{\vartheta}\exp(st)\exp(ikx)$ and derive its evolution as time goes on;

2) adding boundary conditions $\vartheta(0,t) = \vartheta(\ell,t) = 0$, solve the evolution problem formally for a solution starting with an initial condition that can be expanded as a sine series: $\vartheta(x,0) = \sum_{n=1}^{\infty} \vartheta_n \sin(\pi n x/\ell)$; extract its asymptotic solution in the limit $t \to \infty$;

3) going back to the infinite medium, consider functions ϑ that are taken in the form

$$\vartheta(x,t) = \exp(\alpha w(x,t)). \qquad (1.36)$$

Derive the partial differential equation governing $w(x,t)$, and next that governing $v = \partial_x w$; find the value of α bringing this equation in the form

$$\partial_t v + v\partial_x v = \kappa \partial_{xx} v \qquad (1.37)$$

called the *Burgers equation*.[16]

1.5.2 The Brusselator

We consider a simplistic kinetic model introduced by Prigogine and Lefever in 1968, called the Brusselator and accounting for a hypothetical reaction between two components A and B with two end products C and D, four steps and two intermediate compounds X and Y. Concentrations A and B of reactants A and B are the control parameters, whereas those X and Y of the intermediate species X and Y are the variables. The global reaction:
   A + B   →   C + D
can be decomposed into successive steps:
   A   →   X,   B + X   →   Y + C,
   2 X + Y   →   3 X,   X   →   D.

Assuming that the kinetic constants of each reaction is equal to 1, write down the equations for each step and show that the reaction is governed by

$$\tfrac{d}{dt}X = A - (B+1)X + X^2 Y, \qquad \tfrac{d}{dt}Y = BX - X^2 Y$$

This system will be further studied in Exercise 4.4.7.

1.5.3 The SIR model of epidemic

Let us divide the population into three classes: 'Susceptible' individuals (S, variable S), capable of contracting a disease, become 'Infectious' (I, variable

---

[16] J.M. Burgers, *The nonlinear diffusion equation* (Reidel, 1974), see U. Frisch and J. Bec, "Burgulence," in [Lesieur *et al.* (2001)] for a review.

1. Introduction and Overview

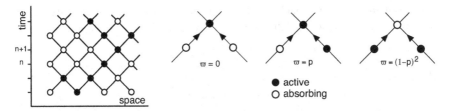

Fig. 1.15 Directed percolation in 1+1 dimensions. Left: Space is along the horizontal and time along the vertical. Right: The state at site j and time $n+1$ depends on the configurations of its two parents at time n and the 'contamination' probability p.

I), thus capable of transmitting the disease. They are next 'Removed' from the system (R, variable R), having recovered and being immunised. The total number $S+I+R$ is constant. Assume that I increases proportionally to the number of contacts IS, with α_i the corresponding proportionality constant and decreases due to recovery at a rate α_r. By applying the "chemical reaction" rules show that one is led to

$$\tfrac{d}{dt}S = -\alpha_i SI\,, \qquad \tfrac{d}{dt}I = \alpha_i SI - \alpha_r I\,, \qquad \tfrac{d}{dt}R = \alpha_r I\,.$$

An epidemic occurs if I increases with time. Discuss its occurrence as a function of the initial condition I_0, assuming $R_0 = 0$ (and thus $S_0 = N - I_0$). Show that $I \to 0$ for t large enough.

More complicated models may involve delays at the contamination and recovery stages or individuals may be susceptible again after having recovered (SIS models), etc.

1.5.4 One-dimensional directed percolation

Directed percolation is a stochastic process defined on a d-dimensional lattice, one direction of which is singled out and will correspond to the time direction. Here we consider the one-dimensional case where $d = 2 = 1 + 1$. Let S_n^j be the state at node j and time n, either 'active' $= 1$ or 'absorbing' $= 0$, see Figure 1.15. The evolution rule $S_{n+1}^j = \mathcal{G}(S_n^j, s_n^{j+1})$ is assumed to be *totalistic*, i.e. the output only depends on $\Sigma_j = S_j + S_{j+1}$. It is defined as follows: (a) $\mathcal{G}(\Sigma = 0) = 0$ with probability 1 (absorbing if both parents are in absorbing state); (b) $\mathcal{G}(\Sigma = 1) = 1$ with probability p (active with probability p if one parent is active).

Verify that $\mathcal{G}(\Sigma = 1) = 0$ with probability $1 - p$, $\mathcal{G}(\Sigma = 2) = 0$ with probability $(1-p)^2$ and $\mathcal{G}(\Sigma = 2) = 1$ with probability $1 - (1-p)^2$.

The mean field approach replaces variables S_j by their spatial averages $\bar{S}$. Show that this yields a discrete time dynamical system

$$\bar{S}_{n+1} = 2p\bar{S}_n + (1 - (1-p^2))\bar{S}_n^2$$

Equilibrium states correspond to fixed points $\bar{S}_{n+1} = \bar{S}_n = \bar{S}_*$ of this iteration. Show that the fraction $\bar{S}_*$ is non-zero provided that $p > p_\mathrm{c} = 0.5$ and that it grows as $(p - p_\mathrm{c})^\beta$ with $\beta = 1$ close to p_c. With the rules defined above, directed percolation in one (space) dimension is a second order phase transition.

The mean-field prediction neglects fluctuations which are however not negligible. The actual values of the critical probability is $p_\mathrm{c} \approx 0.6445$ and exponent β governing the growth of the order parameter is shifted to ≈ 0.277.

Chapter 2

First Steps in Nonlinear Dynamics

In this chapter we consider the dynamics of systems with a very small number of variables, as a prerequisite to the study of chaos, postponed to Chapter 4 after an examination of convection in Chapter 3, which will give us a physical motivation.

2.1 From Oscillators to Dynamical Systems

2.1.1 *First definitions*

Newtonian mechanics is the archetype of deterministic dynamical theories. Governed by an equation of the form

$$m \frac{d^2}{dt^2} X = F, \tag{2.1}$$

where X represents the position of a particle with mass m submitted to a force F, it accounts for processes that are invariant under a change of the arrow of time.

A traditional example of linear system is the *harmonic oscillator* (Fig. 1.1, left, p. 2) that describes the motion of a mass attached to an ideal spring with a restoring force proportional to the extension $F = -kX$ (Hooke's law). This elastic force, of internal origin, derives from a potential:

$$F = -\frac{\partial \mathcal{V}}{\partial X} \qquad \text{with} \qquad \mathcal{V} = \tfrac{1}{2} k X^2$$

depicted in Figure 2.1 (left), next page. Here we have

$$m \frac{d^2}{dt^2} X + k X = 0. \tag{2.2}$$

The evolution is uniquely determined when initial conditions are specified:

$$X = X^{(0)} \qquad \text{and} \qquad \tfrac{d}{dt} X = V^{(0)} \qquad \text{at} \quad t = 0.$$

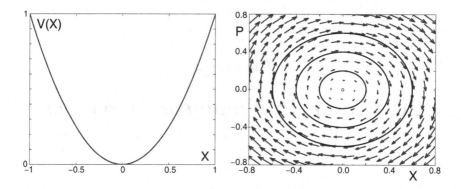

Fig. 2.1 Left: Potential for the harmonic oscillator. Right: Vector field and trajectories in the *phase space*, the plane of variables $X_1 \equiv X$ and $X_2 \equiv P$. Here the vector field is obtained using the MATLAB macro 'quiver', explicitly: `X=[-0.8:0.1:0.8]; U=ones(1,size(X,2)); P=X'; FX=P*U; FP=-U'*X; quiver(X,P,FX,FP)`.

The solution then reads:
$$X(t) = X^{(0)} \cos(\omega_0 t) + \left(V^{(0)}/\omega_0\right) \sin(\omega_0 t) \quad \text{with} \quad \omega_0^2 = k/m\,.$$

Right now, it turns out advantageous to substitute a geometrical description to this analytical formulation. So, let us consider *trajectories* in the phase space where, as shown in Figure 2.1 (right), coordinates are the *position* X and the *momentum*[1]
$$P = mV = m\tfrac{\mathrm{d}}{\mathrm{d}t} X\,. \tag{2.3}$$

Equation (2.2) now reads
$$\tfrac{\mathrm{d}}{\mathrm{d}t} P = -kX\,. \tag{2.4}$$

One can trace back the origin of this representation from the need to turn a high order differential equation into a first order differential system. On general grounds, this is done by introducing the successive derivatives as intermediate variables. For example, equation:
$$\mathrm{d}^n X/\mathrm{d}t^n = \mathcal{F}\!\left(\mathrm{d}^{n-1}X/\mathrm{d}t^{n-1}X, \ldots, \tfrac{\mathrm{d}^2}{\mathrm{d}t^2}X, \tfrac{\mathrm{d}}{\mathrm{d}t}X, X; t\right) \tag{2.5}$$

is canonically reduced to a first order system by setting $X_1 \equiv X$, $X_2 \equiv \tfrac{\mathrm{d}}{\mathrm{d}t}X, \ldots, X_n \equiv \mathrm{d}^{n-1}X/\mathrm{d}t^{n-1}$, which yields:
$$\tfrac{\mathrm{d}}{\mathrm{d}t}X_1 = X_2\,, \quad \tfrac{\mathrm{d}}{\mathrm{d}t}X_2 = X_3\,, \ldots, \quad \tfrac{\mathrm{d}}{\mathrm{d}t}X_n = \mathcal{F}(X_n, \ldots, X_3, X_2, X_1; t)\,.$$

---

[1] See §2.1.2 for a more general but brief introduction to the formalism of analytical mechanics.

This operation is preliminary to any numerical implementation in view of simulations (Appendix B).

There is a slight disadvantage to start from mechanics to introduce basic concepts of the theory of dynamical systems since Newton's equations endow the phase space with a specific structure and give it an even number of dimensions, $d = 2n$. As seen above, this space is indeed constructed as the product of the *configuration space* (variables X_i, $i = 1, \ldots, n$) and of the *momentum space* (variables P_i, $i = 1, \ldots, n$), each pair of *conjugate variables* (X_i, P_i) forming a *degree of freedom*. In the general case when the state of a system is specified using a supposedly sufficient number of variables, no longer grouped by pairs, we shall avoid the expression 'degree of freedom' and call any of these variables a *state variable*. Accordingly, the *state space* will then be the space where these variables live, and the dimension of this space will just be their number. A way to avoid ambiguities is thus to restrict the use of terms "phase space" and "degree of freedom" to mechanics in a strict sense and to use *state variables* and *state space* in all other cases, a practice we shall try to stick to.

So, let us consider the general case of a state space $\mathbb{X}$ with dimension d spanned[2] by d variables $\{X_1, \ldots, X_d\} \equiv \mathbf{X}$. The evolution of the so-defined variables is governed by a system that symbolically reads

$$\tfrac{\mathrm{d}}{\mathrm{d}t}\mathbf{X} = \boldsymbol{\mathcal{F}}(\mathbf{X}; t), \qquad (2.6)$$

where $\boldsymbol{\mathcal{F}}$ is a set of d functions representing the components of a *vector field* defined on $\mathbb{X}$ that specifies the "velocity" of the point representing the system in that space. In agreement with the intuitive concept of *determinism*, the dimension d is also exactly the number of conditions necessary to specify any evolution uniquely. When the properties of $\boldsymbol{\mathcal{F}}$ guarantee the existence and uniqueness of the solution to the initial value problem, in practice when the vector field $\boldsymbol{\mathcal{F}}$ is $\mathcal{C}^1$ (differentiable with continuous first derivatives), one says that it defines a *flow*.

When t is explicitly absent from the definition of $\boldsymbol{\mathcal{F}}$, the system is said to be *autonomous*, otherwise it is *forced*. In practice, among forced systems, only periodically forced systems will be of interest to us, i.e. systems such that $\boldsymbol{\mathcal{F}}(\mathbf{X}; t+T) \equiv \boldsymbol{\mathcal{F}}(\mathbf{X}; t)$ for some minimal time interval T called the *period*. Within this class one often distinguishes *parametric forcing*, for which the expression of $\boldsymbol{\mathcal{F}}$ changes in time, e.g. the parametric linear oscillator (Mathieu equation):

$$\tfrac{\mathrm{d}^2}{\mathrm{d}t^2} X + (1 + a\sin(\omega t))X = 0, \qquad T = 2\pi/\omega,$$

---

[2] Space $\mathbb{X}$ can thus be simply $\mathbb{R}^d$ or a more complicated d-dimensional set embedded in $\mathbb{R}^{d'}$ with $d' > d$.

from *external forcing* where an otherwise autonomous system is submitted to a periodic force independent of its state, i.e.

$$\tfrac{d^2}{dt^2} X + X = f \sin(\omega t) \, .$$

Noisy systems can be understood as particular forced systems with a *random* forcing. In case of additive noise, this defines the so-called *Langevin equation*

$$\tfrac{d}{dt} \mathbf{X} = \mathcal{F}(\mathbf{X}) + \mathbf{\Xi}(t) \, ,$$

where $\mathbf{\Xi}(t)$ is a random vector function. Quite different tools of statistical essence are then required which will not be introduced here since we mostly want to stick to the deterministic point of view.

During its evolution, the system follows a state space *trajectory* starting at $\mathbf{X}^{(0)}$ when $t = t^{(0)}$ and obtained by integration of (2.6):

$$\mathbf{X}(t) = \mathbf{X}^{(0)} + \int_{t^{(0)}}^{t} \mathcal{F}(\mathbf{X}(t'); t') \, dt' \, . \tag{2.7}$$

The *orbit* is the set of points in $\mathbb{X}$ visited by the system in the course of a given trajectory. The description of a system's dynamics in terms of sets of orbits, called its *phase portrait*, as a function of its *control parameters*, is the field of *qualitative dynamics*.

Relation (2.7) allows us to define a map of $\mathbb{X}$ onto itself. Upon specifying an integration time τ, we get a *time-τ map*

$$\mathbf{X}(t+\tau) \equiv \mathbf{\Phi}_\tau(\mathbf{X}(t)) = \mathbf{X}(t) + \int_{t}^{t+\tau} dt' \, \mathcal{F}(\mathbf{X}(t'); t') ,$$

and, starting with an initial condition $\mathbf{X}^{(0)}$, we obtain a discrete sampling of the trajectory, $\mathbf{X}_0 \equiv \mathbf{X}^{(0)}$, $\mathbf{X}_1 = \mathbf{\Phi}_\tau(\mathbf{X}_0)$, $\mathbf{X}_2 = \mathbf{\Phi}_\tau(\mathbf{X}_1), \ldots, \mathbf{X}_{k+1} = \mathbf{\Phi}_\tau(\mathbf{X}_k) \ldots$

The time-τ map is really interesting only when τ corresponds to some characteristics of the system. The most important case corresponds to periodic forcing with period $T = \tau$, in which case $\mathbf{\Phi}_T$ performs a *stroboscopic analysis* of the dynamics, i.e. takes pictures of the system at the period of a strobe signal with a specific phase relation with the forcing. Later we shall see another related way to arrive at such *discrete-time* systems as already mentioned in Chapter 1, p. 3. The concepts of trajectories and phase portraits transpose immediately to systems written as first-order iterations $\mathbf{X}_{k+1} = \mathbf{\Phi}(\mathbf{X}_k; k)$.

From a practical viewpoint the determination of the trajectory issued from some initial condition $\mathbf{X}^{(0)}$ by numerical integration can be viewed as resulting from the iteration of a map Φ_τ integrating the field $\mathcal{F}$ over a time interval $\tau = \Delta t$ chosen for accuracy considerations specific to the numerical scheme (cf. Appendix B).

For the moment, let us consider autonomous systems and especially the simple case of the harmonic oscillator. Its state space is the phase space (X, P) isomorphic to $\mathbb{R}^2$. The right panel of Figure 2.1, p. 40, displays the corresponding vector field. In this representation, trajectories follow elliptic orbits.

On general grounds it turns out useful to scale the variables as much as possible in order to cast the system into its most universal form, hiding its specificity inside the details of the variable change. Here such a transformation simply yields

$$\tfrac{\mathrm{d}}{\mathrm{d}t} X_1 = X_2, \qquad \tfrac{\mathrm{d}}{\mathrm{d}t} X_2 = -X_1. \tag{2.8}$$

Reversibility is one of the fundamental characteristics of ideal mechanical systems (i.e. without friction). The change '$t \mapsto -t$' indeed leaves the Newton equations invariant. This property is associated to energy conservation. Defining $E = \tfrac{1}{2} X_1^2 + \tfrac{1}{2} X_2^2$ and computing $\mathrm{d}E/\mathrm{d}t$ using (2.8) one can check that E is a constant of motion. Things are different for a damped system that *dissipates* its energy. At a linear stage, the introduction of a *viscous friction* proportional to the rate of change of the variable leads to

$$\tfrac{\mathrm{d}^2}{\mathrm{d}t^2} X + 2\eta \tfrac{\mathrm{d}}{\mathrm{d}t} X + X = 0, \tag{2.9}$$

where $\eta > 0$ measures the strength of the damping. Here the change $t \mapsto t/\omega_0$ has been performed in order to normalise the period of the ideal oscillator to 2π.

Trajectories are obtained in parametric form as

$$X(t) = \bar{X} \exp(-\eta t) \cos(\omega t - \varphi),$$

where $\bar{X}$ and φ can be computed from the initial conditions $X^{(0)}$ and $\tfrac{\mathrm{d}}{\mathrm{d}t} X^{(0)}$ at $t = 0$ by identification, and where $\omega = (1 - \eta^2)^{1/2}$ corresponds to the angular frequency of a damped oscillation only when the friction is sufficiently weak, i.e. when $\eta < 1$ (the over-damped case $\eta > 1$ will be considered as an exercise). In the phase space, equation (2.9) reads

$$\tfrac{\mathrm{d}}{\mathrm{d}t} X_1 = X_2, \qquad \tfrac{\mathrm{d}}{\mathrm{d}t} X_2 = -2\eta X_2 - X_1, \tag{2.10}$$

and the orbits now take the aspect of spirals converging towards the origin, see Figure 2.2 (left) that displays the corresponding vector field. The instantaneous dissipation rate of the total energy, still defined as $E = \tfrac{1}{2}(X_1^2 + X_2^2)$, is now given by $\mathrm{d}E/\mathrm{d}t = -2\eta X_2^2 < 0$.

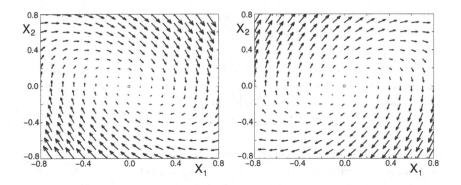

Fig. 2.2 Vector field of the non-ideal linear oscillator in its phase space (X_1, X_2) for $\eta = 0.5$. Left: Damped oscillator ($\eta > 0$). Right: Driven oscillator ($\eta < 0$).

For an excited system that would receive energy from the exterior world by viscous *driving*, thus governed by (2.9) but with $\eta < 0$, the integration of the vector field would give diverging spiral trajectories easily imagined from the vector field depicted in Figure 2.2 (right).

In fact, when defining a state physically, one never considers a mathematical point in state space but rather a small, physically infinitesimal domain around this point. This remark suggests we focus on the future of sets of systems that at a given time belong to volume elements in state space (surface elements in the specific two-dimensional case of interest here). For a general dynamical system (2.6) defined on a d-dimensional state space $\mathbb{X}$, one then shows that the local evolution rate of the volume Δ of an in-

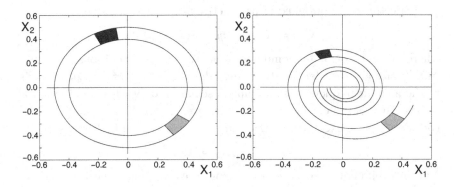

Fig. 2.3 Evolution of volumes in state space. Left: Conservative system. Right: Dissipative system.

finitesimal domain around some point **X** is given by the divergence of the vector field $\mathcal{F}$ computed at this point (cf. Exercise 2.5.1):

$$\frac{1}{\Delta}\frac{d\Delta}{dt} = \text{div}\,\mathcal{F} = \sum_{j=1}^{d} \frac{\partial \mathcal{F}_j}{\partial X_j}. \qquad (2.11)$$

For an ideal oscillator (2.8), one immediately obtains that the divergence is zero: the surface of a small surface element is thus conserved (Fig. 2.3, left). This property can be generalised to all frictionless mechanical systems, that are called *conservative* for that reason. The same calculation for the damped oscillator leads to div $\mathcal{F} = -2\eta < 0$, thus to an indefinite erosion of areas (Fig. 2.3, right). This reduction is characteristic of *dissipative* systems whose permanent regimes, asymptotic states at the limit $t \to \infty$ after the damping out of transients, are described by *attractors*.

As suggested above, the volumes in state space give a measure of the number of accessible states. Asymptotically the damped oscillator always ends at its rest position $X_1 = X_2 \equiv 0$, a single state whatever the initial energy. The attractor is here a *fixed point* in state space, 'fixed point' because the orbit of the trajectory starting there is just reduced to that point. In anticipation, we can say that it is a *stable* fixed point since trajectories starting in its vicinity converge to it as t tends to infinity. In contrast, the fixed point at the origin of the state space associated to an driven oscillator is not an attractor but a *repeller*, it is *unstable* and trajectories depart from it.

General systems are not conservative. They can be fuelled with energy in some regions of their state space while dissipating it in other regions. This feature, that cannot be achieved in the framework of linear systems with constant coefficients (sign of div $\mathcal{F}$ fixed once for all), will turn out to be essential to the existence of self-sustained oscillations and more complicated time behaviour.

2.1.2 *Formalism of analytical mechanics*

In order to ease the solution to some exercises, we give here an introduction to the analytical formalism of classical mechanics that allows one to pass from the Newton equations (second order in time) to the Hamilton equations (first order in time) giving to the intermediate variables so-introduced their status of conjugate momenta to the generalised coordinates.

For a system of n_N (subscript 'N' for Newton) material points with masses m_i and positions $\mathbf{X}_i \equiv (x_i, y_i, z_i)$ submitted to forces $\mathbf{f}_i$ deriving from a potential $\mathcal{V}(\{\mathbf{X}_i\};t)$, the Newton equations read:

$$m_i \frac{\mathrm{d}^2}{\mathrm{d}t^2}\mathbf{X}_i = \mathbf{f}_i = -\frac{\partial \mathcal{V}}{\partial \mathbf{X}_i}, \quad (i=1,\ldots,n_\mathrm{N}). \tag{2.12}$$

The kinetic energy is defined by:

$$\mathcal{T} = \frac{1}{2}\sum_{i=1}^{n_\mathrm{N}} m_i \left(\tfrac{\mathrm{d}}{\mathrm{d}t}\mathbf{X}_i\right)^2,$$

and one easily checks that the total energy

$$E = \mathcal{T} + \mathcal{V}$$

is conserved. But this so-called *Newtonian formulation* makes the treatment of systems with constraints somehow awkward. The *Lagrangian formalism* answers this problem by introducing n_L ('L' for Lagrange) *generalised coordinates*.[3] Let

$$\mathbf{q} \equiv \{q_j; j=1,\ldots,n_\mathrm{L}\}, \quad q_j = q_j(\{\mathbf{X}_i; i=1,\ldots,n_\mathrm{N}\};t), \tag{2.13}$$

be the change of variables from the $\mathbf{X}_i$s to the q_js expressing the constraints. The Lagrangian is then defined by

$$\mathcal{L} = \mathcal{L}(\mathbf{q},\dot{\mathbf{q}}) = \mathcal{T} - \mathcal{V},$$

so that (2.12) can be rewritten as a set of Lagrange equations:

$$\frac{\mathrm{d}}{\mathrm{d}t}\frac{\partial \mathcal{L}}{\partial(\tfrac{\mathrm{d}}{\mathrm{d}t}q_j)} - \frac{\partial \mathcal{L}}{\partial q_j} = 0, \quad j=1,\ldots,n_\mathrm{L}. \tag{2.14}$$

At this stage, the system is still governed by second order differential equations. One then defines the *momenta* $\mathbf{p} \equiv \{p_j\}$ conjugated to the coordinates $\mathbf{q} \equiv \{q_j\}$ by:

$$p_j \equiv \frac{\partial \mathcal{L}}{\partial(\tfrac{\mathrm{d}}{\mathrm{d}t}q_j)}, \tag{2.15}$$

and the Hamiltonian by:

$$\mathcal{H}(\mathbf{q},\mathbf{p}) = \sum_j p_j \tfrac{\mathrm{d}}{\mathrm{d}t} q_j - \mathcal{L}. \tag{2.16}$$

The dynamical equations (2.14) then turn into the *Hamilton equations*:

$$\tfrac{\mathrm{d}}{\mathrm{d}t} q_j = \frac{\partial \mathcal{H}}{\partial p_j}, \quad \tfrac{\mathrm{d}}{\mathrm{d}t} p_j = -\frac{\partial \mathcal{H}}{\partial q_j}, \quad j=1,\ldots,n_\mathrm{L}. \tag{2.17}$$

---

[3]$n_\mathrm{L} < n_\mathrm{N}$ and sometimes $\ll n_\mathrm{N}$, think of a solid body with many (n_N) rigidly linked particles and characterised just by the position of its centre of mass (three coordinates) and its orientation (three Euler angles).

The phase space $\mathbb{X}$ is the Cartesian product of the configuration space with coordinates q_j and the momentum space with coordinates p_j, thus with dimension $d = 2n_\mathrm{L}$. The vector field governing the dynamics then reads

$$\mathcal{F}_{q_j} \equiv \frac{\partial \mathcal{H}}{\partial p_j} \quad \text{and} \quad \mathcal{F}_{p_j} \equiv -\frac{\partial \mathcal{H}}{\partial q_j}, \tag{2.18}$$

expressions that provide it with a so-called 'symplectic' structure ensuring the conservation of phase space volumes automatically (div $\mathcal{F} \equiv 0$, the Liouville theorem).

2.1.3 Gradient systems

Above, we spoke of "forces deriving from a potential" in a strictly mechanical framework. Unfortunately, there is a risk of confusion when using the word "derive" and "potential" without care, because in some branches of dynamical systems theory, they have a somewhat different meaning that now warrants specification.

To do so, let us first consider an autonomous systems with a single real variable X:

$$\tfrac{\mathrm{d}}{\mathrm{d}t} X = \mathcal{F}(X). \tag{2.19}$$

Defining $\mathcal{G}$ from $\mathcal{F}$ by

$$\mathcal{G}(X) = -\int \mathcal{F}(X)\,\mathrm{d}X,$$

we can rewrite (2.19) as:

$$\tfrac{\mathrm{d}}{\mathrm{d}t} X = -\mathrm{d}\mathcal{G}/\mathrm{d}X, \tag{2.20}$$

where $\mathcal{G}$ presents itself as a potential from which the vector field $\mathcal{F}$ can be derived. In higher dimensions, the natural extension of (2.20) for a function of several variables $\mathcal{G}(X_1,\ldots,X_d)$ through

$$\tfrac{\mathrm{d}}{\mathrm{d}t} X_j = \mathcal{F}_j(X_1,\ldots,X_d) = -\partial\mathcal{G}/\partial X_j, \quad j = 1,\ldots,d, \tag{2.21}$$

defines a large class of systems called *gradient flows*. It is also said that such systems are relaxational. The origin of the latter term is to be found in the remark that (2.20) immediately leads to $\mathcal{G}$ being a monotonically decreasing function of time:

$$\tfrac{\mathrm{d}}{\mathrm{d}t}\mathcal{G} = (\mathrm{d}\mathcal{G}/\mathrm{d}X)\tfrac{\mathrm{d}}{\mathrm{d}t} X = -(\mathrm{d}\mathcal{G}/\mathrm{d}X)^2 \leq 0. \tag{2.22}$$

Accordingly, the system evolves from almost all initial conditions so as to relax toward one of the local minima of $\mathcal{G}$ where its stops asymptotically.

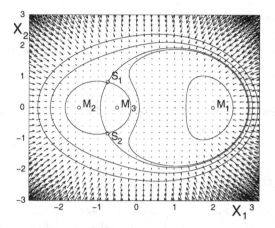

Fig. 2.4 For a gradient flow, the vector field $\mathcal{F}$ is everywhere perpendicular to the level curves of the potential $\mathcal{G}$ from which it derives.

In the d-dimensional case, the vector field $\mathcal{F}$ is seen from (2.21) to be everywhere perpendicular to the level curves of $\mathcal{G}$ so that:

$$\tfrac{\mathrm{d}}{\mathrm{d}t}\mathcal{G} = \sum_j (\partial\mathcal{G}/\partial X_j)\tfrac{\mathrm{d}}{\mathrm{d}t}X_j = -\sum_j (\partial\mathcal{G}/\partial X_j)^2 \le 0, \qquad (2.23)$$

which again expresses the relaxation toward one of its local minima.

Figure 2.4 displays the level lines of a two-dimensional potential[4] $\mathcal{G} = -aX_1 - \tfrac{1}{2}(bX_1^2 + cX_2^2) + \tfrac{1}{4}(X_1^2 + X_2^2)^2$, with $a = 3/2$, $b = 13/4$, $c = 5/4$. The absolute minimum $\mathcal{G} = -5.5$ is at $\mathbf{M}_1 = (2,0)$. There is a relative minimum $\mathcal{G} \simeq -0.14$ at $\mathbf{M}_2 = (-3/2,0)$, a relative maximum $\mathcal{G} \simeq 0.36$ at $\mathbf{M}_3 = (-1/2,0)$, while points $\mathbf{S}_1$ and $\mathbf{S}_2$ at $(-3/4, \pm 0.83)$, later called *saddle points*, belong to the level line $\mathcal{G} = 0.171875\dots$.

In mathematics, gradient flows appear in the theory of elementary *catastrophes*, see p. 137 and [Poston and Stewart (1978)]. In thermodynamics, they offer a good framework for the Landau theory of *phase transitions*, see §1.4.3, especially p. 33, and also [Stanley (1988)].

Now, if $\tfrac{\mathrm{d}}{\mathrm{d}t}\mathbf{X} = \mathcal{F}(\mathbf{X})$ derives from a potential $\mathcal{G}$ in the sense of (2.21), the components of $\mathcal{F}$ fulfil the relations:

$$\partial\mathcal{F}_j/\partial X_{j'} = \partial\mathcal{F}_{j'}/\partial X_j, \qquad \forall j, j', \qquad (2.24)$$

as results from the Schwartz identity:

$$\partial^2\mathcal{G}/\partial X_j \partial X_{j'} = \partial^2\mathcal{G}/\partial X_{j'}\partial X_j.$$

---

[4] For more detail, see P. M. and L. Tuckerman, "Phenomenological modeling of the first bifurcations of the spherical Couette flow," J. Physique **48** (1987) 1461–1469.

Conditions (2.24) are consequently necessary to the existence of a potential. In the general case, the components of the vector field $\mathcal{F}$ do not fulfil such conditions and one can expect an evolution that is richer than a mere relaxation towards a local minimum of some putative $\mathcal{G}$. This is of course the case of ideal mechanical systems for which the total energy is a constant of the motion so that the vector field given by the Hamilton equations is everywhere parallel to the surfaces of constant total energy in phase space (Figure 2.1, right). For profound reasons to be discussed later, we will have to wait for the study of three-dimensional systems before any dynamics more complicated than a relaxation towards fixed points or periodic oscillations can be observed.

2.2 Stability and Linear Dynamics

The first piece of information of interest about the regime attained by a given system under specific conditions relates to its stability, i.e. the way it reaches the state and responds to perturbations. In practice one can only exceptionally determine the response to perturbations of arbitrary amplitude. Except for Lyapunov functions generalising the potentials introduced in §2.1.3 (see also Exercise 2.5.2), we have no tools to attack the problem in full generality and we must restrict ourselves to a study of the evolution of infinitesimal perturbations, for which all the resources of *linear* analysis are available.

2.2.1 *Formulation of the linear stability problem*

In what follows, we are mainly interested in specific time-independent states of autonomous systems:

$$\tfrac{\mathrm{d}}{\mathrm{d}t}\mathbf{X} = \mathcal{F}(\mathbf{X}). \tag{2.25}$$

These states are thus solutions to:

$$\mathcal{F}(\mathbf{X}_\mathrm{f}) = 0 \tag{2.26}$$

and are represented by *fixed points* in state space, hence the subscript 'f' serving to denote them. Equation (2.26) is the formal writing of a system of d nonlinear equations with d unknowns that has a discrete and finite set of solutions in general.

In out-of-equilibrium macroscopic systems, especially continuous media, it is natural to first consider the solution that belongs to the *thermodynamic branch*, defined as the branch of solutions that can be followed from thermodynamic equilibrium by continuity, but what will be said also holds for any other time-independent solution, even if it is more difficult to obtain.

Let $\mathbf{X}_f$ be the fixed point of interest. Inserting $\mathbf{X} = \mathbf{X}_f + \mathbf{X}'$ in (2.25), one expands that equation in powers of $\mathbf{X}'$. Noting that order 0 is identically fulfilled by the fixed point condition and keeping only first order terms, one gets:[5]

$$\tfrac{\mathrm{d}}{\mathrm{d}t}\mathbf{X}' = \mathcal{L}\mathbf{X}'. \tag{2.27}$$

The (Jacobian) operator $\mathcal{L}$ resulting from this linearisation is represented by a matrix with elements

$$a_{ij} = \partial \mathcal{F}_i / \partial X_j \big|_{\mathbf{X}_f},$$

all partial derivatives being evaluated at the fixed point $\mathbf{X}_f$ as indicated by the notation. In this context, the linear stability problem amounts to an integration of equation (2.27) and thus to the evaluation of the action of operator $\exp(t\mathcal{L})$ on some initial condition $\mathbf{X}'^{(0)}$. Infinitesimal perturbations are said to live in the *tangent space* at the fixed point. Their evolution is governed by the *tangent dynamics* (2.27).

A simple way to justify the introduction of the exponential of operator $\mathcal{L}$ consists in considering the scalar equation $\tfrac{\mathrm{d}}{\mathrm{d}t}X = aX$ with $X(0) = X^{(0)}$, in writing recursively

$$X(t) = X^{(0)} + \int_0^t aX(t')\mathrm{d}t'$$
$$= X^{(0)} + \int_0^t a\left[X^{(0)} + \int_0^{t'} aX(t'')\mathrm{d}t''\right]\mathrm{d}t' = \ldots,$$

and in integrating explicitly all what can be integrated. This yields:

$$X(t) = X^{(0)} + atX^{(0)} + \frac{1}{2}a^2 t^2 X^{(0)} + \cdots + \frac{1}{n!}a^n t^n X^{(0)} + \cdots = \exp(at)X^{(0)}.$$

The extension of this approach to solving (2.27) involves what is precisely defined as the exponential $\exp(t\mathcal{L})$, i.e. the limit of a power series. In general, the solution of (2.27) rests on turning $\mathcal{L}$ to its diagonal form, or more precisely to its Jordan normal form. See §A.2, for a reminder.

---

[5] In contrast with the case of nonlinear operator written as $\mathcal{F}(\mathbf{X})$, for linear operator we use the notation $\mathcal{L}\mathbf{X}$. Furthermore, we do not distinguish the operator from the matrix that represents it in a given basis. With little ambiguity $\mathcal{L}$ thus denotes either the operator or the matrix with elements $a_{jj'}$. In developed form, the equation $\mathbf{Y} = \mathcal{L}\mathbf{X}$ then read: $Y_j = \sum_i a_{ji} X_i$. Some elements of linear algebra are recalled in Appendix A.

2.2.2 Two-dimensional linear systems

We first turn to the case of two variables since it contains the essentials of nontrivial aspects of the problem and allows us to introduce the core of the terminology. The extension to dimension d will be alluded to in the next subsection. Thus consider the linear dynamical system:

$$\tfrac{\mathrm{d}}{\mathrm{d}t} X_1 = a_{11} X_1 + a_{12} X_2, \tag{2.28}$$

$$\tfrac{\mathrm{d}}{\mathrm{d}t} X_2 = a_{21} X_1 + a_{22} X_2, \tag{2.29}$$

with initial condition, $X_j = X_j^{(0)}$, $j = 1, 2$, at $t = 0$ as a generalisation of the harmonic oscillator studied previously. Since this system can be studied for itself and not as a tangent problem, we drop the primes indicating that its variables originally measured the departures from the fixed point of some primitive nonlinear system and forger that the matrix elements a_{ij} derive from the evaluation of some Jacobian operator $\mathcal{L}$. Solutions to (2.28, 2.29) are searched in the form:

$$X_i = \bar{X}_i \exp(st).$$

By mere substitution, this leads to an *eigenvalue problem*:

$$s\bar{X}_1 = a_{11}\bar{X}_1 + a_{12}\bar{X}_2,$$
$$s\bar{X}_2 = a_{21}\bar{X}_1 + a_{22}\bar{X}_2,$$

that has nontrivial solutions $\bar{\mathbf{X}} \neq 0$ only if the the growth rate s satisfies the compatibility condition:

$$s^2 - (a_{11} + a_{22})s + a_{11}a_{22} - a_{12}a_{21} = 0, \tag{2.30}$$

also called the *characteristic equation*. In full generality, this quadratic equation has two roots, distinct or not, real or complex conjugate[6] (see Exercise A.5, p. 395).

Eigensolutions corresponding to eigenvalues either positive or complex with positive real part depart exponentially fast from the origin. In the negative case they converge to it.

The general solution is a superposition of eigensolutions. Hence when both roots are negative or complex with negative real parts, the origin is said to be *linearly stable* whereas it is sufficient that one of the roots be real and positive or complex with positive real part to render the origin *unstable*, which immediately yields a universal classification.

---

[6]Let $x + iy = z$ be a complex number, with real part x and imaginary part y, we shall denote its complex conjugate $x - iy$ as z^*.

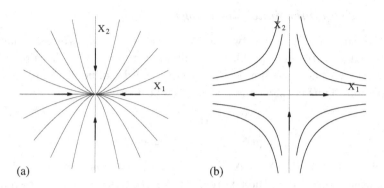

Fig. 2.5 Real roots s_1 and s_2. (a) Stable node, $s_2 < s_1 < 0$. (b) Saddle, $s_2 < 0 < s_1$.

2.2.2.1 Two real distinct roots

Without changing notations, in the eigenbasis we have:
$$\tfrac{\mathrm{d}}{\mathrm{d}t} X_j = s_j X_j \quad \Rightarrow \quad X_j(t) = X_j^{(0)} \exp(s_j t), \qquad j = 1, 2,$$
and we can distinguish two sub-cases:

i) The two roots have the same sign ($s_1 s_2 > 0$): the origin is called a *node*, stable when s_1 and s_2 are negative, unstable in the opposite case. Orbits around the origin have a parabolic shape obtained by elimination of t between the different components of the solution, here between $X_1(t) = X_1^{(0)} \exp(s_1 t)$ and $X_2(t) = X_2^{(0)} \exp(s_2 t)$, which yields $X_2/X_2^{(0)} = (X_1/X_1^{(0)})^{s_2/s_1}$ with a positive exponent s_2/s_1. They open in the direction of X_1 or X_2 according to the relative magnitude of s_1 and s_2, see Figure 2.5(a).

ii) The two roots have opposite signs ($s_1 s_2 < 0$): the origin is a *saddle*, stable along the direction of the negative root and unstable along the other one and thus always unstable. Orbits have a hyperbolic shape as depicted in Figure 2.5(b).

2.2.2.2 A pair of complex conjugate roots

Let us write the roots as $s = \sigma \pm i\omega$. In such a case, there are no real eigenvectors but by a linear variable change one can cast the system into the form
$$\tfrac{\mathrm{d}}{\mathrm{d}t} X_1 = \sigma X_1 + \omega X_2, \tag{2.31}$$
$$\tfrac{\mathrm{d}}{\mathrm{d}t} X_2 = -\omega X_1 + \sigma X_2, \tag{2.32}$$

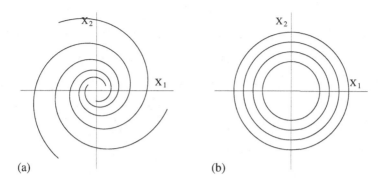

Fig. 2.6 Complex roots $s = \sigma \pm i\omega$. (a) Stable focus, $\sigma < 0$. (b) Centre, $\sigma = 0$.

which, by integration, leads to

$$X_1(t) = \exp(\sigma t)\left[X_1^{(0)} \cos(\omega t) + X_2^{(0)} \sin(\omega t)\right], \qquad (2.33)$$

$$X_2(t) = \exp(\sigma t)\left[-X_1^{(0)} \sin(\omega t) + X_2^{(0)} \cos(\omega t)\right]. \qquad (2.34)$$

Trajectories thus spiral around the origin that is called a *spiral point*, or a *focus*, stable or unstable according to the sign of the real part σ, see Figure 2.6(a).

In the *marginal* case $\sigma = 0$, the fixed point is called a *centre* or an *elliptic point*, Figure 2.6(b). This is of course the case of the harmonic oscillator considered previously. One immediately observes that this property is not *robust*.

The introduction of a damping or a driving, as weak as they could be, converts the centre into a focus, stable or unstable [cf. (2.9): eigenvalues $s = -\eta \pm i(1-\eta^2)^{1/2}$, pure imaginary only for $\eta = 0$]. In such a point of the space of control parameters, the system is said to be *structurally unstable*.

2.2.2.3 Double roots

The last case is when the characteristic equation has (real) double roots, which happens when $(a_{11} - a_{22})^2 + 4a_{12}a_{21} = 0$. Usually, the system has only one eigen-direction so that is cannot be cast into diagonal form by a linear variable change but only into what is called its *Jordan normal form*, here:

$$\tfrac{\mathrm{d}}{\mathrm{d}t} X_1 = sX_1 + X_2, \qquad (2.35)$$

$$\tfrac{\mathrm{d}}{\mathrm{d}t} X_2 = sX_2. \qquad (2.36)$$

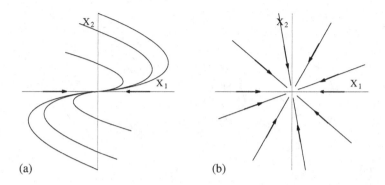

Fig. 2.7 Double real root. (a) Improper node. (b) Star (diagonalisable case).

Integrating (2.36) one gets $X_2(t) = X_2^{(0)} \exp(st)$, which is further inserted in (2.35) which now reads:

$$\tfrac{\mathrm{d}}{\mathrm{d}t} X_1 - s X_1 = X_2^{(0)} \exp(st), \qquad (2.37)$$

which points out the *resonant* character of the right hand side since it evolves at the same rate as that defined by the left hand side. Equation (2.37) is easily integrated by the Lagrange method of *variation of the constant*: solving the homogeneous problem, one gets $X_1 = \tilde{X} \exp(st)$, where $\tilde{X}$ is an integration constant. Assuming that this "constant" is now a function of time, $\tilde{X}(t)$ and introducing this expression into (2.37) one gets

$$\tfrac{\mathrm{d}}{\mathrm{d}t} \tilde{X} = X_2^{(0)},$$

which leads to

$$\tilde{X} = X_2^{(0)} t + X_1^{(0)},$$

and thus to

$$X_1(t) = \left(X_1^{(0)} + X_2^{(0)} t \right) \exp(st) \qquad (2.38)$$

where the second, sub-dominant, term inside the parentheses is called a *secular term*.[7]

This fixed point is called an *improper node*. The corresponding phase portrait is displayed in Figure 2.7(a) which shows that it is indeed intermediate between a node and a focus.

---

[7]The word originates from the Nineteenth Century studies of the period of planets by perturbation methods, when corrections to the relative position of the planets were found to show up on time scales of the order of hundreds of years.

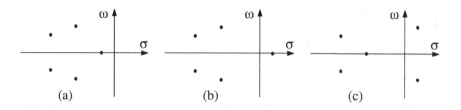

Fig. 2.8 Stability of the fixed point $\mathbf{X}_f$ of a hypothetical 5-dimensional system $\frac{d}{dt}\mathbf{X} = \mathcal{F}(\mathbf{X})$: The operator $\mathcal{L}$ obtained by linearisation of $\mathcal{F}$ around $\mathbf{X}_f$ has 5 real or complex eigenvalues and $\mathbf{X}_f$ can be linearly stable (a), unstable against a stationary mode (b), unstable against an oscillatory mode (c). Instability against several modes is also possible but one is usually interested in the first instability of a state, i.e. the transition from (a) to (b) or (c), which can be achieved by varying just one control parameter in a sufficiently limited range.

When the eigenvalue is double but the system still diagonalisable, which occurs when the operator $\mathcal{L}$ is symmetric, all directions in the plane (X_1, X_2) are eigen-directions, which is another limit case of a node called a *star*, see Figure 2.7(b).

2.2.3 Stability of a time-independent regime

Let us come back to a d-dimensional system linearised around one of its fixed points $\mathbf{X}_f$. At the linear stage, the general solution of the perturbation problem can be expressed as a *superposition* of solutions corresponding to each eigenvalue s_j, with dominant exponential behaviour $\exp(s_j t)$. These eigenvalues can be ordered by decreasing value of their real part, and while in the short term one may observe a complicated evolution of generic perturbations made of an arbitrary superposition of eigenmodes, due to linear interference between them (cf. §A.3, p. 387), in the long term only the contribution corresponding to the eigenvalue with largest real part survives.

The considered base state associated to the fixed point is thus *linearly stable* if all the eigenvalues have negative real parts, Figure 2.8(a), and *linearly unstable* if at least one of the eigenvalues has positive real part, Figure 2.8(b, c). When the unstable root is real, the corresponding unstable mode is said to be *stationary*, Figure 2.8(b), and when it is complex, one speaks of an *oscillatory* mode, Figure 2.8(c).

Here we have made use of the concept of *asymptotic stability*, the amplitude of the perturbation tending to zero as t increases. When the eigenvalue is zero or has zero real part, the corresponding mode is *marginal* or *neu-*

tral. Linear theory then does not allow us to draw any conclusion about the stability of the base state and nonlinearity has to be taken into account. Methods with a more *global* flavour have to be used, in the spirit of the *energy method* introduced in Exercise 2.5.2. In the long time limit, perturbations evolve more slowly than exponentially, generally as some power of time. They may relax, in which case the base state is still asymptotically stable, or grow, in which case it is unstable. This type of situation can be dealt with using the concept of *orbital stability* which is weaker than that of asymptotic stability since it only requires that the perturbed state can permanently depart, but in a controlled way, from the base state, thus remaining in its vicinity. A typical example is that of an elliptic point in a mechanical system, locally equivalent to a harmonic oscillator: trajectories circle around the fixed point without approaching it. An elliptic point with purely imaginary eigenvalue, marginal at a linear level, can thus be orbitally stable or unstable at the nonlinear stage.

The persistent occurrence of purely imaginary eigenvalues often results from symmetry conditions and especially the invariance of the dynamics upon time reversal, which is characteristic of mechanics. One can indeed observe that reversibility implies an exchange '$s \leftrightarrow -s$' upon the change '$t \mapsto -t$', and thus either to complex conjugation in case of purely imaginary roots or an exchange 'stable $\leftrightarrow$ unstable' within a pair of real eigenvalues.[8]

Otherwise the presence of neutral modes must be considered as accidental: since the system usually depends on control parameters, this circumstance only occurs at specific locations in the parameter space. At such points the system is *structurally unstable* since a slight modification of its definition can turn the considered fixed point from stable to unstable. It is ready for a *bifurcation* associated with a qualitative change of its phase portrait.

The breadth of possibilities increases with the dimension of the system but the terminology introduced for two-dimensional systems can be extended straightforwardly. For example, one still speaks of a node when all the eigenvalues are real and have the same sign. In the same way, a fixed point can be called a saddle-focus if it has a pair of complex conjugate eigenvalues with real parts of one sign and its other eigenvalues real with the opposite sign. The case of a three-dimensional system is considered in Exercise 2.5.3.

---

[8]The existence of quadruples $\pm \sigma \pm i\omega$ is not ruled out in the general case, but the symmetry does not change the fact that the system is unstable since, within a quadruple, there is always a pair of roots with positive real parts, see Exercise 2.5.4.

2.3 Two-dimensional Nonlinear Systems

In this section we study the dynamics of autonomous nonlinear systems that evolve in a two-dimensional state space. In a first instance, we take an essentially qualitative view point and use what precedes to draw phase portraits. Then we attack the problem of the explicit quantitative determination of the period of oscillators, presenting several methods in a computational perspective.

2.3.1 Two examples of oscillators

2.3.1.1 The rigid pendulum

The first classical example of nonlinear oscillator that we consider is the simple *rigid pendulum* already introduced in Section 1.1, see Figure 1.1 (right) on p. 2. The potential energy from which the external gravity force can be derived, $\mathcal{V}(\theta) = mg(1 - \cos\theta)$, is illustrated in Figure 2.9 (left). The angular frequency of small oscillations is given by $\omega_0^2 = g/l$. Performing the change $t \mapsto t/\omega_0$, one can rewrite (1.2), p. 2, as a two-dimensional first order system:

$$\tfrac{\mathrm{d}}{\mathrm{d}t}\theta = \varphi, \qquad \tfrac{\mathrm{d}}{\mathrm{d}t}\varphi = -\sin\theta. \tag{2.39}$$

The corresponding state space $\{\theta; \varphi\}$ is a cylinder $\mathbb{S}^1 \times \mathbb{R}$, where $\mathbb{S}^1$ is the unit circle (the one-dimensional sphere, hence the letter 'S') parameterized by the cyclic variable θ, i.e. $\theta + 2\pi \equiv \theta$. On the other hand $\varphi \in \mathbb{R}$ can take

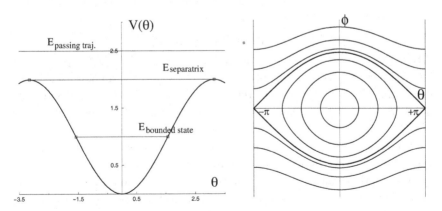

Fig. 2.9 Left: Potential energy of the rigid pendulum. Right: Phase portrait in reduced representation with indication of bounded states, the separatrix, and passing orbits.

its values from $-\infty$ to $+\infty$. This cylinder being open along the generatrix $\theta = \pi$, one obtains a reduced representation of the state space as a band of width 2π along θ and infinite length along φ, with identification of the sides at $\theta = \pm\pi$.

As already noticed, the geometrical constraint fixing the distance of the mass to the rotation axis generates the nonlinearity. The global character of this constraint is reflected in the topological structure of the state space. Nonlinearity reveals itself when large amplitude motions are considered, whereas linearisation remains legitimate close to the origin. Classically, the distinction is made between *passing trajectories* at sufficiently high energy and *bounded trajectories* at low energy, see Figure 2.9 (right).

Fixed points of the system are given by $\varphi = 0$ and $\theta = k\pi$, $k \in \mathbb{N}$. The study of the dynamics close to these points is a straightforward application of the linear approach developed in previous sections. So, it appears that point $(\theta = 0, \varphi = 0)$ is a centre with eigenvalues $\pm i$ and corresponds to the small-oscillation regime. In contrast, the point $(\theta = \pm\pi, \varphi = 0)$ is a saddle with eigenvalues ± 1. Specific trajectories called *separatrices* link one of these points to the other, thus separating the domain of bounded orbits from that of passing orbits.

Breaking the Hamiltonian character of the dynamics, we now consider the effect of viscous friction. The second equation of (2.39) is therefore completed by a term proportional to the angular velocity $\varphi = \frac{d}{dt}\theta$:

$$\tfrac{d}{dt}\varphi = -\eta\varphi - \sin\theta\,.$$

This case is further illustrated in Figure 2.10 using an extended representation that no longer takes advantage of the limitation to the periodized interval $[-\pi, \pi]$. As a matter of fact, this representation makes it easier to understand how the pendulum returns to its rest position after a certain number of complete turns around the axis, a number that depends on the initial energy.

It can be observed that the centres persist but are converted into stable spiral points. On the other hand unstable points at $\theta = \pm(2k+1)\pi$ do not change their nature but trajectories that emerge from them along their unstable direction miss the next point and spiral towards the foci, whereas trajectories that arrive to them along their stable direction need a slightly larger energy than in the conservative case, as expected. Other similar systems are proposed in Section 2.5.

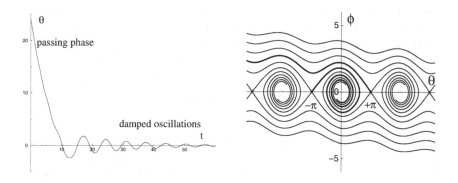

Fig. 2.10 Damped pendulum. Left: Decay of a high energy trajectory. Right: Phase portrait in extended representation.

2.3.1.2 Van der Pol oscillator

Examples of oscillatory processes are numerous in fields other than mechanics, from electronics (the case we consider now) to ecology (prey–predator system, Exercise 2.5.12) or economics (expansion-recession cycles, see [Anderson et al. (1988)]).

In the RLC circuit described in Figure 2.11, the Joule effect in resistor R, accounting for dissipation, is described by a standard ohmic voltage-

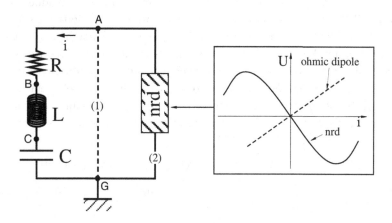

Fig. 2.11 Sketch of the RLC circuit modelling the van der Pol oscillator. Connection (1) is a simple short circuit. Connection (2) is through an active dipole with a voltage-intensity characteristic displaying a range of negative resistance (negative-resistance dipole = nrd, the gray line corresponds to an ordinary ohmic resistor).

intensity relation, $U_A - U_B = RI$. The other elements, the coil with inductance L and the capacitor with capacity C introduce the equivalent of an inertia at the origin of oscillations by setting the intensity and the voltage out of phase: the charge of the capacitor is given by $Q = C(U_C - U_G)$, with $Q = \int I \, dt$ while, when submitted to a varying intensity, the coil responds by building a voltage difference $U_B - U_C = L \frac{d}{dt} I$.

For a series circuit we get:

$$U = U_A - U_G = \frac{Q}{C} + R \frac{d}{dt} Q + L \frac{d^2}{dt^2} Q. \tag{2.40}$$

The capacitor discharge after closing the loop along (1), i.e. $U_A = U_G$, is then given by:

$$\frac{d^2}{dt^2} Q + \frac{R}{L} \frac{d}{dt} Q + \frac{1}{LC} Q = 0.$$

This equation governs a damped linear oscillator $\propto \exp(-i\omega t)$ with a complex angular frequency ω solution to:

$$\omega^2 + i\gamma\omega - \omega_0^2 = 0,$$

where the resonance angular frequency ω_0 is given by $\omega_0^2 = 1/LC$ and the damping factor γ by $\gamma = R/L$.

The energy initially stored in the capacitor is dissipated in the resistor, at the origin of the damping. If one succeeds in injecting energy in the system so as to compensate the losses, one can obtain self-sustained oscillations. To achieve this aim, the circuit is closed on a negative-resistance dipole [loop (2)]. This "active element" is concretely implemented with an operational amplifier that draws its energy from an external electric supply maintaining the whole system in a permanent out-of-equilibrium state.

The voltage-intensity relation accounting for the active element is supposed to be ohmic but with a negative resistance coefficient. In practice, saturation effects come and limit the validity of the "anomalous" Ohm law so that we may take (cf. Fig. 2.11, right):

$$U_{G'} - U_{A'} = -R_0 I + b I^3 \quad \text{with} \quad R_0 > 0 \quad \text{and} \quad b > 0.$$

Using $I = \frac{d}{dt} Q$ as a variable rather than Q itself, the equation governing the circuit then reads:

$$\left[L \frac{d}{dt} I + RI + (1/C) \int I \, dt \right] + \left[-R_0 I + b I^3 \right] = 0,$$

or, upon differentiation with respect to time:

$$L \frac{d^2}{dt^2} I + \left[(R - R_0) + 3 b I^2 \right] \frac{d}{dt} I + I/C = 0. \tag{2.41}$$

As long as $R > R_0$ the coefficient of $\frac{d}{dt}I$ is positive and dissipation plays a normal role: oscillations are damped. This is no longer the case when $R < R_0$: a small perturbation (i.e. such that $3bI^2$ is negligible when compared to $|R - R_0|$) is amplified and the oscillation develops. As soon as the amplitude of the oscillation is large enough, the nonlinear dissipation term plays a normal role and stops the divergence. Decreasing R, we can therefore control the bifurcation from a time-independent steady state (oscillatory perturbations are damped) towards the regime of self-sustained oscillations.

Performing the changes $I \mapsto X$ and $t \mapsto t/\omega_0$ in equation (2.41) we get one of the forms of the van der Pol model

$$\frac{d^2}{dt^2}X - (r - gX^2)\frac{d}{dt}X + X = 0, \qquad (2.42)$$

where $r \propto R_0 - R$ is the control parameter and where $g > 0$ is a measure of the strength of nonlinearity that could have been suppressed by a rescaling of X, ending with $g = 1$. Figure 2.12 presents the results of the numerical integration of (2.42) for two different initial conditions in the quasi-harmonic regime (top line) or strongly anharmonic regime (middle line). In both cases one can observe that orbits spirals toward a closed curve

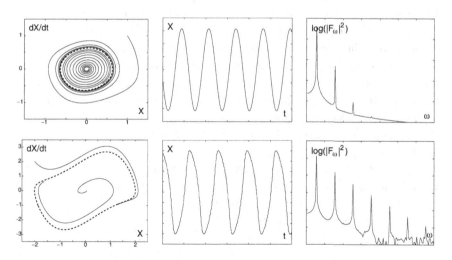

Fig. 2.12 Van der Pol oscillator (2.42) for $r = 0.1$ (quasi-harmonic, top) and $r = 1.0$ (strongly anharmonic, bottom), both with $g = 1$. Left: Phase portraits showing the convergence toward a limit cycle indicated by a dashed line. Middle: Corresponding time series of the intensity signal. Right: Corresponding Fourier spectra (lin–log plot of the modulus of the Fourier amplitudes F_ω squared).

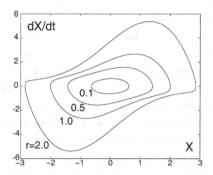

Fig. 2.13 Limit cycles of the van der Pol oscillator for increasing values of r.

called a *limit cycle*, either from the inside or from the outside depending on the initial condition.

This special orbit is nearly elliptical in the first case and rather quadrangular in the second. To them correspond nearly sinusoidal or on the contrary highly anharmonic oscillations, which is also illustrated in the Fourier spectra that have a higher level of harmonics in the second case than in the first. The graphs in Figure 2.13 illustrate the deformation of the limit cycles as the parameter r is varied from 0.1 to 2.0.

The van der Pol limit cycle is an example of attractor that is not trivially reduced to a single point (as was the case for the damped pendulum). It should be noted that, in contrast with ideal mechanical oscillators that do not have attractors and for which the amplitude of the motion is fixed by the total energy (kinetic+potential) in the initial condition, here it is the competition between energy injection and dissipation that fully determines the characteristics (amplitude and period) of the regime achieved beyond the instability threshold.

In Chapter 4 we will come back to the description of this bifurcation when r changes from negative to positive values. For the moment let us determine in an approximate way the amplitude of the cycle in the quasi-harmonic regime, close to the threshold.

Inserting $X \simeq X_{\mathrm{m}} \cos(t)$ in equation (2.42) we get

$$-X_{\mathrm{m}} \cos(t) + \left(r - g X_{\mathrm{m}}^2 \cos^2(t)\right) X_{\mathrm{m}} \sin(t) + X_{\mathrm{m}} \cos(t) = 0\,.$$

Let us restrict to a *first harmonic approximation*. It consists in demanding that the equation be identically fulfilled for terms in $\sin(t)$ and $\cos(t)$, without worrying about higher harmonics generated by the nonlinearities. It is easily observed that the compensation is automatic for the cosine terms.

For the sine terms, using the classical formulas $\cos^2(t) = \frac{1}{2}(1+\cos(2t))$ and $\cos(2t)\sin(t) = \frac{1}{2}(\sin(3t) - \sin(t))$, neglecting the $\sin(3t)$ term, we get:
$$\left(r - \tfrac{1}{4}gX_{\mathrm{m}}^2\right)\sin(t) = 0\,,$$
which leads to:
$$r - \tfrac{1}{4}gX_{\mathrm{m}}^2 = 0 \qquad \text{that is} \qquad X_{\mathrm{m}} = 2\sqrt{r/g}\,. \tag{2.43}$$
This relation fixes the amplitude of the cycle and it can further be checked that the amplitude of term in $\sin(3t)$ generated by the nonlinearity is of order X_{m}^3, hence $\sim r^{3/2} \ll r^{1/2}$ for r small, thus justifying the expression 'quasi-harmonic', see Figure 2.12 (top).

Plotting X_{m} as a function of r, we get exactly the same bifurcation diagram as for A_* in Figure 1.4 (right), p. 13, but restricted to its upper branch, which makes sense since the change $t \mapsto t + \pi$ (legitimate since the system is autonomous) brings the branch $X_{\mathrm{m}} < 0$ on top of the other one.

Getting the result this way is rather crude. In order to improve the solution, one should try to fulfil the equation harmonic by harmonic (a special case of the so-called Galerkin method). This would lead to an infinite nonlinear algebraic system, the lowest order consistent truncation of which is precisely (2.43). Here, an additional implicit assumption has been that the nonlinearity did not change the angular frequency. This property turns out to be correct at lowest order in r for the van der Pol model but not necessarily in other cases, which opens the problem of the general determination of the period of nonlinear oscillators to be examined now.

2.3.2 Amplitude and phase of nonlinear oscillators

2.3.2.1 The Duffing oscillator: period from a direct computation

Let us come back to the special case of the harmonic oscillator and note that the linear relation between the restoring force and the elongation may not stay indefinitely valid. In general the microscopic properties of the elastic forces induce some nonlinearity. If the spring is *hard* the force necessary to obtain a specific elongation grows faster than just being proportional to it and we can assume $F = -kX(1 + cX^2)$ with $c > 0$. (The opposite case of a *soft* spring with $c < 0$ is studied as part of Exercise 2.5.7.) In the absence of friction, choosing the time scale so that the angular frequency of the unperturbed oscillator is equal to 1, we get:
$$\tfrac{\mathrm{d}^2}{\mathrm{d}t^2}X + X + cX^3 = 0\,, \tag{2.44}$$

which called the *Duffing oscillator*.

As long as the amplitude is small, the nonlinear term is negligible and the harmonic approximation is satisfactory. When the amplitude increases, the oscillator feels the effects of the nonlinearity and the motion becomes anharmonic. The period and the shape of the orbits change progressively. However, as long as the amplitude remains small enough, corrections to the linear solution may be obtained using *perturbation methods*.

Here the case is particularly simple. Let us multiply (2.44) by $\frac{d}{dt}X$ and rewrite the result as

$$\frac{d}{dt}\left[\frac{1}{2}(\frac{d}{dt}X)^2 + \frac{1}{2}X^2 + \frac{1}{4}cX^4\right] = 0. \qquad (2.45)$$

The quantity between the brackets is clearly the total energy, the sum of the kinetic energy and the potential energy $\mathcal{V}(X) = \frac{1}{2}X^2 + \frac{1}{4}cX^4$ from which the elastic restoring force derives. Integrating (2.45) we obtain

$$\frac{1}{2}\frac{d}{dt}X^2 + \mathcal{V}(X) = E,$$

called a *first integral* of (2.44). The differential order of the problem has indeed decreased by one since we can rewrite this equation as

$$\frac{d}{dt}X = \pm\sqrt{2(E - \mathcal{V}(X))}, \qquad (2.46)$$

with the implicit assumption that $(E \geq \mathcal{V}(X)$. The condition $\mathcal{V}(X_t) = E$ defines the *turning points* of the problem, which correspond to points with maximal elongation and zero velocity, whereas the region $|X| > X_t$ is forbidden since it corresponds to a negative kinetic energy.

In the case of the nonlinear spring considered here, we get

$$\frac{1}{2}X_t^2 + \frac{1}{4}cX_t^4 = E,$$

which, for c small enough, yields:

$$X_t \approx \sqrt{2E}\left(1 - \frac{1}{2}cE\right),$$

so that the maximum elongation for a hard spring is reduced when compared to that of a harmonic oscillator with the same energy.

The system oscillates between its two turning points. The period can thus be computed by integration of (2.46) between them:

$$\frac{1}{2}T = \int_{-X_t}^{X_t} \frac{dX}{\sqrt{2(E - \mathcal{V}(X))}}. \qquad (2.47)$$

The value of this integral, analytically defined as a complete elliptic integral of the first kind, can be found in tables or numerically computed. Here it is more interesting to find its expression at low energy when c small, and thus for $X_{\mathrm{t}} \ll 1$. Performing the variable change $X = X_{\mathrm{t}} \sin(\varphi)$ and using the parity of the quantity to be integrated, through an expansion truncated at first order we get:

$$T \simeq 4 \int_0^{\pi/2} \left[1 - \tfrac{1}{4}cX_{\mathrm{t}}(1 + \sin^2 \varphi)\right] d\varphi,$$

hence

$$T \simeq 2\pi \left(1 - \tfrac{3}{4}cE\right) \quad \text{or} \quad \omega = \frac{2\pi}{T} \simeq 1 + \tfrac{3}{4}cE. \tag{2.48}$$

For a hard spring the period therefore decreases when the energy increases, which is easily understood by writing $X + cX^3 = X\left(1 + cX^2\right)$ and observing that $(1+cX^2)$ plays the role of an effective elastic constant, the average value of which is always larger than that of the reference linear oscillator.

After this example of direct computation (see also Exercise 2.5.10) let us examine three methods for obtaining the value of the period in cases where nonlinear and/or dissipative effects can be considered as perturbations to harmonic oscillations with intensities scaled by some small parameter.

2.3.2.2 Averaging method

In this first method, it is assumed that the amplitude of the oscillations is modulated on a long time scale that allows one to determine an effective equation for the modulation by *averaging*. Its intuitive simplicity makes it a reasonable first choice but it cannot easily be improved beyond lowest order. As a result we re-obtain the approximate solution to the van der Pol problem previously derived, but here with a bonus.

To explain the method let us come back to the weakly damped linear oscillator (2.9) rewritten here with $\varepsilon = 2\eta$:

$$\tfrac{\mathrm{d}^2}{\mathrm{d}t^2}X + X = -\varepsilon \tfrac{\mathrm{d}}{\mathrm{d}t}X, \tag{2.49}$$

with initial conditions

$$X^{(0)} = 1 \quad \text{and} \quad \tfrac{\mathrm{d}}{\mathrm{d}t}X^{(0)} = 0.$$

The solution reads

$$X(t) = A^{(0)} \exp\left(-\tfrac{1}{2}\varepsilon t\right) \cos\left(\omega t - \varphi^{(0)}\right) \tag{2.50}$$

with $\omega^2 = 1 - \tfrac{1}{4}\varepsilon^2$, $A^{(0)} = 1/\omega$, $\varphi^{(0)} = \arctan(\varepsilon/2\omega)$. This expression can be written as $X(t) = A(t)\cos(t + \varphi(t))$ where $A(t) \propto \exp(-\varepsilon t/2)$ appears as the average of the amplitude over a pseudo-period, which varies slowly provided that $\varepsilon \ll 1$. On the other hand, the argument of the cosine,

$\omega t - \varphi^{(0)}$, can be written as $t + \varphi(t)$ with $\varphi(t) = (\omega - 1)t - \varphi^{(0)}$, a quantity that is also slowly varying since $\omega - 1 \sim \frac{1}{8}\varepsilon^2$.

The problem is now to generalise these notions of average amplitude $A(t)$ and phase $\varphi(t)$, supposed to be sufficiently *slowly varying*, when the system presents itself as a second order differential equation close to that governing the ideal linear oscillator, i.e. when it can be written as:

$$\tfrac{d^2}{dt^2} X + X = -\varepsilon f\bigl(X, \tfrac{d}{dt} X\bigr), \qquad \varepsilon \ll 1. \tag{2.51}$$

The solution is searched in the form:

$$X(t) = A(t)\cos(t + \varphi(t)), \tag{2.52}$$

where $A(t)$ and $\varphi(t)$ are two unknown functions of time. Differentiating this expression we get:

$$\tfrac{d}{dt} X = -A\sin(t + \varphi) + \bigl[\tfrac{d}{dt} A \cos(t + \varphi) - A \tfrac{d}{dt}\varphi \sin(t + \varphi)\bigr]. \tag{2.53}$$

In the absence of modulation ($\tfrac{d}{dt} A \equiv 0$, $\tfrac{d}{dt}\varphi \equiv 0$), we would have:

$$\tfrac{d}{dt} X = -A\sin(t + \varphi),$$

so that it appears natural to reduce the freedom introduced in replacing the original unknown X by the two unknowns A and φ by forcing the quantity between the brackets to cancel identically. The first equation linking A to φ is therefore:

$$\tfrac{d}{dt} A \cos(t + \varphi) - A \tfrac{d}{dt}\varphi \sin(t + \varphi) = 0. \tag{2.54}$$

Differentiating (2.52) once more and taking (2.54) into account we get

$$\tfrac{d^2}{dt^2} X = -A\cos(t + \varphi) - \bigl[\tfrac{d}{dt} A \sin(t + \varphi) + A \tfrac{d}{dt}\varphi \cos(t + \varphi)\bigr]$$

which we insert in (2.51) to obtain

$$\tfrac{d}{dt} A \sin(t + \varphi) + A \tfrac{d}{dt}\varphi \cos(t + \varphi) = \varepsilon g(A, \varphi), \tag{2.55}$$

where $g(A, \varphi) \equiv f(X, \tfrac{d}{dt} X) = f(A\cos(t + \varphi), -A\sin(t + \varphi))$. In order to isolate $\tfrac{d}{dt} A$ and $\tfrac{d}{dt}\varphi$ we can combine (2.54) and (2.55) using the usual trigonometric relations: computing $(2.54) \times \cos(t + \varphi) + (2.55) \times \sin(t + \varphi)$ and $(2.55) \times \cos(t + \varphi) - (2.54) \times \sin(t + \varphi)$, we obtain:

$$\tfrac{d}{dt} A = \varepsilon f\bigl(A\cos(t + \varphi), -A\sin(t + \varphi)\bigr) \sin(t + \varphi), \tag{2.56}$$
$$A \tfrac{d}{dt}\varphi = \varepsilon f\bigl(A\cos(t + \varphi), -A\sin(t + \varphi)\bigr) \cos(t + \varphi). \tag{2.57}$$

Up to now, everything is exact and one can notice that (2.56, 2.57) would result from a change to cylindrical coordinates $X = A\cos(\theta)$, $Y = A\sin(\theta)$, $\theta = -t - \varphi$, in a phase space where the unperturbed second-order equation

$\frac{d^2}{dt^2}X + X = 0$ would be replaced by a system of two first order equations[9] $\frac{d}{dt}X = Y$, $\frac{d}{dt}Y = -X$.

The approximation comes in as soon as one assumes that $A(t)$ and $\varphi(t)$ are slowly variable on the short time scale $T \simeq 2\pi$. If this is the case, we can integrate the above equations over an approximate period while considering A and φ as constants, which leads to

$$\frac{d}{dt}A = \frac{\varepsilon}{2\pi}\int_0^{2\pi} f(A\cos(t), -A\sin(t))\sin(t)\,dt, \qquad (2.58)$$

$$A\frac{d}{dt}\varphi = \frac{\varepsilon}{2\pi}\int_0^{2\pi} f(A\cos(t), -A\sin(t))\cos(t)\,dt. \qquad (2.59)$$

Let us apply these formulas to the van der Pol oscillator, a non-conservative system here taken in the form

$$\frac{d^2}{dt^2}X - \varepsilon(1-X^2)\frac{d}{dt}X + X = 0. \qquad (2.60)$$

This equation is slightly different from (2.42) where the scaling of variable X was more adapted to the problem of the bifurcation when r goes through zero. The scaling chosen here is such that the nonlinearity contributes to the solution when $X \sim \mathcal{O}(1)$ and is valid only for $\varepsilon > 0$. With $f = -(1-X^2)\frac{d}{dt}X$, we get:

$$\frac{d}{dt}A = \frac{\varepsilon}{2\pi}\int_0^{2\pi}(1-A^2\cos^2(t))\,A\sin^2(t)\,dt = \tfrac{1}{2}\varepsilon A\left(1-\tfrac{1}{4}A^2\right),$$

$$A\frac{d}{dt}\varphi = \frac{\varepsilon}{2\pi}\int_0^{2\pi}(1-A^2\cos^2(t))\,A\sin(t)\cos(t)\,dt = 0.$$

The second equation shows that there is no correction to the angular frequency (at least at this order) which implicitly justifies the choice made when developing the first harmonic approximation on p. 62. In contrast, the first equation governing the amplitude (the bonus alluded at the beginning of the section) is non-trivial and may serve us to study the convergence towards the limit cycle corresponding to $A_* = 2$ which is nothing but (2.43) using the scales turning (2.42) in the form (2.60).

2.3.2.3 The Poincaré–Lindstedt method

The second method considered here is called the *Poincaré–Lindstedt* expansion. The solution is now assumed to be periodic with some unknown

---
[9]The minus signs in the definition of θ arise from the fact that the so-defined two-dimensional vector field generates trajectories that rotate clockwise, opposite to the trigonometric convention.

period close to that of the reference oscillator, and both the period and the solution are searched through an expansion in powers of ε. For simplicity, it is explained in the linear case but can easily be applied when nonlinearity is present (Exercise 2.5.11).

Let us consider two harmonic oscillators with nearly equal angular frequencies, the first one governed by $\frac{d^2}{dt^2}Y + Y = 0$, with angular frequency $\omega_Y = 1$, and the second one by $\frac{d^2}{dt^2}X + (1-\varepsilon)X = 0$, with angular frequency $\omega_X = (1-\varepsilon)^{1/2}$. Starting with identical initial conditions are $Y = X = 1$, $\frac{d}{dt}Y = \frac{d}{dt}X = 0$, they will progressively drift out of phase. Choosing the first oscillator as a reference, one can interpret this phase shift as the result of *secular terms* that appear already in the first order expansion of the solution for the second oscillator $X(t) = \cos(\omega_X t)$. From $\omega_X \approx 1 - \frac{1}{2}\varepsilon$, one gets $\cos(\omega_X t) = \cos(t)\cos(\frac{1}{2}\varepsilon t) - \sin(t)\sin(\frac{1}{2}\varepsilon t)$, so that for sufficiently short times such that $\frac{1}{2}\varepsilon t \ll 1$, with $\cos(\frac{1}{2}\varepsilon t) \sim 1$ and $\sin(\frac{1}{2}\varepsilon t) \sim \frac{1}{2}\varepsilon t$, one finds $\cos(\omega t) = \cos(t) + \frac{1}{2}\varepsilon t \sin(t)$, which points out the secular term correction explicitly (cf. p. 54). It is not difficult to obtain this solution directly from the equation by a perturbation expansion to which we now turn.

For the second oscillator, the problem reads:

$$\frac{d^2}{dt^2}X + X = \varepsilon X$$

and the solution is searched for as a power expansion in ε:

$$X = X_0 + \varepsilon X_1 + \varepsilon^2 X_2 + \dots \tag{2.61}$$

We are led to a series of simple linear problems:

$$(\tfrac{d^2}{dt^2} + 1)X_0 = 0, \tag{2.62}$$

$$(\tfrac{d^2}{dt^2} + 1)X_1 = X_0, \tag{2.63}$$

$$(\tfrac{d^2}{dt^2} + 1)X_2 = X_1, \tag{2.64}$$

$$\dots = \dots \tag{2.65}$$

The solution to (2.62) with initial conditions $X = 1$, $\frac{d}{dt}X = 0$ is nothing but $X_0 = \cos(t)$ so that the inhomogeneity in the problem for X_1 just contains a term that is resonant with the left hand side. A special solution to (2.63) is obtained by identification as $X_1 = \frac{1}{2}t\sin t$, which is indeed responsible for the secular growth of the phase shift at the considered order.

Any expansion obtained in this way has thus a limited validity in time. One says that it is *non-uniform*. The aim of the rest of this section and the next one is to obtain a *uniformly valid* expansion. Here the origin of

2. First Steps in Nonlinear Dynamics

the discrepancy is obvious: we must correct the clock and pass from the original time t to a new one τ for which secular terms would be absent. This is precisely the essence of the *Poincaré–Lindstedt method*.

Let us set $\tau = \omega t$ and look for the relation between τ and t in the form of an expansion of ω in powers of the small perturbation parameter ε:

$$\omega = 1 + \varepsilon\omega_1 + \varepsilon^2\omega_2 + \ldots \qquad (2.66)$$

Denoting differentiation with respect to variable τ as $\frac{d}{d\tau}$, we get $\frac{d}{dt}X \equiv \omega\frac{d}{d\tau}X$ and $\frac{d^2}{dt^2}X \equiv \omega^2\frac{d^2}{d\tau^2}X$, and thus

$$\left[(1 + \varepsilon\omega_1 + \varepsilon^2\omega_2 + \ldots)^2 \tfrac{d^2}{d\tau^2} + 1\right](X_0 + \varepsilon X_1 + \varepsilon^2 X_2 + \ldots)$$
$$= \varepsilon(X_0 + \varepsilon X_1 + \varepsilon^2 X_2 + \ldots)$$

leading to a new series of linear problems

$$(\tfrac{d^2}{d\tau^2} + 1)X_0 = 0, \qquad (2.67)$$
$$(\tfrac{d^2}{d\tau^2} + 1)X_1 = X_0 - 2\omega_1\tfrac{d^2}{d\tau^2}X_0, \qquad (2.68)$$
$$(\tfrac{d^2}{d\tau^2} + 1)X_2 = X_1 - (\omega_1^2 + 2\omega_2)\tfrac{d^2}{d\tau^2}X_0 - 2\omega_1\tfrac{d^2}{d\tau^2}X_1, \qquad (2.69)$$
$$\ldots = \ldots \qquad (2.70)$$

The structure of (2.67–2.70) is similar to that of (2.62–2.65), except that the right hand side now contain free parameters that can be fixed so as to "kill" all the resonant terms that generate secular terms at the origin of the time non-uniformity in the initial expansion. This operation is an application of the *Fredholm alternative* stipulating that, when the kernel of a linear operator $\mathcal{L}$ is non-trivial, the problem $\mathcal{L}\mathbf{X} = \mathbf{F}$ has solutions only if the right hand side $\mathbf{F}$ is orthogonal to the kernel of the adjoint operator $\mathcal{L}^\dagger$ of $\mathcal{L}$ (see §A.3.2 for a reminder).

Here the unperturbed problem (2.67) is self-adjoint. Its kernel is generated by the trigonometric functions $\cos(\tau)$ and $\sin(\tau)$. Inserting $X_0 = \cos(\tau)$ in (2.68) we get on the right hand side $F_1 = (1 + 2\omega_1)\cos(\tau)$. The Fredholm alternative imposes us to cancel the coefficient of $\cos(\tau)$, i.e. $\omega_1 = -\tfrac{1}{2}$. At second order we get $F_2 = (\omega_1^2 + 2\omega_2)\cos(\tau)$ which yields $\omega_2 = -\tfrac{1}{2}\omega_1^2 = -\tfrac{1}{8}$, and so on. In this way, by reconstructing ω from its expansion we obtain $\omega = 1 - \tfrac{1}{2}\varepsilon - \tfrac{1}{8}\varepsilon^2$, which is the beginning of the Taylor expansion of $\omega = (1 - \varepsilon)^{1/2}$, as expected from the direct calculation. In Exercise 2.5.11 the method is applied to the Duffing oscillator as a typical example of a nonlinear system.

2.3.2.4 *The method of multiple scales*

The Poincaré–Lindstedt method does not allow for amplitude and phase modulations which were essential in the averaging method. We are thus led to the last and most general approach called the *method of multiple scales* that lifts the restrictions of both previous methods by introducing a hierarchy of time scales.

The natural relevance of the Poincaré–Lindstedt method is to Hamiltonian systems with a single degree of freedom in the mechanical sense, i.e. two-dimensional dynamical systems for which energy conservation implies the periodicity of bounded states, a key feature of the direct calculation above, p. 64. Difficulties appear for non-conservative systems since the occurrence of strict periodicity is then a much less trivial matter. In order to understand how to escape this problem, let us consider again the damped oscillator.

Solution (2.50) $\propto \mathcal{R}e\left\{\exp[(-\frac{1}{2}\varepsilon + i\omega)t]\right\}$ can of course be written as $\propto \mathcal{R}e\left\{\exp[i(\omega + i\frac{1}{2}\varepsilon)t]\right\}$, where $\omega + i\frac{1}{2}\varepsilon$, though a complex quantity, can be understood as the angular frequency of some oscillation. In much the same way as turning to amplitude and phase in the averaging method comes to a change to cylindrical coordinates, the natural extension of the Poincaré–Lindstedt method thus suggests to change from real Cartesian coordinates in the state space to cylindrical coordinates best parameterized using complex variables. As a matter of fact, using (2.66) and applying the previous procedure to (2.49) without assuming that ω is a real quantity, one gets at first order

$$\left(\tfrac{d^2}{d\tau^2} + 1\right)X_1 = -2\omega_1 \tfrac{d^2}{d\tau^2} X_0 - \tfrac{d}{d\tau} X_0 ,$$

instead of (2.68). Injecting the complex solution $X_0 = A\exp(it)$ in this equation, one finds the compatibility condition $2\omega_1 - i = 0$, i.e. $\omega = 1 + \omega_1 \varepsilon = 1 + \frac{1}{2}i\varepsilon$, which correctly accounts for the expected damping at this order.

However, already in this simple problem, we may note the existence of two different time scales: the damping time of order $1/\varepsilon$, and another time scale, the inverse of the frequency shift, of order $1/\varepsilon^2 \gg 1/\varepsilon$ when $\varepsilon \ll 1$. This suggests to base the most general method on a hierarchy of time scales, suitable to correct the lack of synchronisation between a reference linear oscillator and the nonlinear system at hand. The aim of this strategy is to make the approximation uniformly valid in time, much like the introduction of leap years helps us to adjust the calendar according to a complicated algorithm that is more and more complicated as longer and longer periods of time are considered.

Let us look for the solution of the problem in the form
$$X(t) = X(t_0, t_1, t_2, \ldots) \quad \text{with} \quad t_0 = t, \quad t_1 = \varepsilon t, \quad t_2 = \varepsilon^2 t, \ldots$$
For ε small enough, the time scale measured by t_1 is indeed slow with respect to that measured by t_0 since when t_0 varies by a quantity $\mathcal{O}(1)$, the arguments of X in $t_1, t_2, \ldots$ vary by $\mathcal{O}(\varepsilon)$, $\mathcal{O}(\varepsilon^2)$,...

The differentiation with respect to time is then given by
$$\tfrac{d}{dt} = \tfrac{d}{dt} t_0 \partial_{t_0} + \tfrac{d}{dt} t_1 \partial_{t_1} + \tfrac{d}{dt} t_2 \partial_{t_2} + \ldots$$
$$= \partial_{t_0} + \varepsilon \partial_{t_1} + \varepsilon^2 \partial_{t_2} + \ldots,$$
since $dt_0/dt \equiv 1$, $dt_1/dt \equiv \varepsilon$,.... In the same way the second derivative reads:
$$\tfrac{d^2}{dt^2} = \partial_{t_0}^2 + 2\varepsilon \partial_{t_0} \partial_{t_1} + \varepsilon^2 \left(\partial_{t_1}^2 + 2 \partial_{t_0} \partial_{t_2} \right) + \ldots.$$
The solution is taken in the form
$$X(t) = X_0(t_0, t_1, t_2, \ldots) + \varepsilon X_1(t_0, t_1, t_2, \ldots) + \ldots,$$
and we have to insert these expansions in the motion equation and isolate the different orders in ε. Here we consider the specific cases of the Duffing and van der Pol oscillators as two complementary illustrations of the method.

• Duffing oscillator:

We have:
$$\tfrac{d^2}{dt^2} X + X = -\varepsilon X^3,$$
so that at order ε^0 we find
$$(\partial_{t_0}^2 + 1) X_0 = 0,$$
the solution of which reads:
$$X_0(t_0, t_1, t_2, \ldots) = A_0(t_1, t_2, \ldots) \cos\left(t_0 + \varphi_0(t_1, t_2, \ldots)\right).$$
At order ε^1 we obtain:
$$(\partial_{t_0}^2 + 1) X_1 = -X_0^3 - 2 \partial_{t_0} \partial_{t_1} X_0.$$
The strategy is to find conditions on A_0 and φ_0 such that the r.h.s., which expands as:
$$-X_0^3 - 2\partial_{t_0}\partial_{t_1} X_0$$
$$= -A_0^3 \cos^3(t_0 + \varphi_0) + 2(\partial_{t_1} A_0) \sin(t_0 + \varphi_0)$$
$$\quad + 2 A_0 (\partial_{t_1} \varphi_0) \cos(t_0 + \varphi_0)$$
$$= \left(-\tfrac{3}{4} A_0^3 + 2 A_0 (\partial_{t_1} \varphi_0)\right) \cos(t_0 + \varphi_0) + 2(\partial_{t_1} A_0) \sin(t_0 + \varphi_0)$$
$$\quad - \tfrac{1}{4} A_0^3 \cos(3(t_0 + \varphi_0)),$$

contains no term in resonance with the l.h.s. Clearly only the $\cos(3(t_0+\varphi_0))$ term on the last line is naturally non-resonant, while the terms on the previous line are. We thus get two conditions, one from cosine, the other from the sine (the kernel is two-dimensional):
$$-\tfrac{3}{4}A_0^3 + 2A_0\partial_{t_1}\varphi_0 = 0\,,$$
$$\partial_{t_1}A_0 = 0\,,$$
that can both be integrated. The second condition means that A_0 does not depend on t_1 and thus, at best, on $t_2, t_3, \ldots$, i.e. $A_0 = A_0(t_2, \ldots)$. The first equation in turn gives the lowest order correction to the phase:
$$\varphi_0 = \tfrac{3}{8}A_0^2 t_1 + \bar\varphi_0(t_2, \ldots)\,.$$
The non-resonant part can then be computed, if one wants to stop at this order, the full solution can be reconstructed by setting $t_1 = \varepsilon t$ in the result (compare with the output of the Poincaré–Lindstedt method used in Exercise 2.5.11).

- van der Pol oscillator:

The first steps of the computation do not change. At order ε^1 we get:
$$(\partial_{t_0^2} + 1)X_1 = (1 - X_0^2)\partial_{t_0}X_0 - 2\partial_{t_0}\partial_{t_1}X_0\,,$$
the right hand side of which expands as
$$\bigl(1 - A_0^2\cos^2(t_0+\varphi_0)\bigr)\bigl(-A_0\sin(t_0+\varphi_0)\bigr) + 2(\partial_{t_1}A_0)\sin(t_0+\varphi_0)$$
$$+ 2A_0(\partial_{t_1}\varphi_0)\cos(t_0+\varphi_0)$$
$$= \bigl(-A_0(1 - \tfrac{1}{4}A_0^2) + 2\partial_{t_1}A_0\bigr)\sin(t_0+\varphi_0) + 2A_0(\partial_{t_1}\varphi_0)\cos(t_0+\varphi_0)$$
$$+ \tfrac{1}{4}A_0^3\sin\bigl(3(t_0+\varphi_0)\bigr)\,.$$
The conditions that annihilate the resonant terms now read:
$$-A_0\bigl(1 - A_0^2/4\bigr) + 2\partial_{t_1}A_0 = 0\,,$$
$$2A_0\partial_{t_1}\varphi_0 = 0\,.$$
The second condition, $\partial_{t_1}\varphi_0 = 0$, hence $\varphi_0 = \bar\varphi_0(t_2, \ldots)$, shows as before that there is no correction to the angular frequency at this order. On the other hand, the first equation is seen to govern the evolution of A_0, like in the averaging method, except for the presence of the factor ε that is incorporated in the definition of t_1 and would reappear if we were to come back to the independent variable t.

The method of multiple scales can be pursued at higher order in a completely systematic way. The price to be paid is increasingly heavy computations that are greatly eased by the use of formal algebra software such as MAPLE or MATHEMATICA. It can be adapted to treat the problem of periodically forced nonlinear oscillators, consult e.g. [Nayfeh and Mook (1979)] for further reference.

2.4 What Next?

The notion of determinism implemented in time-continuous dynamical systems considered up to now implies that a unique trajectory goes through any regular point in state space (trajectories do not intersect since a given state cannot have several pasts and futures). The consequence is of utmost importance for two-dimensional systems which experience a strong topological constraint, so that their behaviour remains "simple" in a sense to be re-examined in Chapter 4. In higher dimensions, in contrast, trajectories have enough space to wind one around each other, which is ultimately at the origin of "complicated" behaviour. Before considering this problem, we examine in the next chapter how high dimensional systems, especially continuous media, may behave as effective low dimensional systems making the approach just introduced of great practical use.

2.5 Exercises

2.5.1 *Evolution of volumes in state space*

Consider a continuously differentiable ($\mathcal{C}^1$) two-dimensional map

$$Y_1 = \mathcal{G}_1(X_1, X_2),$$
$$Y_2 = \mathcal{G}_2(X_1, X_2).$$

1) Determine the transform of a state space element $[X_1, X_1 + \delta X_1] \times [X_2, X_2 + \delta X_2]$ under map $\mathcal{G}$ and compute its surface at lowest significant order [Hint: Figure 2.14, next page].

2) Assuming that $\mathcal{G}$ is the time-τ map of some continuous-time system

$$\tfrac{\mathrm{d}}{\mathrm{d}t} X_1 = \mathcal{F}_1(X_1, X_2),$$
$$\tfrac{\mathrm{d}}{\mathrm{d}t} X_2 = \mathcal{F}_2(X_1, X_2),$$

over an infinitesimal time interval $\tau = \delta t$, compute $\mathcal{G}$ from $\mathcal{F}$ and infer that the variation of state space volumes is locally given by the divergence of the vector field $\mathcal{F}$, formula (2.11).

2.5.2 *The energy method and its application*

The energy method is an example of *global* method for studying stability problems without reference to specific perturbations, e.g. infinitesimal ones.

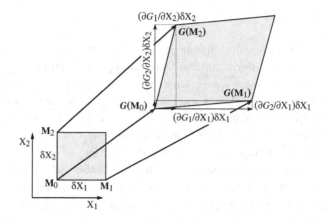

Fig. 2.14 Transformed domain $[X_1, X_1 + \delta X_1] \times [X_2, X_2 + \delta X_2]$ under map $\mathcal{G}$. Exercise 2.5.1, first part.

1) As a preliminary, consider the linear system $\frac{d}{dt}V = -\mu V$ and check that the "kinetic energy" $E = \frac{1}{2}V^2$ is governed by:
$$\frac{d}{dt}E = -\mu V^2 \leq 0,$$
which implies a monotonic return to equilibrium, illustrating the concept of *asymptotic* stability.

2) Consider next the oscillator defined as
$$\frac{d}{dt}X = Y + X\sqrt{X^2 + Y^2}, \qquad \frac{d}{dt}Y = -X + Y\sqrt{X^2 + Y^2},$$
where the nonlinear terms come and modify a marginally stable dynamics in the neighbourhood of the origin (a centre) and for which the weaker notion of *orbital* stability has been introduced.

2a) Define $E = \frac{1}{2}(X^2 + Y^2)$, compute $\frac{d}{dt}E$ and conclude that the trivial solution is orbitally unstable. [Answer: $\frac{d}{dt}E = (X^2 + Y^2)^{3/2} > 0$ for any $(X, Y) \neq (0, 0)$).]

2b) Determine the evolution of $Z = \sqrt{2E}$ (distance to the origin) starting from some initial condition $Z_0 > 0$ at $t = 0$ and show that that Z diverges at $t_* = 1/Z_0$, i.e. the corresponding trajectory spirals away to infinity in a finite time.

3) We now seek to extend the concept of energy used up to now, in such a way that the study of its variations allows one to decide about stability or instability of a given state, conveniently taken as the origin of coordinates.

- *Definition*: Let $\mathcal{G}(\mathbf{X})$ be a function of point $\mathbf{X}$ in state space $\mathbb{X}$, taking its values in $\mathbb{R}^+$, *definite positive*, i.e. such that $\mathcal{G}(\mathbf{0}) = 0$ and $\mathcal{G}(\mathbf{X} \neq \mathbf{0}) > 0$ (in practice a definite positive quadratic form). A sufficient condition of asymptotic stability, is that the amplitude of the perturbation, as measured by $\mathcal{G}$ decreases in the course of time and tends to zero as t goes to infinity. This will be the case if $\frac{d}{dt}\mathcal{G} = \sum_j \partial_{X_j}\mathcal{G} \frac{d}{dt}X_j = \sum_j \partial_{X_j}\mathcal{G}\, \mathcal{F}_j$ is *negative definite* as a function of $\mathbf{X} \in \mathbb{X}$ taking its values in $\mathbb{R}$, such that $\frac{d}{dt}\mathcal{G}(\mathbf{0}) = 0$ and $\frac{d}{dt}\mathcal{G}(\mathbf{X} \neq \mathbf{0}) < 0$. Such a $\mathcal{G}$ is called a *Lyapunov function*.[10]

- *Application*: Consider the system defined as:

$$\frac{d}{dt}X = XY - X^3,$$
$$\frac{d}{dt}Y = -Y - 2X^2.$$

Show that the origin is the unique fixed point and further that it is marginally stable. Since no conclusion can be drawn at the linear stage, by generalising the energy E to a function in the form:

$$\mathcal{G}(X,Y) = \tfrac{1}{2}\left(\alpha X^2 + \beta Y^2\right),$$

determine the conditions under which $\mathcal{G}$ is a Lyapunov function, thus proving the asymptotic stability of this state.

2.5.3 Linear stability and bifurcation in dimension three

As a useful preliminary to Exercise 3.3.3 in Chapter 3, consider a real, three-dimensional, linear dynamical system in the form

$$\frac{d}{dt}X_i = \sum_{j=1}^{3} a_{ij} X_j, \quad i = 1, 2, 3.$$

1) Recall the relations between the coefficients and the roots of its characteristic equation written as $s^3 + as^2 + bs + c = 0$. Further identify all the possible non-degenerate or degenerate cases (real or complex nature of the roots, their sign or the sign of their real part, simple and multiple roots). Write down the Jordan normal form of the operator corresponding to each situation (see §A.2 for a reminder). [Answer: see Fig. A.1(b).]

2) When the control parameters vary, the eigenvalues of the system move in the complex plane. Determine the remarkable relations fulfilled by the coefficients of the characteristic equation in the marginal cases and sketch

---
[10] This definition can be extended to the case of continuous media described in terms of fields (velocity, temperature, concentration,...), in which case the Lyapunov function becomes a functional of the fields.

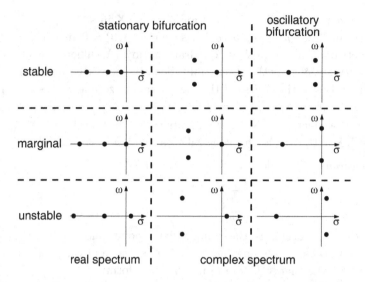

Fig. 2.15 Typical bifurcations in the three-dimensional case, Exercise 2.5.3.

the spectrum of the operator in the complex plane before, at, and after a stationary bifurcation ($s = 0$) or an oscillatory bifurcation ($\mathcal{R}e(s) = 0$). [Answer: see Fig. 2.15.]

2.5.4 Coupled linear oscillators

Consider the following system of two coupled oscillators

$$\tfrac{\mathrm{d}^2}{\mathrm{d}t^2} X + X = Y,$$
$$\tfrac{\mathrm{d}^2}{\mathrm{d}t^2} Y + \Omega^2 Y = -cX.$$

Find the equation governing the angular frequencies of the eigenmodes taken as $(X, Y) = (X_0, Y_0) \exp(i\omega t)$, observe that the solutions appear as pairs of opposed eigenvalues. Determine the domain in the plane of parameters Ω^2 and c where the eigenmodes are stable. [Hint: remember that for a mechanical system, stability means orbital stability, with purely imaginary eigenvalues.]

2.5.5 Logistic equation

Re-examine the limited growth model (logistic equation) introduced in Chapter 1, p. 26. Study the stability of its two fixed points and draw its phase portrait.

2.5.6 Dynamical systems and solitons

1) Consider a dynamical system with a single degree of freedom in the sense of analytical dynamics (a pair of conjugate variables) with a force deriving from the potential

$$V(X) = -\frac{X^2}{2} + \frac{X^3}{3}. \tag{2.71}$$

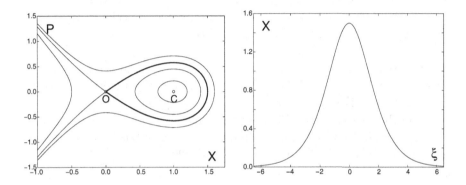

Fig. 2.16 Left: Phase portrait of the system with a force deriving from potential (2.71) Right: Profile of the soliton corresponding to the separatrix on the left part.

1a) Write down the second order differential equation governing X (mass $m = 1$) and then the first order system for X and $P = \frac{d}{dt}X$. Find the fixed points, compute their eigenvalues and eigenvectors (if any). Sketch the phase portrait in the phase space (X, P). [Answer: Figure 2.16, left.]

1b) Draw the graph of the potential $V(X)$ and determine the energy corresponding to the separatrix, the special trajectory that limits the domain of orbits bounded around the minimum of the potential. Find the turning point and the equation accounting for this trajectory. Determine its solution by identification with

$$X(t) = X_0/\cosh^2(t/\tau).$$

Why is this form "natural"?

1c) Sketch the phase portrait of the system when perturbed by the introduction of a weak viscous friction, taking the weakly damped pendulum as an example.

2) The Korteweg–de Vries equation reads:
$$\partial_t h + h\partial_x h + \beta \partial_{xxx} h = 0 \,.$$
This is a non-dissipative equation with a nonlinearity of hydrodynamic type, identical to that of the Burgers equation considered in Chapter 1, Exercise 1.5.1. It is integrable and its solution can be expressed in terms of a superposition of interacting solitons. Here we are interested in a solution with a single soliton such that $h \to 0$ when $x \to \pm\infty$ moving without deformation at speed c, and thus only function of the combination $\xi = x-ct$.

2a) Determine the equation in ξ governing such a moving solution (make the substitutions $\partial_x \mapsto d/d\xi$ and $\partial_t \mapsto -c\,d/d\xi$). Integrate this equation and find the value of the integration constant that corresponds to the soliton solution.

2b) By appropriate variable changes put this equation in the form considered in the first part of the exercise and deduce the analytical expression of the soliton. Discuss the relation between the speed, the height, and the width of the soliton. [Hint: Figure 2.16, right.]

For a in-depth introduction to solitons in physics and their mathematical properties consult e.g. [Dauxois and Peyrard (2006)].

2.5.7 Variants of the Duffing oscillator

1) Consider the system
$$\tfrac{d^2}{dt^2}X + X(a + X^2) = F \,,$$
where a is a coefficient with unspecified sign for the moment, and F a quantity playing the role of an external force.

1a) Find the potential from which the dynamics can be derived in the sense of mechanics, i.e. $\mathcal{V}$ such that $m\tfrac{d^2}{dt^2}X = -\partial \mathcal{V}/\partial X$, and the time independent response to a force $0 < F \ll 1$ (observe that the roots of an equation in the form $x^3 - px + q = 0$ with $q \ll 1$ and $p \sim 1$ are approximately given by $px \sim q$ for x small and $x^3 \sim px$ for x large.

1b) Discuss the dynamics in the phase space $(X, Y \equiv \tfrac{d}{dt}X)$ as a function of the sign of parameter a. Locate the fixed points and determine their stability properties (when solving the linearised problems, neglect the nonlinearity for the small solution and the constant term for the large one). Draw a few typical orbits. In which sense can one speak of the system as of an oscillator?

2) Consider now
$$\tfrac{d^2}{dt^2}X + X + cX^3 = 0 \qquad \text{with} \quad c < 0 \quad \text{(soft spring)}.$$

2a) Find the elastic potential from which the dynamics can be derived and discuss the behaviour of the system in its phase space as above.

2b) For which set of initial conditions is the system a physically well posed one (bounded orbits)? How should one correct the model in order to avoid the divergence of some trajectories? Interpret this limitation by considering the theory leading to the approximate expression (2.48) for the angular frequency of small oscillations when $c > 0$ and assuming that it is still valid for large amplitude orbits when $c < 0$.

2.5.8 *Carriage with a spring*

Consider the mechanical system built with an ideal spring of equilibrium length ℓ_0 fixed by one of its ends at point P as described in Figure 2.17(a). Its other end is attached at M to a small carriage with mass m sliding along a horizontal line forming the x axis. Point P is at a distance ℓ_1 from this axis. The intensity of the restoring force is proportional to the elongation $|\mathbf{F}| = k(\ell - \ell_0)$.

1) Determine the components of the forces exerted on the carriage at abscissa X and the corresponding equation of motion. Taking ℓ_0 as length unit, $\sqrt{m/k}$ as time unit, and setting $\ell_1 = \lambda \ell_0$, rewrite this equation in dimensionless form as a system of two equations for X and $Y = \frac{d}{dt}X$.

2) Compute the potential $\mathcal{V}(X)$ from which the force $F(X)$ derives, i.e. $F(X) = -\partial \mathcal{V}/\partial X$. Find the fixed points as a function of λ, study their stability by linearising the system.

3) Sketch the phase portraits in the two cases $\lambda < 1$ and $\lambda > 1$. What can be said from the trajectory issued from the vicinity of the origin along the unstable direction in the case $\lambda < 1$? Using the fact that the system is frictionless, determine the corresponding turning point quantitatively.

2.5.9 *Carriage sliding on a rotating hoop*

Consider now a carriage sliding along a rail in the form of a hoop with radius a, itself rotating around a vertical diameter (cf. Fig. 2.17b). The rotation period is $T = 2\pi/\omega$. The motion is governed by the following equation:

$$a\frac{d^2\theta}{dt^2} = -g\sin(\theta) + a\omega^2 \cos(\theta)\sin(\theta)$$

where g is the gravitational acceleration.

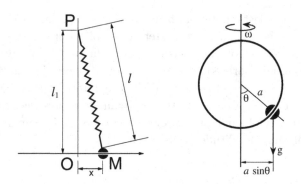

Fig. 2.17 Left: Spring and carriage sliding along a horizontal rail. Right: Carriage sliding along a rail in the form of a hoop further rotating around a vertical diameter.

1) Justify this equation by taking advantage of the elements of analytical mechanics recalled in §2.1.2. To this aim, compute first the gravitational potential energy and then the kinetic energy resulting from the superposition of the two independent rotation motions, around the axis and within the plane of the hoop. Set the equation in the form of a two-dimensional differential system for θ and $\varphi = \frac{d}{dt}\theta$.

2) Find the fixed points and study their stability as a function of the angular speed ω. Sketch the phase portraits in the different cases. [Answer: Figure 2.18.]

3) Examine the effects of a slight viscous friction proportional to φ.

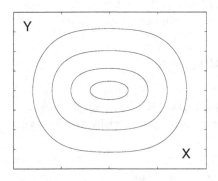
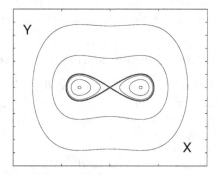

Fig. 2.18 Phase portrait of the hoop system for $\omega^2 = 0.8$ (left) and $\omega^2 = 1.25$ (right).

2.5.10 Period of an oscillator in a quartic potential

Consider a strongly anharmonic oscillator with a restoring force deriving from the potential

$$V(X) = V_0 + \tfrac{1}{4}X^4.$$

From equation (2.47) giving the period of an oscillator, show without explicit calculations that one gets $T \propto E^\alpha$ with an exponent α to be determined (perform the change of variables that applies the interval between the turning points onto $[-1, 1]$).

2.5.11 Averaging and Poincaré–Lindstedt methods

The Duffing oscillator is taken in the form

$$\left(\tfrac{d^2}{dt^2} + 1\right)X = -\varepsilon X^3.$$

1) Extract the relation between the amplitude and the angular frequency of the oscillator by making use of equation (2.59) for the phase. What is the meaning of equation (2.58) for the amplitude?

2) Following the Poincaré–Lindstedt method (§2.3.2.3), derive the complete solution at first order in ε starting from the initial condition $X^{(0)} = A$, $\tfrac{d}{dt}X^{(0)} = 0$ at $t = 0$. Using expansions (2.61–2.66), show first that at the relevant order the problem simply reads

$$\left(\tfrac{d^2}{d\tau^2} + 1\right)X_1 = 2\omega_1 X_0 - X_0^3,$$

and next that the elimination of resonant terms leads to the same result as the first order Taylor expansion of the solution obtained by a direct calculation (2.48). [Observe that, at lowest order, the averaging method and the Poincaré–Lindstedt method involve exactly the same computations.]

3) Find the complete solution at first order, once the Fredholm alternative is fulfilled.

2.5.12 Prey–predator system

Consider the system introduced in Chapter 1, p. 26:

$$\tfrac{d}{dt}X = \mathcal{F}_X(X,Y) = (\alpha'Y - \alpha_0)X, \qquad (2.72)$$

$$\tfrac{d}{dt}Y = \mathcal{F}_Y(X,Y) = (\beta - \gamma X)Y, \qquad (2.73)$$

in the physical quadrant $X \geq 0$, $Y \geq 0$ (X and Y are population counts).

1) Draw a qualitative phase portrait by studying the vector field along

lines defined by $\mathcal{F}_X(X,Y) = 0$ and $\mathcal{F}_Y(X,Y) = 0$ (vertical and horizontal isoclines respectively). Search and study the characteristics of the fixed points (roots of $\mathcal{F}_X(X,Y) = 0 = \mathcal{F}_Y(X,Y)$).

2) Show that the quantity
$$H(X,Y) = \alpha' Y - \alpha_0 \log(Y) + \gamma X - \beta \log(X)$$
is conserved along a trajectory and that H displays a minimum at the non-trivial fixed point.

3) Consider the successive intersections of a trajectory with line $Y = \alpha_0/\alpha'$ and conclude from the behaviour of H that all trajectories are periodic orbits. Notice that they are not limit cycles and explain why the elliptic character of the non-trivial fixed point is a deficiency of the model. Try to correct it by considering the most general model of the form $\mathcal{F}_X = X\tilde{\mathcal{F}}_X(X,Y)$, $\mathcal{F}_Y = Y\tilde{\mathcal{F}}_Y(X,Y)$, at most quadratic in X and Y. In particular find the conditions to be fulfilled in order to reproduce the qualitative features characteristic of a prey–predator system.

Chapter 3

Life and Death of Dissipative Structures

In this transition chapter we focus on the emergence of convection and how patterns that have developed further desegregate. This rather intuitive example helps us to introduce a few general ideas and techniques to analyse instabilities (§3.1). The theory that allows us to interpret their disorganisation, presented here from a purely phenomenological perspective (§3.2) will be re-examined in Chapters 4 and 6.

3.1 Emergence of Dissipative Structures

3.1.1 *Qualitative analysis of the instability mechanism*

Let us go back in more detail on the argument previously sketched in relation with Figure 1.3, p. 11, to explain the onset of convection. The two parts of the mechanism, instability due to differential buoyancy and stability through dissipation (viscous relaxation and heat diffusion) will first be qualitatively analysed in terms of *characteristic times*.

Let us consider a horizontal layer (height h) of fluid heated from below (Figure 3.1):
$$T_{\rm b} = T_{\rm t} + \Delta T > T_{\rm t}$$
('b' for 'bottom' and 't' for 'top'). The fluid is initially at rest in a regime of *pure conduction*. The temperature profile is linear
$$T_0(z) = T_{\rm b} - \beta z \quad \text{with} \quad \beta = \Delta T/h\,,$$
and notations imply that the temperature gradient β is positive in case of heating from below.

The corresponding density distribution is given by the equation of state which, in a first approximation, reads:
$$\rho(T) = \rho_{\rm ref}\left(1 - \alpha\left(T - T_{\rm ref}\right)\right) \tag{3.1}$$

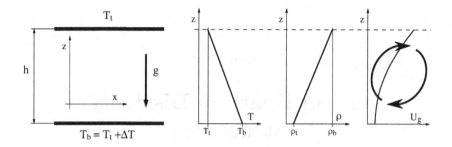

Fig. 3.1 Left: Geometry of the convection experiment. Right: Profiles of temperature $T(z)$, density $\rho(z)$ and gravitational potential energy $U_g(z)$ for a fluid particle at altitude z, indicating the tendency to restore a stable density stratification with heavy fluid at the bottom.

where T_{ref} is a reference temperature, $\rho_{\text{ref}} = \rho(T_{\text{ref}})$, and α is the thermal expansion coefficient (1/273 for an ideal gas), hence $\rho_0(z) = \rho(T_0(z))$.

A first characteristic time, a transport time τ_b, can be defined from the buoyancy (hence subscript 'b'). Assume a fluid particle experiencing a temperature fluctuation θ at some height z, i.e. $T(z) = T_0(z) + \theta$, from (3.1) the differential force to which it is submitted is $\rho g \alpha \theta$. The quantity $g \alpha \theta$ is thus an acceleration, homogeneous to a length divided by the square of a time. Natural length and temperature scales are h and ΔT, respectively. We can thus define the time τ_b through:

$$\frac{h}{\tau_b^2} = g \frac{\Delta \rho}{\rho} \sim g \alpha \Delta T.$$

Physically, τ_b is the typical time a hot (cold) bubble would take to move up (down) over a distance h with a constant acceleration due to thermal expansion.

Dissipative processes, viscous friction (Stokes law, kinematic viscosity $\nu = \mu/\rho$) and thermal conduction (Fourier law, thermal diffusivity $\kappa = \chi/C$ where χ is the thermal conductivity and C the specific heat) are diffusive in essence. The relaxation times associated with these processes can be deduced from the form of a diffusion equation, $\partial_t q \propto \nabla^2 q$, in which the proportionality coefficient is the diffusivity, homogeneous to $[\ell]^2 [t]^{-1}$, hence here:

$$\nu = h^2/\tau_v, \qquad \kappa = h^2/\tau_\theta.$$

(See also Exercise 1.5.1.)

3. Life and Death of Dissipative Structures

The result of the competition between the destabilising mechanism and the stabilising processes can be estimated by forming the ratio:

$$R = \frac{\tau_v \tau_\theta}{\tau_c^2} = \frac{\alpha g \Delta T h^3}{\kappa \nu} \tag{3.2}$$

called the *Rayleigh number*. By construction, it is a dimensionless number. Convection develops when the buoyancy is more effective (τ_b short) than the dissipative processes (τ_v and τ_θ long) and thus when R is large. This strictly dimensional analysis is qualitative and cannot help us to determine the value of ΔT necessary to induce convection. It is the reason why we go one step further and develop a more quantitative model of the instability. The detailed analysis is the subject of Exercise 3.3.2.

3.1.2 Simplified model

In order to study the stability of the base flow, here the fluid at rest ($\mathbf{v}_0 \equiv 0$) in a regime of pure conduction with linear temperature profile ($T_0(z) \sim -\beta z$, $\beta = \Delta T/h$), we must derive the equations governing small perturbations around this state, a temperature fluctuation θ defined by $T = T_0(z) + \theta$ and a velocity fluctuation $\mathbf{v}$. We thus insert the full solution into the primitive equations (here: continuity + Navier–Stokes + Fourier in a fluid), expand these equations in powers of the perturbations, and finally keep only the first order terms (linear stability theory).

The analysis of the mechanism (Chapter 1, §1.3.1) points out a direct coupling of the horizontal modulations of the temperature fluctuation and the vertical velocity component. Accordingly, we assume that a model involving just θ and v_z, depending only on the horizontal coordinate x and time t, will capture the physics.

- *Equation for the vertical velocity:* The unperturbed temperature field $T_0(z)$ induces a density distribution $\rho_0(z) = \rho(T_0(z))$ through (3.1). The differential buoyancy generated by a temperature fluctuation θ reads:

$$-g(\rho - \rho_0) = -g[\rho(T_0 + \theta) - \rho(T_0)] \approx \rho_0 \alpha g \theta,$$

where the minus sign comes from the fact that the vertical unit vector is oriented up while the force is directed down. It appears as an external force term in the z component of the Navier–Stokes equation:

$$\partial_t v_z = \nu \partial_{x^2} v_z + \alpha g \theta, \tag{3.3}$$

where the first term on the r.h.s. corresponds to viscous diffusion (only along x). The term $\mathbf{v} \cdot \boldsymbol{\nabla} \mathbf{v}$ is of higher order since there is no velocity at order zero. Also, one can notice that $\theta > 0$ implies $\partial_t v_z > 0$, i.e. an upward acceleration as guessed intuitively.

- *Heat equation*: The Fourier equation must be written for a fluid particle (cf. p. 7) since, as already mentioned, it is its advection in a spatially varying temperature field that plays the essential role in the feedback loop, hence:

$$\tfrac{\mathrm{d}}{\mathrm{d}t} T = \partial_t T + \mathbf{v} \cdot \boldsymbol{\nabla} T = \kappa \partial_{x^2} T\,.$$

Expanding the term $\mathbf{v} \cdot \boldsymbol{\nabla} T$ to first order (linearisation) we find:

$$v_z \partial_z [T_0(z) + \theta] = v_z \partial_z T_0(z) = -v_z \beta\,,$$

which comes from the temperature field at order zero (the second-order term $v_z \partial_z \theta$ is neglected). This leads to:

$$\partial_t \theta = \kappa \partial_{x^2} \theta + \beta v_z\,. \tag{3.4}$$

Remark: Lateral boundary conditions have not yet been specified. Here, we tacitly assume that we deal with a horizontally unbounded layer, or at least that the horizontal dimensions are large when compared to the sole characteristic length in the problem, the height h of the layer. Moreover, the horizontal velocity component and the pressure are absent from the problem at this stage. They are only indirectly coupled to v_z and θ by the need to ensure the continuity of the fluid and the closing of flow lines. The model is thus highly simplified. This deficiency will impede us to determine the critical wavelength that will thus be fixed from the outside by dimensional considerations. Anyway, we shall now illustrate the extension to continuous media of the stability analysis introduced at the beginning of the previous chapter using the simplified model (3.3, 3.4).

3.1.3 Normal mode analysis, general perspective

The system formed by (3.3, 3.4) is typical of linear stability problems in continuous media, i.e. a system of linear partial differential equations (with constant coefficients in the simplest case). It presents itself as an *initial value problem* for the perturbations that we write formally as:

$$\partial_t \mathbf{V} = \boldsymbol{\mathcal{L}}_r(\partial_x, \dots) \mathbf{V}\,, \tag{3.5}$$

where $\mathbf{V}$ represents the set of perturbations. The linear operator $\boldsymbol{\mathcal{L}}_r$ contains spatial partial derivatives $(\partial_x, \dots)$ and also depends on a set of *control parameters* denoted as r. In general, the instability can be controlled using a single quantity that can be varied from the outside (for convection, it is simply the applied temperature gradient β), all other parameters of the system being kept fixed.

Problem (3.5) is linear. Its solution can therefore be searched by means of a *superposition*:

$$\mathbf{V}(\mathbf{x},t) = \sum_n A_n \mathbf{X}_n(\mathbf{x},t), \tag{3.6}$$

to be introduced in the differential problem. Setting:

$$\mathbf{X}(\mathbf{x},t) = \exp(st)\hat{\mathbf{X}}(\mathbf{x}), \tag{3.7}$$

and inserting this assumption in (3.5), we get

$$s\hat{\mathbf{X}}(\mathbf{x}) = \mathcal{L}_r \hat{\mathbf{X}}(\mathbf{x}). \tag{3.8}$$

The stability study then comes to an *eigenvalue problem*. The states $\hat{\mathbf{X}}(\mathbf{x})$ are called the *normal modes* of the problem. These modes have spatial structures that mathematically express the physical *coherence* of the processes at work in the system. On general grounds, the eigenvalues s are complex since the spectrum is entirely real only when the operator can be made self-adjoint for some well-chosen scalar product (see §A.3).

The nature of the spectrum of $\mathcal{L}_r$ depends on the applied *confinement* conditions (Exercise 3.3.1). When the system is unbounded in some directions of space, the spectrum is formed with continuous branches, indexed by as many continuous parameters as there are unbounded directions. For example if the mechanism singles out a specific direction ('vertical' in the case of convection), and if the system is invariant under translations in two complementary directions ('horizontal'), performing a Fourier transform, one looks for normal modes in the form:

$$\hat{\mathbf{X}}_n(x,y,z) = \exp(i(k_x x + k_y y))\tilde{\mathbf{X}}_n(z), \tag{3.9}$$

which defines the *wavevector* $\mathbf{k}_\perp = (k_x, k_y)$ as an indexing parameter. Under the substitution $\nabla_\perp \equiv (\partial_x, \partial_y) \mapsto i\mathbf{k}_\perp$, the operator $\mathcal{L}_r = \mathcal{L}_r(\partial_x, \ldots)$ is transformed into an ordinary differential operator in z, the only remaining independent variable, i.e. $\mathcal{L}_r(ik_x, ik_y, \mathrm{d}/\mathrm{d}z)$. The wavevector $\mathbf{k}_\perp$ thus labels the eigenmode branches in addition to a discrete index related to the confinement direction z. Equation (3.8) has nontrivial solutions provided that the growth rate and the wavevector $\mathbf{k}_\perp$, introduced in (3.7) and (3.9) fulfil a compatibility condition called the *dispersion relation*:

$$s = s_n(r, \mathbf{k}_\perp). \tag{3.10}$$

Let us examine a few questions relative to confinement effects.

- As for convection between unbounded horizontal plates, when the system is rotationally invariant in the plane orthogonal to the direction in which the mechanism is operating, then s_n can only depend on $k = |\mathbf{k}_\perp|$ and not on its orientation.

- When the system is bounded in two space directions, e.g. y and z, and translationally invariant in the last one x, the continuous component k_y is replaced by a discrete index.

- When the system is bounded in all three directions, i.e. when its size is of the order of the scale over which the instability mechanism is operating in all directions, the spectrum loses its last continuous dependence on k and becomes fully discrete. The eigenvalues are all distinct, except for degeneracy linked to physical symmetries. The spatial structure of the normal modes is specific to the geometry considered and mirrors spatial resonance properties of the mechanism with the shape of the set-up. The situation is then the closest to the one considered in the previous chapter, but with infinite series of eigenvalues and normal modes to acknowledge the fact that we deal with continuous media with infinitely many degrees of freedom.

Considering the case of a system unbounded in just one direction (single continuous parameter k), we can write

$$s_n(k,r) = \sigma_n(k,r) - i\omega_n(k,r).$$

Given r and k, the modes can then be ordered by decreasing values of their real part $\sigma_n(k,r)$, which, from definition (3.7), allows the distinction between *stable* normal modes with $\sigma_n < 0$ (damped) and *unstable* modes with $\sigma_n > 0$ (amplified). The dissipative character of the medium implies that modes with the shortest wavelengths are strongly damped, i.e. $\sigma_n(k) \to -\infty$ when $k \to \infty$.

A given mode n with wavevector k bifurcates when on its way from stable to unstable it becomes *neutral*, upon variation of the control parameter. The corresponding *marginal* condition, superscript '(m)', is defined thus by the condition $\sigma_n(k,r) = 0$. Once solved for r, this condition reads:

$$r = r_n^{(\mathrm{m})}(k)$$

Let us assume that, as in convection, increasing the stress corresponds to increasing the control parameter r, the marginal curve for some mode n usually reaches its minimum for some $k = k_n^{(\mathrm{c})}$ called the *critical wavevector* for that mode. The quantitiy $r_n^{(\mathrm{m})}(k_n^{(\mathrm{c})}) = r_n^{(\mathrm{c})}$ is the corresponding *threshold*.

Now, according to the general discussion about stability in the previous chapter, *linear instability* takes place as soon as one normal mode becomes unstable. The linear instability threshold r_c is therefore the minimum over n of all the so-defined $r_n^{(c)}$, achieved for some $n = n_c$. The wavevector corresponding to that mode is called the critical wavevector of the instability. The latter is hence characterised by the set n_c, $r_c = r_{n_c}^{(c)}$, $k_c = k_{n_c}^{(c)}$.

Apart from its growth properties, the rest of the time dependence of a mode depends on the value of $\omega_n(k, r)$. When $\omega_n = 0$, the mode is said to be *stationary*, whereas when $\omega_n \neq 0$, one speaks of an *oscillatory* mode. The value of the angular frequency at threshold, $\omega_c = \omega_{n_c}(k_c, r_c)$, thus allows one to distinguish stationary from oscillatory instabilities. The classification of instabilities according to the spatiotemporal structure of their critical mode will be re-examined in §3.1.6.

3.1.4 Back to the model

Let us come back to the simplified model (3.3, 3.4). We assume that the fluid layer is unbounded in the x direction so that, according to (3.7–3.9), solutions are searched in the form $\{v_z, \theta\} = \{V, \Theta\} \exp(st) \exp(ikx)$. We obtain:

$$sV = -\nu k^2 V + \alpha g \Theta,$$
$$s\Theta = -\kappa k^2 \Theta + \beta V,$$

which is in fact a homogeneous algebraic system of two equations for two unknowns:

$$(s + \nu k^2)V - \alpha g \Theta = 0,$$
$$-\beta V + (s + \kappa k^2)\Theta = 0.$$

The system has non-trivial solutions only if its determinant cancels:

$$(s + \nu k^2)(s + \kappa k^2) - \alpha g \beta = s^2 + (\nu k^2 + \kappa k^2)s + \kappa \nu k^4 - \alpha g \beta = 0. \quad (3.11)$$

This *compatibility condition* linking the growth-rate s to the wavevector k of the perturbation is here the expression taken by the dispersion relation (3.10). We get a single branch (and thus no discrete index) since the differential problem in z has been replaced by an algebraic system, due to our neglect of the z-dependence of the fluctuations.

If the real part of $s(k)$ is negative, the mode is damped and the layer is *stable* against a perturbation with wavelength $\lambda = 2\pi/k$. Otherwise the fluctuation is amplified and the mode k is *unstable*.

Let us estimate the threshold from (3.11). Here it is a quadratic equation in s that can have two real or complex roots.[1] The discriminant

$$\Delta = \left(\nu k^2 + \kappa k^2\right)^2 + 4\left[\alpha g \beta - (\nu k^2)(\kappa k^2)\right] = \left(\nu k^2 - \kappa k^2\right)^2 + 4\alpha g \beta$$

can be negative, and the corresponding solutions to (3.11) have non-zero imaginary parts, only when β is sufficiently large and negative, in the case of strong heating from above. But in this case the modes are always damped since the sum of the roots

$$S = -\tfrac{1}{2}\left(\nu k^2 + \kappa k^2\right) \qquad (3.12)$$

is then negative. This analysis thus confirms the intuition according to which heating must be from below in order to have an instability. Moreover, if there is an instability, it can only be *stationary*, with two real roots, one positive, the other negative since their sum is negative. In order to determine the sign of the roots we have just to consider their product

$$P = (\nu k^2)(\kappa k^2) - \alpha g \beta. \qquad (3.13)$$

The change of sign takes place at $\beta = \beta^{(\mathrm{m})}(k)$ with

$$\beta^{(\mathrm{m})}(k) = \frac{\nu \kappa k^4}{\alpha g}. \qquad (3.14)$$

As long as $\beta < \beta^{(\mathrm{m})}$, the product is positive and the two roots negative, mode k is stable. When $\beta > \beta^{(\mathrm{m})}$, it becomes negative and one of the roots is positive, the mode is unstable, convection sets in. The value $\beta = \beta^{(\mathrm{m})}(k)$ of the applied temperature gradient thus defines the *marginal stability condition* that makes mode k neutral. One can observe that the negativity of the sum and the positivity of the product, the two stability factors, come from stabilising dissipative processes and that the instability factor involves a term in β arising from the advection of the fluctuation through the layer submitted to the temperature gradient.

The marginal stability condition (3.14) implies an increase of the marginal temperature gradient as k^4 for k large, which expresses the growing efficiency of the stabilising mechanisms as the scale of the fluctuations decreases (Figure 3.2, left). According to this relation, the longer the wavelength, the lower the threshold. However one should not conclude that the fluid layer is unstable at $k = 0$ for $\beta = 0$, i.e. $\Delta T = 0$. because this low-k behaviour is an artefact of the one-dimensional character of the model that

---
[1] Let $s_{1,2}$ be the two roots, one has $0 = (s-s_1)(s-s_2) = s^2 - (s_1+s_2)s + s_1 s_2 = s^2 - Ss + P$ where S is the sum of the roots and P their product. The discriminant is $\Delta = S^2 - 4P$ and the roots are $s_{1,2} = s_\pm = \tfrac{1}{2}(S \pm \sqrt{\Delta})$.

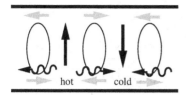 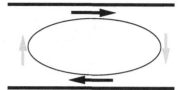

Fig. 3.2 Left: For $\lambda = 2\pi/k \ll h$, viscous dissipation associated with the horizontal shear $\partial_x v_z$ (black arrows) and thermal diffusion (undulated arrows) combine their effects to prevent convection, while the vertical shear $\partial_z v_x$ (grey arrows) can be neglected. Right: When $\lambda \gg h$ the horizontal shear (grey arrows) becomes negligible while the vertical shear (black arrows) becomes dominant.

neglects the z-dependence of the fluctuations and the associated dissipation processes: The viscous damping by the horizontal component of the flow that closes the streamlines can no longer be neglected as $k \to 0$ (Figure 3.2, right), hence the need for a corrected argument valid at small k.

Keeping v_z as a reference since it is directly involved in the instability mechanism, one can estimate the order of magnitude of v_x from the continuity equation

$$\partial_x v_x + \partial_z v_z = 0.$$

Boundary conditions on v_z are at the horizontal plates, a distance h apart, and imply a z-dependence such that $\partial_z v_z \sim v_z/h$ and therefore $kv_x \sim v_z/h$ or $v_x \sim v_z/kh$. But the presence of v_x imposes us to take the x-component of the Navier–Stokes equation into account. We can simplify it as:

$$-kp/\rho - \nu v_x/h^2 \simeq 0,$$

and introduce the so-evaluated pressure in the equation for v_z, which yields:

$$\partial_t v_z = -\partial_z p/\rho + \nu \left(\partial_{x^2} + \partial_{z^2}\right) v_z + \alpha g \theta. \tag{3.15}$$

A sketchy analysis of the space dependence of the different perturbations then shows that, once expressed in terms of v_z using the continuity equation, $\partial_z p/\rho$ goes as $\nu v_z/k^2 h^4$, so that the pressure term dominates those involving v_z on the r.h.s. of (3.15). As a matter of fact, it diverges as k^{-2} when $k \to 0$, while the second term tends to zero as k^2 and the third one does not vary with k. We thus arrive at an effective equation for v_z replacing (3.3):

$$\partial_t v_z = -\nu(1/h^4 k^2) v_z + \alpha g \theta,$$

only valid in the limit $k \ll 1/h$ (the minus sign expresses the fact that it is indeed a damping term). In this limit, it suffices to replace $-k^2$ by $-1/h^2$

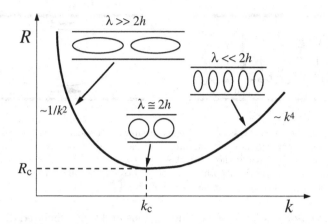

Fig. 3.3 Marginal stability curve from the semi-quantitative argument.

in the heat equation (3.4) to account for the dominant dissipative process. A stability analysis parallel to that leading to (3.14) yields:

$$\kappa\nu/(h^6 k^2) - \alpha g \beta^{(m)} = 0,$$

so that the marginal stability condition for small wavevectors reads:

$$\beta^{(m)}(k) = \frac{\nu\kappa}{\alpha g h^6 k^2} \qquad (3.16)$$

thus showing a divergence as $1/k^2$ for $k \to 0$ that can be demonstrated through a detailed calculation (Exercise 3.3.2, see Figure 3.16, p. 118).

Between the divergence as k^4 for $k \gg 1/h$ and as $1/k^2$ for $k \ll 1/h$, we must find a minimum corresponding to some optimum between the stabilising effects of different origins and the destabilising buoyancy force. This optimum is achieved for some intermediate value of the wavevector that, for dimensional reasons, can only be related to the thickness of the layer (Figure 3.3). Assuming that the diameter of the convection cells is, at threshold, of the order of h, i.e.

$$k_c = \frac{2\pi}{\lambda_c} = \frac{\pi}{h},$$

and inserting this value of the *critical wavevector* in the expression of β_m we get the *instability threshold* beyond which the convection regime develops, in the form of regular structures with typical wavelength $\lambda_c = 2\pi/k_c \simeq 2h$. Using expression (3.2) for the Rayleigh number, (3.14) yields:

$$R_c \sim \pi^4.$$

The semi-quantitative argument developed so far stresses on the physics of the processes at stake. The threshold value turns out to be grossly underestimated because a large part of the dissipating processes is badly evaluated, but it remains reasonable as an order of magnitude. We should notice that the model, however simple it may be, reproduces the two main characteristics of the instability: its stationary character and the general shape of the marginal stability curve with a correct asymptotic behaviour for $k \ll k_c$ and $k \gg k_c$. Similar simplified analyses will be developed as exercises to study the effects of molecular diffusion on convection in binary mixtures (Exercise 3.3.3), the stability of an angular momentum stratification in a cylindrical shear experiment (Taylor–Couette instability, Exercise 3.3.5), or the emergence of spatial structures in reaction–diffusion systems (Turing instability, Exercise 3.3.4).

3.1.5 Vicinity of the threshold: linear stage

Studying the linear dynamics of fluctuations in the neighbourhood of the threshold is best developed by first turning equations (3.3, 3.4) into dimensionless form. In order to do this, we have to choose length, time, and temperature scales. The thickness h of the layer is the obvious natural length scale. The thermal diffusion time over distance h, $\tau_\theta = h^2/\kappa$, being chosen as the time scale (the alternate possibility would be the viscous time $\tau_v = h^2/\nu$), the velocity scale then reads $h/\tau_\theta = \kappa/h$. Since it is preferable to keep ΔT as the control parameter, the composite quantity[2] $\kappa\nu/\alpha g h^3$ is taken as the temperature scale. Performing the changes $x \mapsto hx$, $t \mapsto \tau_\theta t,\ldots$ in (3.3, 3.4) we obtain

$$\partial_t v_z = P\left(\partial_{xx} v_z + \theta\right), \qquad (3.17)$$

$$\partial_t \theta = \partial_{xx}\theta + R v_z. \qquad (3.18)$$

A second dimensionless number has been introduced:

$$P = \frac{\nu}{\kappa} = \frac{\ell^2/\kappa}{\ell^2/\nu} = \frac{\tau_\theta}{\tau_v}. \qquad (3.19)$$

Called the *Prandtl number*, it characterises the physical properties of the fluid, specifying which of the viscous diffusion ($\tau_v = h^2/\nu$) or the thermal diffusion (τ_θ defined above) is the dominant relaxation process.

In gases P is of the order of unity and varies little with the nature of the gas since momentum (τ_v) is transported by the molecules themselves at

---

[2] From expression (3.2) for the Rayleigh number, it is easily checked that it is homogeneous to a temperature.

the same rate as energy (τ_θ). In condensed fluids this number can largely vary. For example, in liquid metals (e.g. mercury) it is very small, typically $< 10^{-2}$, since energy is efficiently transported by conduction electrons while atoms must be moved to smooth out velocity fluctuations, hence $\tau_\theta \ll \tau_v$. In insulating fluids, thermal diffusion mainly involves molecular vibrations that keep the same order of magnitude whatever the fluid, while the viscosity can vary by large amounts. P is of the order of 2–10 in water or alcohol, 10^2–10^4 in silicon oils depending on the polymerisation degree (molecular weight), and essentially infinite for the Earth mantle which is extraordinarily viscous and in which convection develops on geological times only.

In the limit $P \gg 1$, the flow adjusts itself to the temperature field instantaneously, which can be understood from the consideration of equation (3.17) written as

$$P^{-1}\partial_t v_z \simeq 0 = \partial_{xx} v_z + \theta,$$

showing that v_z is merely obtained by integrating θ over space. The dynamics is therefore simplified since we have just one relevant scalar field. On the contrary, when P is small, the inertia of the fluid cannot be neglected and a full hydrodynamic problem is recovered, with the vector nature of the velocity field and the incompressibility condition playing a crucial role.

Let us stay in the limit $P \gg 1$ and consider the critical mode $(v_z, \theta) \sim \sin(k_c x)$ with $k_c \simeq \pi$ (π/h if the physical dimension is restored). Within the framework of the simplified model we get:

$$v_z = \theta/\pi^2,$$

and upon insertion in (3.18):

$$\partial_t \theta = -\pi^2 \theta + R v_z = \left(-\pi^2 + R/\pi^2\right) \theta.$$

Dividing both members of this equation by π^2 and defining:

$$\tau_0 = \frac{1}{\pi^2} \quad \text{and} \quad r = \frac{R - R_c}{R_c}, \tag{3.20}$$

with here $R_c = \pi^4$ (but this value is only anecdotal) we simply get:

$$\tau_0 \partial_t \theta = r \theta. \tag{3.21}$$

The coefficient τ_0 therefore presents itself as a characteristic evolution time (τ_θ/π^2 in physical units) for convection, while r measures the relative distance to the threshold and is, of course, our control parameter.

Defining A as the amplitude of the most unstable convection mode and setting:

$$\theta \propto A(t) \sin(k_c x), \tag{3.22}$$

we get from (3.21) the linear evolution equation for A

$$\tfrac{\mathrm{d}}{\mathrm{d}t} A = \sigma A, \tag{3.23}$$

where $\sigma = r/\tau_0$ is its effective growth rate, which substantiates (1.16) introduced on phenomenological grounds in Chapter 1, p. 12. The corresponding time $\tau = 1/\sigma = \tau_0/r$ therefore diverges as r^{-1} close to the threshold ($r \ll 1$), a phenomenon called the *critical slowing-down*. Amplitude A plays the role of an *effective degree of freedom* for the fluid layer as a whole.

Result (3.23) is valid much more generally than suggested by the derivation above on the special case $P \to \infty$, and indeed holds in the vicinity of any linear instability. This is a consequence of the fact that σ is a non-singular function of the parameters, and thus can be expanded in Taylor series. Since the condition that defines the threshold $r_c = 0$ is precisely $\sigma = 0$, generically the expansion begins with its first order term $\sigma = r\, \partial_r \sigma|_c$, hence the observed behaviour of σ as a function of r.

The argument just produced can be repeated for a value of k different from the critical value k_c provided that we replace the threshold R_c by the corresponding marginal value $R^{(m)}(k)$. As long as k stays sufficiently close to k_c, the natural characteristic evolution time has no reason to be very different from τ_0, so that we can write at lowest order

$$\tau_0 \sigma(k) = \frac{R - R^{(m)}(k)}{R^{(m)}(k)}. \tag{3.24}$$

On the other hand, any curve in the vicinity of an *extremum* is generically equivalent to a parabola. The marginal curve close to its minimum at (k_c, R_c) is not an exception so that, for $k = k_c + \delta k$ and $\delta k/k_c \ll 1$, we can write

$$\frac{R^{(m)}(k) - R_c}{R_c} = \xi_0^2 \, \delta k^2, \tag{3.25}$$

where ξ_0^2 presents itself as the square of a characteristic length, the *coherence length*, which accounts for the curvature of the marginal stability curve at threshold.

3.1.6 *Classification of unstable modes*

One can arrange (3.24) and (3.25) together to write down the real part of the dispersion relation in the condensed form:

$$\tau_0 \sigma(k) \simeq r - \xi_0^2 (k - k_c)^2. \tag{3.26}$$

When the minimum of the marginal stability curve is reached for $k_c \neq 0$, case considered up to now of convection in a simple fluid, one says that the instability is *cellular*. Otherwise, it may happen that the most unstable mode is for $k_c = 0$ and the instability is then termed *homogeneous*. This situation, which occurs for example when convection takes place between horizontal plates that are bad thermal conductors, is often difficult to treat since the system is sensitive to lateral boundary conditions and/or any kind of slowly varying perturbations, while when $k_c \neq 0$, each cell with width $\lambda_c/2$ plays its own game, without worrying about lateral boundaries as soon as they are sufficiently far apart, say three or four wavelengths.

From a temporal viewpoint, Rayleigh–Bénard convection in a simple fluid is a *stationary* instability and the imaginary part $\omega(k)$ of the dispersion relation is identically zero. In other cases, the instability may be *oscillatory* with $\omega_c \neq 0$, where ω_c is the angular frequency at threshold. When the instability sets in with $k_c \neq 0$ and $\omega_c \neq 0$, the critical mode is in fact a *wave* propagating at some *phase velocity* c since, factoring out k_c one can write $\exp\bigl(i(k_c x - \omega_c t)\bigr) \equiv \exp\bigl(ik_c(x - ct)\bigr)$, as will be done in Chapter 7.

The real part σ of the eigenvalue of the marginal mode is well approximated by (3.26) in the neighbourhood of the threshold (r_c, k_c). In the same way, its imaginary part $\omega(k)$ can be expanded as:

$$\omega(k) = \omega_c + r\, \partial_r \omega|_c + \delta k\, \partial_k \omega|_c + \tfrac{1}{2} \delta k^2\, \partial_{kk}\omega|_c\,, \tag{3.27}$$

where derivatives with respect to r or k are computed at threshold.

The coefficient of δk (third term on the r.h.s.) corresponds to the *group velocity* of the waves. This can be seen by looking at a wave packet formed by superposition of elementary waves written as $V(x,t) = \int A(k) \exp i(kx - \omega t)\, \mathrm{d}k$, where $A(k)$ is the amplitude of mode k presenting a peak at some wavevector $k = k_0$. Setting $\omega_0 = \omega(k_0)$, we get:

$$V(x,t) = \exp i(k_0 x - \omega_0 t) \int A(k_0 + \delta k) \exp[i\, \delta k(x - \partial_k \omega|_{k_0} t) + \mathcal{O}(\delta k^2)]\, \mathrm{d}\delta k.$$

In the long time limit ($t \gg 1/\omega_0$), V is negligible everywhere except where the argument of the exponential is zero ('stationary phase' approximation) since elsewhere the rapid oscillations of the complex exponential "kill" the signal. This happens when $x/t = \partial_k \omega|_{k_0}$ which shows that this quantity is precisely the velocity of the wavepacket. In the same way, the coefficient of δk^2 in (3.27) accounts for the *dispersion* of the wavepacket, i.e. its smearing out due to changes in phase velocity. In a non-dispersive medium, the phase velocity is independent of the wavevector, i.e. $\partial_k(\omega/k) = 0$, so that $c_g \equiv \partial_k \omega = \omega/k \equiv c$ and of course $\partial_{kk}\omega \equiv 0$.

The Taylor–Couette instability of a fluid sheared between two coaxial cylinders rotating at different angular speeds (Exercise 3.3.5) is also cellular and stationary. In chemistry, the Belousov–Zhabotinsky reaction is an example of homogeneous oscillatory instability. Finally, in some circumstances, convection in binary fluid mixture develops in the form of dissipative waves (Exercise 3.3.3).

3.2 Disintegration of Dissipative Structures

The study of the transition to turbulence of structures generated by an instability mechanism consists of several steps. The first one is the determination of states achieved beyond threshold. The next relates to the destabilisation of such states, and so on. The game is then repeated up to a point where the regime obtained is completely irregular. In this section we begin with a simple modelling of nonlinear effects in convection, §3.2.1. A brief account of experimental observations about the transition is then given in §3.2.2, where we point out the role of geometrical effects. This leads to a fundamental distinction between *confined* systems for which the concept of *temporal chaos* is relevant, §3.2.3, and *extended* systems for which the disorganisation in space is as important as the irregularity in time, i.e. *spatiotemporal chaos*, §3.2.4. We conclude the chapter with a brief presentation of convection in the post-transitional regime where the concept of *developed turbulence* begins to make sense, §3.2.5. Here we mostly stay at a phenomenological level, deferring the introduction of theoretical tools to subsequent chapters. A more in-depth presentation of experiments and related theory is to be found in [Getling (1998)].

3.2.1 *Simplified model of nonlinear convection*

Relation (3.22) defines a variable A measuring the intensity of the perturbation of the base state. At steady state, we thus expect $A \equiv 0$ below threshold and $A \neq 0$ above. In the theory of thermodynamic *phase transitions* A would be called an *order parameter* [Stanley (1988)]. However, (3.23) is valid only as long as A stays infinitesimal and must be completed to account for the range $r > 0$. In order to get (1.19), p. 13, we just postulated heuristically that convection was a self-limiting process and we replaced σ in (3.23) by an effective value,[3]

$$\sigma_{\text{eff}} = \tau_0^{-1}(r - gA^2) \quad \text{with} \quad g > 0.$$

Beyond threshold, for $r > 0$ ($R > R_c$), several processes indeed come and limit the growth of A. First the dissipation increases, and second the destabilising force decreases since part of the heat is transported by the flow, so that the bulk effective temperature gradient that governs the conductive part of the heat flux decreases below its nominal value β.

It is of course possible to derive an accurate model of nonlinear evolution from the primitive equations in a systematic way. Here we rather continue to develop a heuristic formulation, guided by the result to be obtained. We no longer assume that $P \gg 1$ but restrict ourselves to the consideration of the most unstable linear mode. On more general grounds than for (3.22), we then take $\{v_z, \theta\} = \{V(t), \Theta(t)\}\sin(k_c x)$, where V et Θ are two time-dependent amplitudes. Injecting this assumption in (3.17, 3.18) we obtain:

$$\frac{d}{dt}V = P(\Theta - \pi^2 V), \tag{3.28}$$

$$\frac{d}{dt}\Theta = RV - \pi^2 \Theta. \tag{3.29}$$

System (3.28, 3.29) needs to be completed with nonlinear terms arising from the advection of the fluctuations $\mathbf{v} \cdot \nabla \mathbf{v}$ and $\mathbf{v} \cdot \nabla \theta$. In the spirit of a first harmonic approximation now developed in space and not in time as for the van der Pol oscillator, p. 62, we guess that the terms that contribute are those resonating with the postulated dependence in $\sin(k_c x)$. From the continuity equation $\partial_x v_x + \partial_z v_z = 0$, assuming $v_z \propto \sin(k_c x)$ one gets $v_x \propto \cos(k_c x)$, so that, in the equation for v_z, the advection term $v_x \partial_x v_z + v_z \partial_z v_z$ varies as $\sin^2(k_c x) = \frac{1}{2}(1 - \cos(2k_c x))$, i.e. produces nothing in resonance with $\sin(k_c x)$. Averaging over the thickness of the layer and over a wavelength, we thus expect a negligible contribution from these terms to (3.28) which remains unchanged at this order.

The problem is different for (3.29). As a matter of fact, a parallel argument would also imply no complementary term, but this would not reflect the fact that, as indicated above, part of the heat is transported by the convection motion. The corresponding flux is easily identified with the product $v_z \theta$. As discussed in Sec. 3.1.2, below threshold, the destabilising part of the convection mechanism relies on the advection, due to differential buoyancy, of temperature fluctuations in a purely conductive temperature gradient. Above threshold, since part of the heat is transported

---
[3] Notation g introduced to measure the intensity of nonlinear couplings is traditional. In the context of convection it should not be mistaken with the gravitational acceleration, but the risk is limited.

by convection, differential buoyancy has to be appreciated with respect to a conductive temperature gradient β_{eff} which is decreased by the contribution of convection from its nominal value β evaluated from the applied temperature. That contribution, $v_z\theta \propto \sin^2(k_c x)$, produces: (i) a second harmonic component $\propto \cos(2k_c x)$ that averages to zero over a fluctuation wavelength, but also (ii) a term at $k = 0$ that corresponds to a correction brought to the averaged temperature profile.

Considering this correction as a variable in itself, let us call it Ψ and look for its governing equation. A simple calculation then yields:

$$\tfrac{d}{dt}\Psi = V\Theta - b\Psi, \tag{3.30}$$

where the first term on the right hand side is the source term issued from space-independent part of $v_z\theta$ – contribution (ii) above – and the second term accounts for its diffusive relaxation according to the Fourier law, at a decay rate b that could be computed explicitly.

The argument about the effective intensity of the convection mechanism sketched above is simply implemented by subtracting the convective contribution Ψ from the nominal Rayleigh number R to form an effective Rayleigh number R_{eff} replacing it in (3.29). This yields:

$$\tfrac{d}{dt}\Theta = (R - \Psi)V - \pi^2\Theta. \tag{3.31}$$

Equations (3.28, 3.31, 3.30) generalise the linear model derived previously. They form the celebrated *Lorenz model* that played an important role in the development of ideas about chaos since 1963 (Note 6, p. 17), its original expression being recovered by appropriately re-scaling time, variables V, Θ, Ψ, and parameter R. See §B.4.2, p. 420 for a hands-on study.

Let us first show how this model allows one to recover the effective Landau equation (1.19) introduced p. 13, extending (3.23) to the nonlinear regime. As noticed earlier, close to the threshold, the dynamics of the system is very slow. Its evolution rate is proportional to $r = (R - R_c)/R_c \ll 1$. But equation (3.30) shows that the natural relaxation time of correction Ψ remains $\mathcal{O}(1)$. We can thus assume that Ψ rapidly relaxes towards a value $V\Theta/b$, itself slowly varying at a rate $\mathcal{O}(r)$. Let us insert this value in (3.31) and admit that $V = \Theta/\pi^2$ for all times, which, from (3.28), is true only in the limit $P \to \infty$. We get:

$$\tfrac{d}{dt}\Theta = [(R/\pi^2) - \pi^2]\Theta - \Theta^3/\pi^4 b,$$

which we rewrite as:

$$\frac{1}{\pi^2}\frac{d}{dt}\Theta = \frac{R - \pi^4}{\pi^4}\Theta - \frac{1}{b\pi^6}\Theta^3. \tag{3.32}$$

This equation is therefore exactly (1.19), i.e. :

$$\tau_0 \tfrac{\mathrm{d}}{\mathrm{d}t} A = rA - gA^3, \tag{3.33}$$

with Θ, the amplitude of the temperature modulation playing the role of the effective variable A, $\tau_0 = 1/\pi^2$, $r = (R - R_\mathrm{c})/R_\mathrm{c}$, $R_\mathrm{c} = \pi^4$, and $g = 1/b\pi^6$.

Obtaining the effective equation (3.33) is an example of reduction by *adiabatic elimination of enslaved variables*. Here variables V and Ψ are enslaved to Θ: At every instant their values are fixed by that of $\Theta \equiv A$ according to relations $V = \Theta/\pi^2$ and $\Psi = \Theta^2/b\pi^2$. This step plays an essential role in the study of nonlinear dissipative systems.

Equation (3.33) accounts for the bifurcation from the conduction branch corresponding to the trivial solution $A \equiv 0$ towards the convection branch associated to the pair of (here time-independent) nontrivial solutions. These *bifurcated solutions*, $A_{(\pm)} = \pm\sqrt{r/g}$, are given by the condition $\tfrac{\mathrm{d}}{\mathrm{d}t} A = 0$, and therefore correspond to *fixed points* of (3.33). Quantity g is positive, so that they exist for $r > 0$: the bifurcation is *supercritical*. All this has already been presented in the introductory chapter, see Figure 1.4 p. 13, and will be extended to general systems involving a single stationary mode in Chapter 4, especially in Exercise 4.4.4, p. 165.

We have previously stated without justification that the bifurcated solutions corresponding to convection beyond threshold are stable. Let us show how this arises from (3.33) using tools introduced in Chapter 2. Setting $A = +\sqrt{r/g} + A'$, we easily get the equation for perturbation A' by substitution. After simplification we obtain

$$\tau_0 \tfrac{\mathrm{d}}{\mathrm{d}t} A' = -2rA', \tag{3.34}$$

so that A' decays for $r > 0$. In fact this is valid only close enough to the threshold, before new instabilities have any chance to set in.

The argument developed above only holds as long as $R \approx R_\mathrm{c}$. When $R \gg R_\mathrm{c}$, the relaxation time of Θ and V towards their equilibrium values, derived from (3.34), infinite at threshold, shortens as r increases and rapidly becomes of the order of magnitude of relaxation time of Ψ. Adiabatic elimination of the latter is then no longer legitimate: Ψ is less and less enslaved to Θ and V but on the contrary gains a status of genuine degree of freedom.

3.2.2 Transition to turbulence of convection cells

Let us now consider nonlinear convection from an experimental point of view. The actual situation is less transparent than what has just been

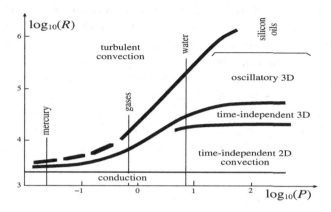

Fig. 3.4 Transition toward turbulence in convection, after Krishnamurti, Note 4. The Prandtl number is varied by changing the fluid. The transition lines are intentionally made thick to indicate orders of magnitude rather that precise thresholds.

described by using the simplified model, but one fact remains: the effective dimension of the problem increases with R. Unfortunately, the physical mechanisms that destabilise the cellular structure to produce the *secondary modes* are much less intuitive than the primary mechanism.

In principle, the method is the same as for the primary mode but, at steady state, the base flow beyond threshold is now made of finite-amplitude time-independent convection cells. The study is considerably more complicated than when we had to deal with the uniform conducting state since the new base flow is periodic along one horizontal direction. Accordingly, the operator obtained through linearisation now explicitly depends on space, which forbids the direct recourse to Fourier transforms to solve the problem. This will be re-examined theoretically later. For the moment, let us describe the cascade towards turbulence from a phenomenological point of view.

The convection threshold was independent of the Prandtl number P, whose value just played some role in the nature of the primary mode, thermal when $P \gg 1$, hydrodynamic when $P \ll 1$. This simple fact has profound consequences on the shape of the secondary modes and the subsequent bifurcation cascade towards turbulence. A compilation of early results adapted from Krishnamurti[4] is displayed in Fig. 3.4.

---

[4] R. Krishnamurti, "Some further studies on the transition to turbulent convection," J. Fluid Mech. **60** (1973) 285.

Upon increasing R, the fluid layer first experiences a transition from pure conduction (fluid uniformly at rest) to two-dimensional time independent convection (2D: fluctuations depend locally on two coordinates, z and one coordinate in the direction perpendicular to the local direction of the convection rolls). At sufficiently large Prandtl number a three-dimensional regime sets in (3D: fluctuations now depend on x, y, z); at first time-independent, next periodic, and eventually turbulent. At small P the domain of 'time-independent two-dimensional' convection is very narrow and an irregular time dependence rapidly sets in, here called "turbulent convection".[5]

3.2.2.1 *Large-Prandtl-number fluids*

When $P \gg 1$ (e.g. , with highly viscous oils), the temperature field drives everything, inducing the vertical velocity component directly and the horizontal component indirectly *via* the continuity condition.

Secondary instabilities specific to this case remain localised within thermal boundary layers close to the horizontal plates. These boundary layers get thinner and thinner as the Rayleigh number is increased and, at some point, they become unstable against the plain Rayleigh mechanism. A stationary secondary instability called *bimodal* sets in, with rolls oriented at right angles with the primary rolls and located in the thermal boundary layers. Since the fluctuations are now modulated in the three directions of space, the regime is labelled "time-independent 3D" in Figure 3.4 on p. 101. Time dependence next manifests itself as a periodic break-down and reformation of thermal boundary layers first analysed by Howard. Strict periodicity is then lost and an irregular dynamics sets in.

3.2.2.2 *Intermediate- and low-Prandtl-number fluids*

At smaller P, the situation is more confused. Busse and his collaborators have identified a large number of possible secondary modes leading to a complicated picture in the (R, P, k) parameter space called the *Busse balloon*, owing to the global shape of the region where straight rolls are stable. When $P \sim 1$ (water, gases) or smaller (liquid metals), the velocity field becomes dominant through specific contributions of the advection term $\mathbf{v} \cdot \nabla \mathbf{v}$. The unstable secondary modes appear close to the convection threshold and occupy the whole thickness of the layer. Cells enter a kind

---

[5]For a review of higher instability modes in convection, consult: F.H. Busse, "Transition to turbulence in Rayleigh–Bénard convection," in [Swinney and Gollub (1985)].

of free-wheel regime where friction on the plates is dealt with inside thin viscous boundary layers. Time dependence enters very early in the form of 'Busse oscillations' that are sorts of waves propagating along the convection rolls due to an inertial call-back of roll axis undulations. Most often it turns out to be difficult to identify a range of Rayleigh numbers over which the periodic behaviour is strictly regular, and the flow is often considered turbulent right at the onset of oscillations.

3.2.2.3 Transition towards turbulence, conceptual problems

At least for $P \gg 1$ there seems to be a small number of well defined steps on the way between the conduction regime and turbulence. This apparently supports the viewpoint advanced by Ruelle et Takens in 1971 (note 5, p. 16) according to whom the stochastic behaviour, a fundamental property of turbulence, generically appears at the end of a short cascade of three or four bifurcations. Previously, Landau (note 4, p. 16) explained his understanding of turbulence as the result of an indefinite superposition of modes, each with its own time-space scale, i.e. quasi-periodicity with an infinite number of incommensurate frequencies. These two interpretations were sketched in Figure 1.7, p. 17.

In fact, neither of the Ruelle–Takens or Landau pictures offer a satisfactory interpretation of experimental results reported above (Figure 3.4). All observations were made in containers that were very wide in order to check theories developed for a laterally unbounded system. The so-obtained convection *patterns* were rarely regularly organised but on the contrary were inhomogeneous with lots of defects, so that the transition thresholds were not defined as sharply as a bifurcation point. Moreover, a slow residual time dependence was often observed.

Having recognised that these interpretation problems were mostly due to spatial disorder, which in turn resulted from the presence of a large number of cells, and that lateral boundaries at large distances were ineffective in maintaining long range order in the patterns, experimentalists have tried to better control the situation by turning to systems with a small number of cells, hence lateral dimensions of containers of basically the same order of magnitude as their heights.

Confinement effects can be appreciated through *aspect ratios* defined as

$$\Gamma = \ell/h, \qquad (3.35)$$

where ℓ represents the typical lateral extension of the system, see Figure 3.5. We shall re-examine their physical role later in Chapter 4, §4.1.

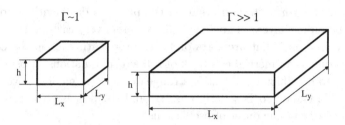

Fig. 3.5 Aspect ratio for closed systems, either confined (left) or extended (right).

Early experiments reported above were performed in the limit $\Gamma \gg 1$ that characterise *extended systems* and for which the concept of *spatiotemporal chaos* to be introduced in Chapter 6 seems more appropriate. In contrast, strongly *confined* systems, characterised by $\Gamma \sim 1$, can be expected to better fit the framework proposed by Ruelle and Takens and their concept of *temporal chaos*. As a matter of fact, guaranteeing strong spatial coherence among a small number of convection cells, confinement effects should be instrumental in restricting the dynamics to couplings within a small set of effective variables.

3.2.3 *Transition toward chaos in confined systems*

The literature about the transition from regular to chaotic time behaviour is sufficiently rich that we can limit ourselves to the presentation of few experimental results obtained at the beginning of the eighties as typical examples of the main scenarios. This sketchy description is given mainly as an illustration of the kind of phenomena to be understood theoretically rather than as a review that would rather be premature at this stage. Consult the general bibliography for more detailed information, especially [Hao (1990); Cvitanović (1989)].

3.2.3.1 *Subharmonic cascade*

The first experiment to be reported here has been performed by Libchaber et Maurer.[6] Liquid helium with $P \sim 1$ is placed in a parallelepipedic container with aspect ratios $\Gamma_x = 2.4$, $\Gamma_y = 1.2$. Stationary convection

---

[6]A. Libchaber and J. Maurer, "Une expérience de Rayleigh–Bénard en géométrie réduite; multiplication, accrochage et démultiplication de fréquences," J. Physique Colloques **41-C3** (1980) 51–56.

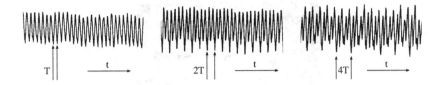

Fig. 3.6 Time series of the temperature signal measured at a given point during the first steps of a subharmonic cascade, after Libchaber and Maurer, Note 6.

sets in beyond some threshold R_c. At $R \simeq 30 R_c$ the system experiences a bifurcation toward an oscillatory regime. Then, at $R \simeq 39.5 R_c$, a second mode with an incommensurate period sets in. This two-periodic regime persists up to $R \simeq 40.5 R_c$ when, while shifting, the second period gets locked to twice the first one, the system is then periodic with a period $2T$. The scenario under study now begins: a second period doubling (period $4T$) at $R \simeq 42.7 R_c$ (Figure 3.6). After several supplementary period doublings (period $8T$, $16T$,... the system enters a chaotic regime for $R > 43 R_c$. A complementary study of Fourier spectra would reveal first fine lines at one frequency and its harmonics, then a second family of lines and many combinations (two-periodic regime), then, after the locking, the return to a simpler spectrum with one fundamental line at $\omega = 2\pi/T$ and its harmonics. The period doubling cascade manifests itself by the growth of *subharmonics* at $\omega/2$, next $\omega/4$, etc. As long as the system is periodic, no matter how long the period, the spectral lines remain narrow but when it becomes chaotic, they get measurably enlarged at their foot.

3.2.3.2 *Chaos on a two-periodic background*

This transition, closely reminiscent of the scenario originally proposed by Ruelle and Takens, has been observed roughly at the same epoch by Dubois and Bergé[7] again in parallelepipedic geometry with similar aspect ratios, $\Gamma_x = 2$, $\Gamma_y = 1.2$, but this time with silicon oil ($P \simeq 130$). In contrast with the previous experiment, visualisation of the structure was possible by means of differential interferometry, Figure 3.7, which made easier the understanding of motions at the origin of the observed fluctuations and the choice of points where to measure the velocity using a LASER Doppler anemometer.

The following sequence was observed: 1) conduction regime up to R_c.

---

[7]M. Dubois and P. Bergé, "Instabilités de couche limite dans un fluide en convection: évolution vers la turbulence," J. Physique **42** (1981) 167.

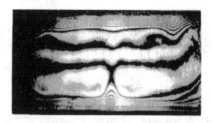

Fig. 3.7 Isotherms in silicon oil can be visualised by differential interferometry; the fringes originate from the variations of the refraction index induced by the local temperature gradients. (Courtesy M. Dubois.)

2) Time-independent convection from R_c to $R \simeq 215 R_c$. 3) Bifurcation towards a periodic regime with period T_1. 4) Two-periodic dynamics with a second period T_2 from $R \simeq 250 R_c$ up. Geographically well separated, the two oscillation modes are weakly coupled, which explains the relative robustness of the two-periodic regime and a characteristic alternation of locking/unlockings when the ratio of the periods, that slightly shifts with R, passes from incommensurate to commensurate values and *vice versa*. 5) For $R > 305 R_c$, temporal chaos enters as an irregular slow modulation of a locked periodic behaviour that gives a series of widened spectral lines and low frequency power in the Fourier spectrum, Figure 3.8.

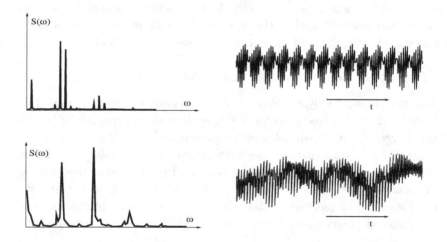

Fig. 3.8 Quasi-periodic convection in silicon oil. Fourier spectra (left) and time series (right) of a velocity component at a given point in the experimental cell. After Dubois and Bergé, Note 7.

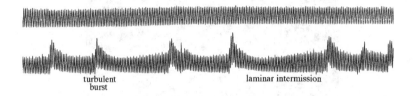

Fig. 3.9 The intermittency scenario: A periodic regime ("laminar" intermissions) is irregularly interrupted by chaotic bursts that become more frequent as R increases beyond the *intermittency threshold*. Upper trace: regular signal observed below threshold. Time is running from left to right. After Bergé et al., Note 8.

3.2.3.3 Intermittency

The third scenario to be described here has also been observed by Bergé and Dubois[8] with the same fluid and the same experimental set-up but with a slightly different initial convection structure. Accordingly, a different transition scenario developed after a single step involving a secondary instability mechanism with a hot droplet transported by the general convection and playing the role of a pacemaker. Convection was time-independent up to $R \simeq 250 R_c$, then periodic with period T. Not far above a subharmonic bifurcation (hence period $2T$), for $R = 290 R_c$, the system experienced a transition to chaos with irregularly distributed "turbulent" bursts interrupting the previously observed regular periodic behaviour forming "laminar" intermissions. When the Rayleigh number was increased, the frequency of the bursts was seen to increase. Time series of the velocity signal before and after the transition are displayed in Figure 3.9, upper trace and lower trace, respectively.

3.2.4 Dynamics of textures in extended systems

In contrast with what has just been described, before 1975 the transition to turbulence was studied in extended systems and focused more on the occurrence of a developed turbulent regime where most of the spatial structure was lost and the time dependence strongly irregular. The low frequency noise often observed at early stages was not recognised as an interesting phenomenon related to the transition process. The study of the emergence of chaos in confined systems has also led to reconsidering the situation for

---

[8] P. Bergé, M. Dubois, P. M., Y. Pomeau, "Intermittency in Rayleigh–Bénard convection," J. Physique Lettres **41** (1980) L341–345.

Fig. 3.10 Texture observed in convection at large Prandtl number as seen from above. (Courtesy V. Croquette.)

extended systems and introducing the notion of *spatiotemporal chaos* as an element of interpretation of the transition process. In practice, the instability mechanisms preserve coherence at the *local* scale (few convection cells) but are unable to maintain it on a *global* scale (the set-up). This can be understood as the result of the possible interference of a large number of neighbouring modes easily excited immediately beyond threshold (§3.1.5 and Chapter 6). Disordered patterns with many defects of all sorts, called *textures*, are generally observed in the absence of any induction process forcing the growth of regularly oriented "clean" convection structures. Here again we have to distinguish between fluids according to their Prandtl number.

3.2.4.1 Textures in fluids with high Prandtl number

When P is large, quasi-stationary convection structures are obtained, which relates to the fact that the dominant field is the temperature, a scalar, and that the fluid layer mostly behaves as a gradient system (see §2.1.3, p. 47). It is then possible to interpret the principal features of the textures observed in terms of a single amplitude field. An example[9] is given in Figure 3.10, where one can easily identify *grains* of convection rolls with nearly uniform orientations, *grain boundaries* along which two grains with different orientations meet, *dislocations* where a pair of rolls suddenly ends. It should also be noted that rolls arrive mostly perpendicular to the lateral boundaries and that the frustration implied by this topological constraint is partly resolved by the presence of a large scale *curvature* of the rolls. The evolution of such textures is very slow, when compared to the velocity

---

[9] A. Pocheau and V. Croquette, "Dislocation motion: a wavenumber selection mechanism in Rayleigh–Bénard convection," J. Physique **45** (1984) 35–48.

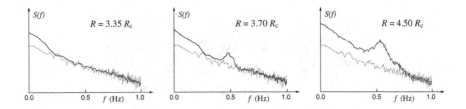

Fig. 3.11 Weak turbulence observed in a cylindrical cell with $\Gamma = D/h = 12$ in liquid helium at low Prandtl number. Typical background noise is the grey line in each figure. For $R = 3.35R_c$ low frequency noise develops first. At $R = 3.70R_c$ noisy Busse oscillations at finite frequency around 0.5 Hz have settled in the system. As seen for $R = 4.50R_c$ and beyond, the level of noise continues to increase gently. After Libchaber and Maurer, Note 10.

of the fluid. The transition from 2D to 3D convection mediated by defects, and the emergence of (weak) turbulence can be understood as a kind of melting of the global pattern, with progressive loss of local order.

3.2.4.2 Transition at small Prandtl number and large aspect ratios

All experimental observations show that turbulence occurs early when P is small. This feature has to be attributed to the fact that the viscosity is low and that Reynolds numbers constructed from the velocity induced by convection and the size of the cells rapidly become large. Like for low-dimensional systems, the complexity of the dynamics is considerably enriched by inertial effects that favour oscillatory behaviour (here mainly Busse oscillations). This explains that mode interactions, even close to the convection threshold, generate a much more "active" behaviour than what is observed at higher Prandtl numbers. Another source of complexity comes from the existence of large scale flows directly generated by curvature and defects in the global texture.

Figure 3.11 illustrates the scenario observed again by Libchaber and Maurer[10] in liquid helium but in a cylindrical container with diameter D and aspect ratio $\Gamma = D/h = 24$. Remarkably enough, a low frequency noise sets in before any trace of secondary instability and when the latter develops (Busse oscillations) the system is already disordered so that the corresponding frequency is not sharply defined. It took some time before this behaviour could be understood.

---

[10] A. Libchaber and J. Maurer, "Local probe in a Rayleigh–Bénard experiment in liquid helium," J. Physique Lettres **39** (1978) L369–L372.

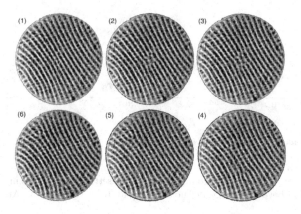

Fig. 3.12 Cyclic evolution from state (1) → (2) → ... → (6) → (1): Mechanism of nucleation–dissociation–migration–annihilation of dislocations pointed by Pocheau et al., Note 11, and here illustrated with pictures kindly provided by V. Croquette from a similar experiment with $\Gamma = D/h = 40$.

In an experiment where helium was replaced by argon under pressure at room temperature, making visualisations possible, Pocheau et al.[11] later showed that the noise developing slightly above threshold was due to the synchronisation loss of an initially periodic process of nucleation, migration, annihilation of dislocations, as illustrated in Figure 3.12. The migration of dislocations was driven by a secondary flow at the scale of the container, the existence of which was predicted by earlier theoretical studies[12] and explicitly confirmed by specific experiments later.[13]

Again for low Prandtl numbers, somewhat above the range of Rayleigh numbers where the convection pattern is made of possibly slowly evolving, essentially straight rolls, a much more disorganised active state is observed in the form of rotating spirals as illustrated in Figure 3.13. This *spiral defect*

---

[11] A. Pocheau, V. Croquette, and P. Le Gal, "Turbulence in a cylindrical container of Argon near threshold of convection," Phys. Rev. Lett. **55** (1985) 1094–1097, later reviewed by V. Croquette, "Convective Pattern Dynamics at Low Prandtl Number. Part I, II," Contemporary Physics **30** (1989) 113-133, 153–171.

[12] (a) E.D. Siggia, A. Zippelius: "Pattern selection in Rayleigh–Bénard convection near threshold," Phys. Rev. Lett. **47** (1981) 835–838. (b) P. Manneville, J.M. Piquemal: "Transverse phase diffusion in Rayleigh–Bénard convection," J. Physique Lettres **43** (1982) L253–L258. (c) M.C. Cross, A.C. Newell: "Convection patterns in large aspect ratio systems," Physica D **10** (1984) 299–328.

[13] V. Croquette et al. "Large-scale flow characterisation in a Rayleigh–Bénard convective pattern," Europhys. Lett. **1** (1986) 393–399.

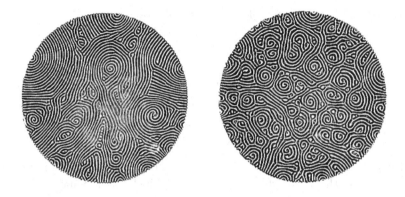

Fig. 3.13 Convection in CO_2 with $\Gamma = D/h \simeq 150$, spiral defect chaos observed at $r = 0.536$ (left) and $r = 0.894$ (right). After Morris et al., Note 14, courtesy G. Ahlers (UCSB).

chaos, observed in particular by Morris et al.[14] permanently evolves in both space and time. It takes place only at large aspect ratios but then does not depend on the lateral shape of the container. Furthermore, it is extensive in the sense that it can be characterised by a surface density of spiral cores seen to increase with increasing Rayleigh numbers beyond a threshold that depends on the Prandtl number. Curvature-induced secondary flows seem essential to its occurrence.

3.2.5 *Turbulent convection*

The local study of convection structures has been, since its very beginning, completed by measurements of the heat flux through the whole experimental container, characterising the global behaviour. Results are usually expressed in terms of the dimensionless *Nusselt number*:

$$N = \frac{\text{total heat flux}}{\text{conduction heat flux}}$$

where the total heat flux is the quantity actually measured and the conduction heat flux is the flux that would be computed from the temperature difference upon assuming that the fluid is at rest in the pure conduction state. Hence one gets $N \equiv 1$ when $R < R_c$, while $N - 1$ measures the

---

[14] S.W. Morris, E. Bodenschatz, D.S. Cannell, G. Ahlers, "The spatio-temporal structure of spiral-defect chaos," Physica D **97** (1996) 164–179.

contribution of convection. Close to threshold, one expects:

$$N - 1 \propto v_z \theta \propto \frac{R - R_\text{c}}{R_\text{c}}, \qquad (3.36)$$

since both θ and v_z vary as $[(R - R_\text{c})/R_\text{c}]^{1/2}$, which is indeed well observed experimentally, see Figure 3.14, p. 113.

Far beyond threshold, from scaling arguments familiar in the theory of turbulence (Chapter 8), a power law behaviour is expected instead:

$$N \sim R^\gamma.$$

Early experiments seemed to support a theory by Malkus predicting $\gamma = 1/3$ but the Rayleigh-number range studied was too narrow, while other studies for $P \ll 1$ suggested rather $\gamma = 1/4$.

At the end of the eighties, the problem became a topic of renewed interest, experimental and theoretical. The heat flux was studied over R-ranges extending up to 10^6, then 10^{12} and even 10^{17} in fluids with various Prandtl numbers and in containers with aspect ratio of order 1/2 or 1. Exponents γ ranging from 1/2 to 1/4, through 1/3, 0.3 or 2/7, have been measured over (sometimes very) limited ranges of Rayleigh numbers.[15]

In Figure 3.14 drawn after the results of Chavanne et al. in liquid helium, Note 15(c), one can identify the linear behaviour close to threshold expected from (3.36), a 'soft turbulence' regime where chaos is still mostly temporal as discussed in §3.2.3, then 'hard turbulence' with an exponent $\gamma \simeq 2/7$ explained by a theory involving thermal transfer through turbulent layers sheared by the general circulation blowing as "wind" along the horizontal walls, and an "ultimate" regime with exponent tending to 1/2. The discussion bears on the existence of asymptotic regimes with a single exponent or rather on a superposition of power laws in the form $N = C_1 R^{\gamma_1} + C_2 R^{\gamma_2}$ and the role of P (Note 15 d, e), of the geometry and nature of lateral walls, etc.

---

[15] We quote here only few references, first the general presentation by (a) E.D. Siggia, "High Rayleigh number convection," Annu. Rev. Fluid Mech. **26** (1994) 137-168, and next specific results by (b) J.J. Niemela et al., "Turbulent convection at very high Rayleigh numbers," Nature **404** (2000) 837–840, err. **406** (2000) 439; (c) X. Chavanne et al., "Turbulent Rayleigh–Bénard convection in gaseous and liquid He," Phys. Fluids **13** (2001) 1300–1320; a theory by (d) S. Grossmann and D. Lohse, "Scaling in thermal convection: a unifying theory," J. Fluid Mech. **407** (2000) 27–56 and "Thermal convection for large Prandtl numbers," Phys. Rev. Lett. **86** (2001) 3316–3319; corresponding experimental work by (e) G. Ahlers and X.-c. Xu, "Prandtl-number dependence of heat transport in turbulent Rayleigh–Bénard convection," Phys. Rev. Lett. **86** (2001) 3320–3323; (f) for a general review, consult: G. Ahlers et al: "Heat transfer and large scale dynamics in turbulent Rayleigh–Bénard convection," Rev. Mod. Phys. **81** (2009) 503–537.

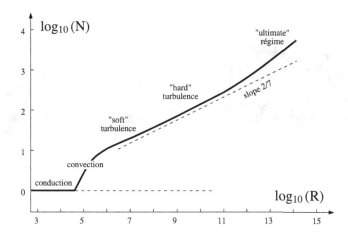

Fig. 3.14 Total heat flux as measured in terms of the Nusselt number as a function of the Rayleigh number (in log-log scale), after Chavanne et al., Note 15(c).

To conclude, convection presents itself as the prototype of stationary cellular instabilities. In this chapter we have described its particularly intuitive mechanism, and its subsequent destabilisation up to turbulence. We have also noted the role of confinement effects on the nature of these steps. In the next three chapters we shall examine in more detail some mathematical aspects of the theory that allows us to interpret these phenomena and, at the same time, to tackle a large class of instabilities in continuous media. We shall not come back to turbulent convection, owing to the limited scope of Chapter 8 devoted to the simpler case of turbulent shear flows.

3.3 Exercises

3.3.1 Simple model of cellular instability

Consider the linear part of the *Swift–Hohenberg model*[16] that reads:
$$\partial_t v = rv - (\nabla_\perp^2 + 1)^2 v\,. \tag{3.37}$$

This model accounts for the emergence of convection cells in a simplified but physically meaningful way and will be used further in §B.4.4–B.4.5.

Variable v may represent the vertical velocity component in the fluid or the departure from the base temperature profile, r is the control parameter

---

[16] J. Swift, P.C. Hohenberg: "Hydrodynamic fluctuations at the convective instability," Phys. Rev. A **15** (1977) 319–328.

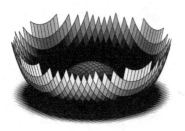 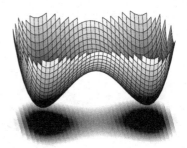

Fig. 3.15 Marginal stability surfaces for the linearised Swift–Hohenberg model in the isotropic (left) and anisotropic (right) cases.

measuring the relative distance to the threshold: $r \propto (T-T_{\rm c})/T_{\rm c}$. In (3.37), $\nabla_\perp^2 \equiv \partial_{xx} + \partial_{yy}$ is the Laplacian operator acting on the space dependence of the fluctuations in the plane of the layer.

1) Laterally unbounded medium.

a) Determine the dispersion relation $s = s(\mathbf{k})$ for Fourier modes taken in the form $\exp(i\mathbf{k} \cdot \mathbf{x})$ where $\mathbf{k} = (k_x, k_y)$ and $\mathbf{x} = (x, y)$. Check that it depends only on $k = |\mathbf{k}_{\rm h}|$.

b) Draw the graph of $s(k)$ for $r < 0$, $r = 0$ and $r > 0$ and conclude that the system bifurcates towards a cellular structure at $r = 0$. Show in particular that the most dangerous modes correspond to $|\mathbf{k}| = 1$.

c) Sketch the marginal stability surface ($s(\mathbf{k}r) = 0$) in the three-dimensional space (k_x, k_y, r); determine the set of marginal modes and the domain of unstable wavevectors $s(\mathbf{k}, r) > 0$ when $r > 0$. [Answer: Figure 3.15 (left).]

d) System (3.37) is isotropic in the $(x\text{–}y)$ plane. Consider now the modified anisotropic model

$$\partial_t v = rv - \left[(\partial_{xx} + 1)^2 - \partial_{yy}\right] v. \qquad (3.38)$$

Sketch the marginal stability surface for this case and conclude that, in contrast with the isotropic case, the linear stability operator selects a non degenerated mode. [Answer: Figure 3.15 (right).]

2) Consider now the same system but restricted to one space dimension:

$$\partial_t v = rv - (\partial_{xx} + 1)^2 v, \qquad (3.39)$$

for a function $v(x,t)$ defined on a finite interval of length ℓ with boundary conditions $v = \partial_{xx} v = 0$ at $x = 0$ and $x = \ell$.

Check that the eigenmodes can be taken in the form $V = A\sin(k_n x)$ with $k_n = n\pi/\ell$. Determine the marginal stability condition $r = r_n^{(\text{m})}(\ell)$ for mode n at given ℓ, and next the instability threshold $r_\text{c}(\ell) = \inf_n r_n^{(\text{m})}(\ell)$, as a function of ℓ.

How does the spatial resonance between the intrinsic length-scale $\lambda_\text{c} = 2\pi/k_\text{c} = 2\pi$ and the size ℓ of the system manifest itself? Find the condition for two neighbouring modes being simultaneously marginal.

The reader is encouraged to perform the same (but much more difficult) study for boundary conditions $v = \partial_x v = 0$ at $x = \pm \ell/2$. This is more typical of the general case since simple trigonometric lines are no longer appropriate. The eigenmodes will be searched for as a superposition of elementary solutions to the 4th order differential equation

$$sv = \left[r - \left(\tfrac{\mathrm{d}^2}{\mathrm{d}x^2} + 1\right)^2\right] v$$

that fulfil the boundary conditions (the scalar s is the eigenvalue). Separate odd from even solutions and find corresponding marginal conditions (given by transcendental equations to be solved numerically by some root-finder program).

This model, completed by appropriate nonlinear terms, will be used again in Chapter 4, Exercise 4.4.2, and for the hands-on numerical experiments in §B.4.4–B.4.5.

3.3.2 Rayleigh–Bénard convection: detailed study

The purpose of the exercise is to go beyond the semi-quantitative approach developed in §3.1.4 and determine the marginal stability condition from the full primitive equations. The theory rests on the *Boussinesq approximation* of moderate heating which supports the idea that the fluid's physical parameters are independent of the temperature, except the density in the term responsible for the differential buoyancy force. The linearised thermo-hydrodynamic equations governing two-dimensional (x, z) perturbations then read[17]:

$$\partial_t (\partial_{xx} + \partial_{zz}) v_z = P\left((\partial_{xx} + \partial_{zz})^2 v_z + \partial_{xx}\theta\right), \tag{3.40}$$

$$\partial_t \theta = R\, v_z + (\partial_{xx} + \partial_{zz})\theta. \tag{3.41}$$

These equations are written here in dimensionless form after elimination of the pressure and horizontal velocity component. The scales are the same as

---

[17] These equations and the boundary conditions are derived below on p. 118.

those leading to (3.17, 3.18). As far as the (x,t)-dependence is considered, their structure is also the same as that of the simplified system, but they now retain the additional z-dependence of the fluctuations explicitly.

Boundary conditions are set at the position of the plates z_p, at top, $z_\mathrm{t} = 1$ or $+1/2$, and bottom $z_\mathrm{b} = 0$ or $-1/2$, one or the other choice making computations more transparent depending on the cases considered.

We consider here infinitely good heat-conducting horizontal plates. Accordingly the temperature is strictly fixed by heat baths so that the fluctuations are zero there:

$$\theta|_{z_\mathrm{p}} = 0. \tag{3.42}$$

For the velocity components, rigid plates imply a *no-slip* condition

$$v_z|_{z_\mathrm{p}} = \partial_z v_z|_{z_\mathrm{p}} = 0. \tag{3.43}$$

In practice, calculations are easier with the somewhat artificial *stress-free* conditions considered initially by Rayleigh, which leads to the replacement of (3.43) by

$$v_z|_{z_\mathrm{p}} = \partial_{zz} v_z|_{z_\mathrm{p}} = 0. \tag{3.44}$$

In the following we consider symmetrical cases where top and bottom conditions are identical but any non-symmetrical condition (good/bad conductor, no-slip/stress-free) can be considered in the same way at the expense of a (much) more cumbersome analysis since parity considerations are no longer useful to classify the solutions.

1) Normal mode analysis in the stress-free case (Rayleigh solution, 1916). Boundary conditions are set at $z_\mathrm{p} = 0$ and 1.

a) For a laterally unbounded layer, check that solutions to (3.40, 3.41) can be taken in the form:

$$(v_z, \theta) = (\bar{V}, \bar{\Theta}) \sin(n\pi z) \exp(ikx) \exp(s_n t).$$

b) Determine the marginal stability condition for the mode n that first bifurcates when R is increased. Find the corresponding critical wavevector and threshold.

[Answer:

$$R_n^{(\mathrm{m})}(k) = (n^2\pi^2 + k^2)^3/k^2, \tag{3.45}$$

$k_\mathrm{c} = \pi/\sqrt{2}$, $R_\mathrm{c} = 27\pi^4/4$. See Figure 3.16, p. 118, curve A.]

c) Show that the linear dynamics close to the threshold is governed by:

$$\tau_0 s = r - \xi_0^2 (k - k_\mathrm{c})^2,$$

with $r \equiv (R - R_{\rm c})/R_{\rm c}$. Compute coefficients τ_0 and ξ_0.

2) *Normal mode analysis in the no-slip case.* The exact solution, obtained by Pellew and Southwell (1940) is presented in [Chandrasekhar (1961)]. The corresponding marginal stability curve is given in Figure 3.16, p. 118, as line B' (fine, solid) with critical conditions $k_{\rm c} \approx 3.11632$, $R_{\rm c} \approx 1707.76$.

Here we look for an approximate solution by a so-called *Galerkin method*, a special case of weighted-residual approximation introduced, e.g. in [Finlayson (1972)].

In a few words, the solution, expanded on a complete basis of functions, is further injected in the equations that are projected on a complementary basis using some scalar product. The Galerkin method comes in when the functions in the original and complementary bases are identical and fulfil the boundary conditions of the problem.

The approximate solution is seen to converge to the exact solution as the number of functions is increased, especially when the problem has an underlying variational structure, which is the case for Rayleigh–Bénard convection, as discussed in [Chandrasekhar (1961)].

Here, the solution of (3.40, 3.41) is searched for in the form:
$$\{v_z(x,z,t), \theta(x,z,t)\} = \{\bar{V}(z), \bar{\Theta}(z)\}\sin(k_x x)\exp(st)\,.$$
The z dependence of $\bar{V}_z$ and $\bar{\Theta}$ is taken as polynomials. Boundary conditions are set at $z = \pm 1/2$.

a) Show that in order to fulfil boundary conditions automatically, one must take:
$$\bar{V}(z) = (\tfrac{1}{4} - z^2)^2 P_v(z)\,,$$
$$\bar{\Theta}(z) = (\tfrac{1}{4} - z^2) P_\theta(z)\,,$$
where $P_v(z)$ and $P_\theta(z)$ are polynomials in z, $P_v(z) = \sum_{n=0}^{\infty} V_n z^n$, $P_\theta(z) = \sum_{n=0}^{\infty} \Theta_n z^n$. The approximation enters when polynomials are truncated beyond some given maximal degree N.

b) The differential problem being formally written as $\mathcal{L}\mathbf{U} = 0$, where $\mathbf{U}$ has two components $\bar{\Theta}$ et $\bar{V}_z$, the projection onto the basis is defined by integrals
$$\int_{1/2}^{1/2} z^n (\tfrac{1}{4} - z^2)^2 (\mathcal{L}\mathbf{U})_{v_z}\, dz = 0\,,$$
$$\int_{1/2}^{1/2} z^n (\tfrac{1}{4} - z^2) (\mathcal{L}\mathbf{U})_{\theta}\, dz = 0\,,$$

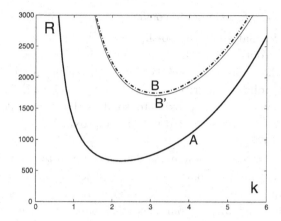

Fig. 3.16 Marginal stability curves of the most dangerous modes in the stress-free case (curve A) and no-slip case (curves B). Thin solid line → exact result. Dot-dashed line → Galerkin approximation at lowest significant order.

for $n = 0, 1, \ldots, N$. This leads to a system of $2(N+1)$ linear equations for the $2(N+1)$ unknown coefficients introduced in the polynomials.

Considering only the stationary case at marginality, i.e. $s = 0$, derive the system at lowest significant order, i.e. $N = 0$ (2 equations for 2 unknowns, Θ_0 and V_0) and the corresponding marginal stability condition; compare the result to the Rayleigh solution (3.45). Then compute the critical wavevector and the threshold; further compare them to the exact result given above.

[Answer:
$$R^{(m)}(k) = \frac{28}{27} \frac{(k^4 + 24k^2 + 504)(k^2 + 10)}{k^2}, \qquad (3.46)$$
threshold: $k_c \approx 3.1165 \simeq$ exact value, $R_c \approx 1750$, 2.5% too high only. See Figure 3.16, curve B (dot-dashed line), but so close an agreement is somehow accidental!]

Derivation of system (3.40–3.44).

In the two-dimensional case (x, z), the continuity, Navier–Stokes and Fourier equations read:

$$\partial_x v_x + \partial_z v_z = 0,$$
$$\rho(\partial_t v_x + v_x \partial_x v_x + v_z \partial_z v_x) = -\partial_x p + \eta(\partial_{xx} + \partial_{zz})v_x,$$
$$\rho(\partial_t v_z + v_x \partial_x v_z + v_z \partial_z v_z) = -\partial_z p + \eta(\partial_{xx} + \partial_{zz})v_z + g\alpha\theta,$$
$$\partial_t \theta + v_x \partial_x \theta + v_z \partial_z \theta = \kappa(\partial_{xx} + \partial_{zz})\theta + \beta v_z.$$

The formally quadratic terms have been dropped owing to the linearisation step. Following the usual procedure, pressure is eliminated by differentiating the equation for v_z with respect to x, the equation for v_x with respect to z and subtracting the two. In order to eliminate v_x, the result is further differentiated with respect to x and the so-obtained $\partial_x v_x$ replaced with $-\partial_z v_z$ using the continuity equation. This leads to:

$$\partial_t(\partial_{xx} + \partial_{zz})v_z = \nu(\partial_{xx} + \partial_{zz})^2 v_z + \alpha g \partial_{xx}\theta\,,$$
$$\partial_t\theta = \kappa(\partial_{xx} + \partial_{zz})\theta + \beta v_z\,,$$

which is finally cast into (3.40, 3.41) by scaling v_z and θ appropriately.

No-slip and stress-free boundary conditions respectively read $v_x(z_\text{p}) = v_z(z_\text{p}) = 0$ and $\partial_z v_x(z_\text{p}) = v_z(z_\text{p}) = 0$. Differentiating the conditions on v_x with respect to x and replacing $\partial_x v_x$ with $-\partial_z v_z$ yields the boundary conditions (3.43, 3.44), expressed in terms of v_z exclusively.

3.3.3 Simplified model of convection in a binary mixture

Thermal convection in the presence of additional molecular diffusion processes is called *thermohaline*, by reference to the diffusion of salt in water. Here we consider the emergence of convection in a binary mixture. The local state of the fluid is thus characterised by the concentration C of a solute, in addition to its temperature T and its velocity $\mathbf{v}$. The temperature gradient is still generated by the temperature difference ΔT, while a concentration gradient is applied by putting the fluid layer in contact with two reservoirs at different concentration through appropriate porous membranes. The solute concentration difference between top and bottom is a second control parameter and molecular diffusion (Fick law, coefficient D) is a supplementary stabilising mechanism, while advection of the solute may contribute to the destabilisation of the layer. We develop a simplified one-dimensional model in the spirit of §3.1.2, i.e. assuming fluctuations that are functions of x and t only, while the driving gradients are imposed along the z direction.

1) Construction of the model:

The fluid layer is supposed to be at rest ($\mathbf{v} \equiv 0$) in contact with two baths at temperatures T_t and $T_\text{b} = T_\text{t} + \Delta T$ (heating from below implies $\Delta T > 0$) and concentrations C_t and $C_\text{b} = C_\text{t} + \Delta C$ (ΔC positive or negative). The purely diffusive temperature and concentration profiles read:

$$T(z) = T_\text{b} - \beta z \quad \text{and} \quad C(z) = C_\text{b} - \beta' z\,,$$

where $\beta = \Delta T/h$ et $\beta' = \Delta C/h$ are the applied gradients. The differential buoyancy force is still induced by variations of the density ρ but it now has two origins, thermal expansion and composition change. Accordingly, the state equation can be taken as

$$\rho = \rho_b(1 - \alpha\theta - \alpha'c),$$

where α (> 0) is the same coefficient as that introduced previously and where the sign of α' depends on the composition of the mixture. Justify the equations governing the linearised model:

$$\partial_t v_z = \nu \partial_{x^2} v_z + g(\alpha\theta + \alpha'c),$$
$$\partial_t \theta = \kappa \partial_{x^2} \theta + \beta v_z,$$
$$\partial_t c = D \partial_{x^2} c + \beta' v_z,$$

and explain the origin of terms βv_z and $\beta' v_z$ and discuss in simple terms the possible instability mechanisms involving each fluctuation.

Take h as length unit, $\tau_\theta = h^2/\kappa$ as time unit, introduce the Lewis number $L = D/\kappa$ in addition to the Prandtl number $P = \kappa/\nu$, define the control parameters

$$R = \frac{\alpha g \Delta T h^3}{\kappa \nu} \quad \text{et} \quad R' = \frac{\alpha' g \Delta C h^3}{D \nu}$$

(thermal and chemical Rayleigh numbers) and, finally, cast these equations in the form

$$\partial_t v_z = P(\partial_{x^2} v_z + \theta + Lc),$$
$$\partial_t \theta = \partial_{x^2} \theta + R v_z,$$
$$\partial_t c = L \partial_{x^2} c + R' v_z.$$

Interpret the Lewis number from a physical viewpoint; what can be its order of magnitude in a liquid, in a gas?

2) Normal mode analysis:

a) Introducing $\{v_z, \theta, c\} = \{V, \Theta, C\} \cos(kx) \exp(st)$, write down the algebraic linear system fulfilled by the amplitudes $\{V, \Theta, C\}$. (Here we assume directly and without justification that $k \sim \pi$ (i.e. in dimensional units $\lambda = 2h$ where λ is the wavelength of the unstable mode).

b) Show that for highly viscous fluids ($P \to \infty$), the resulting system of three equations for three unknowns can be reduced to a system of two equations for two unknowns by eliminating the velocity component v_z. In the following we keep this supplementary assumption but the general case can be treated in the same way using Exercise 2.5.3, p. 75.

c) Derive the compatibility condition of the simplified two-dimensional system and show that, in contrast with ordinary convection, complex roots are possible in the marginal case.

3) Different instability modes:

a) From a discussion of the sign of coefficients of the quadratic equation expressing the compatibility condition above, find the threshold of the stationary instability mode (the product of roots change its sign), then that of the oscillatory mode (the sum of roots change its sign).

b) In the parameter plane of parameters R' (horizontal axis) and R, draw the graph of these threshold conditions. Discuss the nature of the regime expected in each of the regions bound by these lines. Find the coordinates of the point where the system is simultaneously marginal against the two modes. Try to explain the physical origin of the oscillations by returning to the different contributions to the density changes and their respective relaxation times.

[Answer: Figure 3.17. The oscillations result from an interplay of differential buoyancy with two competing dissipation processes, one (thermal diffusion) being much faster than the other (molecular diffusion). This induces delays and phase shifts between the different fluctuations, ending in overshoots and oscillations.]

3.3.4 Turing patterns and reaction-diffusion systems.

In chemistry, mechanisms combining reaction and diffusion may produce dissipative structures called *Turing patterns*, as mentioned in §1.4.2, p. 27. Here we consider a simplified reaction–diffusion system where two species U and V with diffusion coefficients D_U and D_V also react with each other.

In one dimension, with coordinate x, the model reads:

$$\partial_t U = F_U(U, V) + D_U \partial_{xx} U, \qquad (3.47)$$
$$\partial_t V = F_V(U, V) + D_V \partial_{xx} V, \qquad (3.48)$$

where reaction terms F_U and F_V need not be specified at this stage.

1) Neglect diffusion ($D_U = D_V \equiv 0$) and assume that a fixed point solution (U_0, V_0) exists; linearise the governing equations around that point, set

$$a = \partial_U F_U|_0, \quad b = \partial_V F_U|_0, \quad c = \partial_U F_V|_0, \quad d = \partial_V F_V|_0,$$

'$|_0$' meaning computed at (U_0, V_0), next determine the conditions on a, b, c, d that guarantee the stability of solution (U_0, V_0).

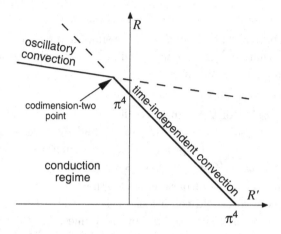

Fig. 3.17 Stability diagram for convection in a binary fluid mixture. When $R' > 0$, the temperature and concentration fluctuations co-operate, whereas when $R' < 0$, they play antagonistic roles. Oscillations occur as their evolutions get sufficiently out of phase. At the intersection of the two lines, called a *codimension-two* point (see p. 129) the two mechanisms are equally efficient to destabilise the layer.

[Answer: the linear stability matrix must have eigenvalues either real and negative or complex with negative real parts, hence negative sum $(a+d < 0)$ and positive product $(ad - bc > 0)$.]

2) Assume that these conditions are fulfilled and add the effect of diffusion ($D_U \neq 0$, $D_V \neq 0$). Determine the dispersion relation of fluctuations around the state (U_0, V_0) supposed uniform in space.

Write down the linearised system governing the amplitudes $\overline{U}(t)$ and $\overline{V}(t)$ of Fourier normal modes in $\exp(ikx)$ and show that the instability, if any, is necessarily cellular and stationary ($k \neq 0$, $\omega = 0$).

Show that the occurrence of the instability requires:

$$aD_V + dD_U > 0 \quad \text{and} \quad 4(ad - bc)D_U D_V \leq (aD_V + dD_U)^2,$$

in addition to the conditions already found. When this is the case, determine the range of unstable wavevectors k.

3.3.5 Taylor–Couette instability

We study the stability of the flow between infinite coaxial cylinders rotating at different angular speeds (Couette flow). This problem was studied by Rayleigh (1916) at the limit of zero viscosity. Taylor (1923) developed the

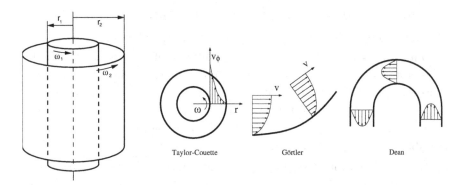

Fig. 3.18 Left: Geometry of the Taylor–Couette experiment. Right: Examples of curved flows.

first theoretical analysis of the viscous case and performed the corresponding experiments.

In the geometry of Figure 3.18 (left), assume that the base flow is purely azimuthal, show that $v_\phi = r\omega_0(r)$ with

$$\omega_0(r) = a + b/r^2, \tag{3.49}$$

obtain a and b from the no-slip condition $\omega(r_i) = \omega_i$ at r_i, $i = 1, 2$.
[Answer: $a = (\omega_2 r_2^2 - \omega_1 r_1^2)/(r_2^2 - r_1^2)$, $b = (\omega_1 - \omega_2) r_1^2 r_2^2/(r_2^2 - r_1^2)$.]

1) Rayleigh instability mechanism (see also [Chandrasekhar (1961)] or [Drazin and Reid (1981)]). The base flow is characterised by the fact that the centrifugal force at distance r from the axis, $\rho r \omega^2(r)$, is compensated by a centripetal pressure gradient.

Consider a rotating fluid particle displaced from distance r to distance $r + \delta r > r$. In the absence of viscous friction, angular momentum $\rho r v_\phi$ is a conserved quantity. From this conservation law, derive its angular speed at the new position and compare it with that of a fluid particle originally at the same place.

2) If the speed of the displaced particle is smaller than that of the surrounding fluid, the local pressure gradient is larger and pushes the particle back to its original position, the purely azimuthal flow is stable. In the opposite situation it is unstable. Show that this stability condition reads

$$\frac{d\omega}{dr}\delta r \geq -2\frac{\omega}{r}\delta r,$$

then turn it to the form:

$$\delta\left(r^2\omega\right) \geq 0$$

and express this *Rayleigh stability criterion* in words.

3) Application to the Couette profile and other curved flows.

a) Coming back to the base flow profile (3.49) identify the different possible cases (rotation direction identical or different, and when the rotation directions are identical, which cylinder is rotating faster) and find the situations that are stable according to the Rayleigh criterion.

b) When cylinders rotate in opposite directions, determine the region which is stable according to Rayleigh.

c) By comparison with the case of convection, and by anticipation of Chapter 7 about shear flows, guess the role of viscosity, especially when the flow is mechanically stable.

d) The same instability mechanism is expected to work when flow lines are curved. Identify the regions of the flow where the centrifugal instability can develop according to Rayleigh, in the boundary layer flow along a concave wall (Görtler instability; what about a convex wall?) or in the flow along a curved channel (Dean instability) depicted in Figure. 3.18 (right).

Chapter 4

Nonlinear Dynamics: from Simple to Complex

We now examine in more detail the theoretical context where the convection experiments described in the previous chapter can be best situated. The interest of this study stems from the fact that most instabilities that develop in bounded continuous media enter the framework of dynamical systems theory. Its foundation rests on an analysis of the effects of lateral confinement, §4.1.1. In a strongly confined system, even far enough from the instability threshold, the space dependence of unstable modes remains frozen and the state of the system is sufficiently well characterised by a few amplitudes playing the role of effective degrees of freedom, which permits the analysis of the transition to turbulence in the spirit of Chapter 2. Here we limit ourselves to a heuristic presentation of nonlinear dynamics without insisting much on mathematical aspects.

Our interest is in characterising sustained regimes of *dissipative dynamical systems* obtained after the decay of transients. These asymptotic regimes reached in the limit $t \to \pm\infty$ are represented in state space by objects called *limit sets*. Stable limit sets are reached in the normal course of time, i.e. for $t \to +\infty$ and unstable limit set in the reverse case $t \to -\infty$. *Fixed points* and *limit cycles* were introduced in Chapter 2 for time-independent and periodic regimes, respectively. More complicated, multi-periodic or chaotic, behaviour is accounted for by objects called *limit tori* or *strange attractors*, which were alluded to in §1.3. The main purpose of this chapter is thus to come back to this problem and describe the growth of complexity observed as the number of instability modes increases when control parameters are varied, with a particular emphasis on the reduction to a low-dimensional effective dynamics §4.1.2. By increasing the order of complexity, we examine the transitions between time-independent states and from time-independent states to regular periodic or multi-periodic regimes, and next to chaos, which enters following *universal scenarios* that are independent of the specific system studied. Section 4.2 is devoted to the first steps, saturation of oscillations and stability of periodic behaviour in the general case, the mathematics of chaotic behaviour being introduced in §4.3.

In contrast to regular regimes, chaotic states are characterised by their *unpredictability*, generally associated with a *fractal* occupation of state space. These different facets will be examined in Chapter 5. However, this approach remains valid only in so far as the projection of the dynamics onto well isolated spatially frozen modes makes sense. In practice, unfreezing takes place when lateral confinement effects are not strong enough to maintain coherence in physical space so that chaos becomes spatiotemporal. This extension is the topic of Chapter 6.

4.1 Reduction of the Number of Degrees of Freedom

4.1.1 *Role of aspect ratios*

Let us come back to eigenmodes obtained from the linear stability analysis of the instability studied, e.g. convection. The problem has been formalised in Section 3.1.3, p. 86. We assume that confinement is specified by *lateral* boundary conditions at a distance ℓ, whereas the instability generates cells with typical scale λ_c. The aspect ratio can be defined as $\Gamma = \ell/\lambda_c$. The difference between the cases "$\Gamma \sim 1$" and "$\Gamma \gg 1$" is illustrated in Figures 4.1 and 4.2. Exercises 4.4.1 and 4.4.2 should be worked out to gain a more analytical understanding of the problem.

We first consider the case when the horizontal dimensions are of the order of the wavelength λ_c so that all modes are physically distinct (Figure 4.1). Only the value of the corresponding thresholds $R_1, R_2, \ldots$ matters, the two-dimensional presentation of the different modes is given just in order to rationalise the classification in terms of the number of substructures in x and z, which is supported by the idea that these states should stay close to the branches that would exist in the absence of lateral boundary conditions (fine solid lines). They account for a kind of spatial resonance between the lateral extension of the system and the length scale originating from the instability mechanism, ending in the formation of an integer number of roughly circular cells, as already illustrated in Figure 3.7 p. 106, displaying silicon oil convection cells in a small aspect ratio box. The lower branch would then correspond to the "fundamental" states with one layer of cells and the upper one to "excited" states with two layers of cells stacked

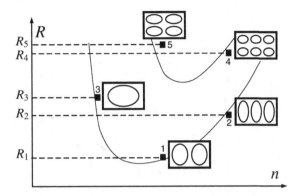

Fig. 4.1 Spectrum of the linearised stability problem for a confined system, $\Gamma \sim 1$; mode 1 is the sole unstable mode for $R_1 < R < R_2$. (n is the number of cells in the horizontal direction; thin lines suggest the marginal stability conditions of modes with different vertical structures as obtained in the absence of lateral boundary conditions.)

one above the other as could be obtained at higher values of the control parameter. In practice, the actual structure of the modes can be obtained at the price of a full three-dimensional calculation, in general difficult and with little supplementary insight into their physical significance. Since the spectrum of the linear stability problem is entirely made of isolated values far from one another, it is in principle easy to limit the number of linearly excited modes ($\sigma_n > 0$ for $R > R_n$) from those that are still damped ($R < R_m$, $\sigma_m < 0$, $|\sigma_m| \gg 1$).

In contrast, when $\Gamma \gg 1$, that is $\ell \gg \lambda_c$ (Figure 4.2), the spectrum is quasi-continuous. It is obtained from the dispersion relation for a laterally unbounded system by adding a quantisation condition on the wavevector, typically $k_n \sim \pi n/\ell$ for n cells of width $\lambda/2$ over an interval of width ℓ. The difference between neighbouring wavevectors for patterns with ± 1 cell is $k_{n+1} - k_n = \pi/\ell \ll k_c$, which is small when ℓ is large. Close to k_c, owing to the quadratic shape of the dispersion relation (3.26), p. 95, one expects $(R_{n_c \pm 1} - R_{n_c})/R_c = \xi_0^2 \pi^2/\ell^2$, where n_c is here the number of cells expected at threshold and given by $n_c \lambda_c/2 \simeq \ell$. Corresponding modes are thus quasi-degenerate. Apart from the precise value of the wavevector, all modes of the lowest branch are expected to have similar spatial structures. Interference between modes then yields modulated patterns, also called *textures*, as illustrated in Figure 3.10, p. 108. The present chapter is entirely devoted to the case $\Gamma \sim 1$ and to the development of *temporal chaos* phenomenologically introduced in §3.2.3, p. 104, and is thus placed in direct continuity with Chapter 2.

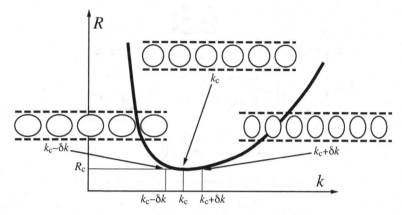

Fig. 4.2 Marginal stability curve for an extended system, $\Gamma \gg 1$; for $R > R_c$ the modes belonging to a range $\mathcal{O}(\sqrt{R - R_c})$ centred at k_c are unstable and may serve to build the nonlinear pattern.

4.1.2 Low-dimensional effective dynamics

4.1.2.1 Framework

It may seem reasonable to take advantage of the work done in solving the linear stability problem to treat the nonlinear problem. This is done by expanding the solution onto the basis of eigenmodes:

$$\mathbf{V} = \sum_n A_n \mathbf{X}_n, \qquad (4.1)$$

and then searching the equations to be fulfilled by the amplitudes A_n of the modes $\mathbf{X}_n$, in order to extend the linear dynamics simply governed by $\frac{\mathrm{d}}{\mathrm{d}t} A_n = s_n A_n$ to the nonlinear range. From the qualitative arguments developed above, we are led to think that the unstable (and neutral) modes play the most active role and that the stable modes evolve passively under the forcing action of the others.

Let us incorporate this feature in the formalism itself and assume that the problem is initially posed in the general form:

$$\tfrac{\mathrm{d}}{\mathrm{d}t} \mathbf{V} = \mathcal{L}\mathbf{V} + \mathcal{N}(\mathbf{V}), \qquad (4.2)$$

where $\mathcal{L}$ accounts for the linear interactions and where $\mathcal{N}(\mathbf{V})$ is at least quadratic in $\mathbf{V}$. An appropriate projection of the equations onto the eigenbasis yields a dynamical system for the amplitudes A_n introduced in (4.1), which formally reads:

$$\tfrac{\mathrm{d}}{\mathrm{d}t} A_n = s_n A_n + \sum_{mp} g_n^{mp} A_m A_p + \ldots, \qquad (4.3)$$

where the coefficient g_n^{mp} describes how the interaction of modes $\mathbf{X}_m$ and $\mathbf{X}_p$ contributes to the dynamics of mode $\mathbf{X}_n$. In fact this projection procedure is rarely achievable without approximation, so that some additional modelling of the physical situation is usually required. A favourable case is when the analytical expression of eigenmodes can be handled by hand (or using some formal algebra software), for example trigonometric functions as illustrated in Exercises 4.4.1 and 4.4.2. Otherwise one can attempt to extend to the nonlinear domain the *Galerkin method* already used at the linear stage, e.g. in the second part of Exercise 3.3.2, p. 117.

Replacing the primitive problem expressed in terms of partial differential equations by (4.3), i.e. a differential system of infinite order, may not seem an important progress. However, this is very powerful when combined with the idea of separating unstable from stable modes to truncate the system and eliminate the stable modes, the *slaves*, in order to obtain an effective system involving only the unstable *driving* modes. This is the *adiabatic elimination procedure* advanced as the founding principle of *synergetics*, the science of nonlinear co-operative phenomena in out-of-equilibrium systems [Haken (1983)].

Let us consider a system on the verge of bifurcating or having just bifurcated ($R \approx R_c$). This comes to the assumption that, among all linear eigenmodes, we can isolate a small subgroup of those that are "dangerous," nearly neutral, either unstable or stable but slightly, i.e. with the real parts σ of their growth rates either positive or negative but small. Strictly speaking, the *centre subspace*, subscript 'c', is the space spanned by eigenmodes having eigenvalues with $\sigma = 0$ exactly, and we assume that we have n_c such modes. On general grounds, having n_c strictly critical modes is a problem of codimension[1] n_c, since n_c conditions $\sigma_n(\mathbf{r}) = 0$ have to be fulfilled, $\mathbf{r}$ being the set of control parameters. In practice, we can vary only one control parameter while remaining in the vicinity of a manifold of codimension n_c in parameter space and we just require $\sigma_c \approx 0$ for n_c driving modes.

On the contrary, all other modes are supposed to remain *stable* (subscript 's') with a related condition $\sigma_s < 0$ meaning more precisely that $|\sigma_c| \ll |\sigma_s|$, so that the driving modes have relaxation rates well separated

---

[1] *Codimension* is a term used to characterise the difference between the dimension of a set and the dimension of a larger set that contains it: in a d-dimensional space, a point, with dimension 0, has codimension d, its position can be defined as a root of a system of d equations; a line, with dimension 1 has codimension $d - 1$, it can be parametrically described using d equations depending on a parameter so that a system of $d-1$ equations is obtained by elimination of this parameter.

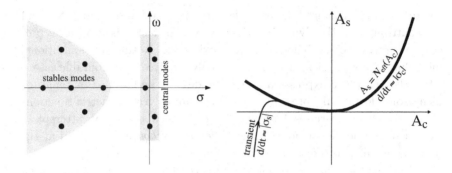

Fig. 4.3 Left: Spectrum of a linear operator with a group of *central modes* separated from *stable modes* by a wide *spectral gap*. Right: Adiabatic elimination of stable modes. The whole sets of centre-mode and stable-mode amplitudes are featured by axes $\mathbf{A}_c$ and $\mathbf{A}_s$, respectively. A trajectory starting anywhere in state space rapidly migrates towards a *centre manifold* with equation (4.6) which is the nonlinear extrapolation of the centre subspace at the origin ($\mathbf{A}_s = 0$), then slowly evolves along it according to the effective dynamics (4.7).

from those of the slave modes. The linear spectrum is therefore supposed to display a wide *spectral gap*, see Figure 4.3 (left).

The effective dynamics is obtained by elimination of the enslaved modes. For simplicity we consider a d-dimensional case, $d < \infty$, with n_c driving modes and $n_s = d - n_c$ stable modes. The considerations developed now can be, at least at a heuristic level, easily extended to the infinite dimensional case corresponding to instabilities in continuous media, with an infinite number of stable modes and the series of eigenvalues with real parts extending down to $-\infty$. The adaptation of this elimination procedure to the case of systems with $\Gamma \gg 1$ will be examined in Chapter 6.

4.1.2.2 *Heuristic approach*

Let us explicitly separate driving modes with amplitudes collectively denoted as $\mathbf{A}_c$ from enslaved ones with amplitudes $\mathbf{A}_s$ in (4.3). For the driving modes we get:

$$\tfrac{\mathrm{d}}{\mathrm{d}t}\mathbf{A}_c = \mathcal{L}_c \mathbf{A}_c + \mathcal{N}_c(\mathbf{A}_c, \mathbf{A}_s), \qquad (4.4)$$

where $\mathcal{L}_c$ is the restriction of the linearised operator $\mathcal{L}_r$ to the space spanned by the $\mathbf{A}_c$ and $\mathcal{N}_c$ accounts for the nonlinear interactions between driving and enslaved modes. Similarly we write:

$$\tfrac{\mathrm{d}}{\mathrm{d}t}\mathbf{A}_s = \mathcal{L}_s \mathbf{A}_s + \mathcal{N}_s(\mathbf{A}_c, \mathbf{A}_s) \qquad (4.5)$$

for the amplitudes of stable modes. The idea of the reduction is to solve the problem for the stable modes as if they were submitted to a forcing from the (slowly evolving) driving modes, then to insert the solution back into (4.4) to obtain the effective dynamical system we are looking for.

In order to simplify the application of this strategy, let us assume that the s_s are real and non degenerate, and that $\mathcal{N}_s$ involves only the $\mathbf{A}_c$ so that we can rewrite (4.5) in a basis such that $\mathcal{L}_s$ is in diagonal form, which leads to:

$$\tfrac{d}{dt}\mathbf{A}_s + |s_s|\mathbf{A}_s = \mathcal{N}_s(\mathbf{A}_c(t)).$$

It is easily checked that the complete solution of this non-homogeneous linear problem then reads

$$\mathbf{A}_s(t) = \exp(-|s_s|t)\mathbf{A}_s(0) + \int_0^t \exp\bigl(-|s_s|(t-t')\bigr)\mathcal{N}_s(\mathbf{A}_c(t'))\,dt'$$

where $\mathbf{A}_s(0)$ is specified by the initial conditions at $t = 0$. According to the assumption made about the stable part of the spectrum, we see that the first term on the right hand side contributes to the solution only during a brief transient of duration $\mathcal{O}(1/|s_s|)$ and that, in the integral, the exponential kernel has a short range, so that we may approximate the solution as

$$\mathbf{A}_s(t) \approx \frac{1}{|s_s|}\mathcal{N}_s(\mathbf{A}_c(t)). \qquad (4.6)$$

In the d-dimensional state space, these $n_s = d - n_c$ relations define a n_c-dimensional manifold parameterized by the n_c amplitudes of the driving modes. Inserting them into the evolution equations for the $\mathbf{A}_c$, we get:

$$\tfrac{d}{dt}\mathbf{A}_c = \mathcal{L}_c\mathbf{A}_c + \mathcal{N}_c\bigl(\mathbf{A}_c, \mathcal{N}_s(\mathbf{A}_c)\bigr) = \mathcal{L}_c\mathbf{A}_c + \mathcal{N}_{\text{eff}}(\mathbf{A}_c), \qquad (4.7)$$

which defines an *effective dynamics* on this manifold. An explicit example is treated in Exercise 4.4.2.

4.1.3 Centre manifolds and normal forms

4.1.3.1 Reduction to the centre manifold

The result illustrated in Figure 4.3 (right) can be derived in a mathematically rigorous way, asymptotically valid in the long time limit, i.e. $t \gg 1/\min(|\sigma_s|)$. This approach is called the *centre manifold reduction*.[2]

---
[2] See for example: J.D. Crawford, "Introduction to bifurcation theory" Rev. Mod. Phys. **63** (1991) 991–1037.

Assuming that, as suggested by the heuristic approach, the stable modes "live" on a slow manifold defined through

$$\mathbf{A}_s = \mathcal{H}(\mathbf{A}_c), \tag{4.8}$$

one can determine the functional equation governing $\mathcal{H}$ by inserting (4.8) in (4.5) and replacing $\frac{d}{dt}\mathbf{A}_c$ by its expression from (4.4). This yields

$$\partial \mathcal{H}(\mathbf{A}_c) \cdot \left[\mathcal{L}_c \mathbf{A}_c + \mathcal{N}_c\!\left(\mathbf{A}_c, \mathcal{H}(\mathbf{A}_c)\right) \right] = \mathcal{L}_s \mathcal{H}(\mathbf{A}_c) + \mathcal{N}_s(\mathbf{A}_c, \mathcal{H}(\mathbf{A}_c)),$$

where $\partial \mathcal{H}$ denotes the Jacobian matrix of $\mathcal{H}$ with elements $\mathcal{H}_{ij} = \partial \mathcal{H}_i / \partial A_{cj}$. This functional equation for $\mathcal{H}$ is then solved by representing $\mathcal{H}(\mathbf{A}_c)$ as a formal series in powers of the components of $\mathbf{A}_c$:

$$\mathcal{H}(\mathbf{A}_c) \equiv \sum_{n \geq 2} \mathcal{H}_n(\mathbf{A}_c),$$

where $\mathcal{H}_n(\mathbf{A}_c)$ is a polynomial formed with a series of homogeneous increasing-degree monomials in the form:

$$\prod_{m=1}^{d_c} A_{cm}^{n_m}, \qquad n_m \geq 0, \qquad \sum_m n_m = n,$$

with coefficients to be determined by identification. The so-obtained expression is further inserted in (4.4), which leads to the effective dynamics one is looking for.

4.1.3.2 Reduction to the normal form

In practice, the physical contents of the expression just obtained for the effective dynamics is obscured by the presence of a large number of terms that can be eliminated by nonlinear changes of variables, in much the same way as linear changes of variables allow one to represent a linear operator in its Jordan normal form (§A.2, p. 378). This supplementary step, called *normal form reduction* leads to the elimination of all *non-resonant* terms in the effective dynamical system. The term 'resonant' is here understood in the same way as in the study of the improper node, Chapter 2, p. 54.

For example, let us consider the case of a two-dimensional centre manifold parameterized using (A_1, A_2) or rather the pair of complex conjugate modes $(A = A_1 + iA_2, A^*)$, with eigenvalues $\pm i\omega_c$, $\omega_c \neq 0$,[3] exactly marginal at some specific value $\mathbf{r} = \mathbf{r}_c$ of the set of control parameters, i.e. $\sigma(\mathbf{r}_c) = 0$.

---
[3] $\omega_c \neq 0$ means $\omega_c \sim \mathcal{O}(1)$ since $\omega_c \ll 1$ would rather correspond to a reference situation with a double root $s = 0$, perturbed so as to display real and imaginary parts that would be simultaneously small.

Consider one of these variables, say A, the previous reduction procedure generically leads to an effective dynamics

$$\tfrac{\mathrm{d}}{\mathrm{d}t} A = -i\omega_\mathrm{c} A + \sum_{m=0}^{2} g_m^{(2)} A^{2-m} A^{*m} + \sum_{m=0}^{3} g_m^{(3)} A^{3-m} A^{*m} + \ldots, \qquad (4.9)$$

where the coefficients $g_m^{(n)}$ are complex *a priori*. The equation governing the other variable A^* is of course the complex conjugate of (4.9). It is however easily seen that none of the quadratic terms present in these equations is resonant.

Looking for a solution to (4.9) as an expansion in powers of A, at first order we indeed get $A^{(1)} \propto \exp(-i\omega_\mathrm{c} t)$. The correction brought by the first quadratic term is then a solution to:

$$\tfrac{\mathrm{d}}{\mathrm{d}t} A^{(2)} + i\omega_\mathrm{c} A^{(2)} \propto (A^{(1)})^2 \sim \exp(-2i\omega_\mathrm{c} t),$$

admitting $\exp(-2i\omega_\mathrm{c} t)$ as special solution with a non-singular coefficient. Terms

$$AA^* \sim 1 \quad \text{and} \quad (A^*)^2 \sim \exp(2i\omega_\mathrm{c} t)$$

are non-resonant for the same reason and, among cubic terms only the monomial

$$A^* A^2 = |A|^2 A \sim \exp(-i\omega_\mathrm{c} t)$$

turns out to be resonant. It is just a little long and tedious to find the change of variables:

$$A = \bar{A} + \sum_{m=0}^{2} a_m^{(2)} \bar{A}^{2-m} (\bar{A}^*)^m + \sum_{m=0}^{3} a_m^{(3)} \bar{A}^{3-m} (\bar{A}^*)^m + \ldots \qquad (4.10)$$

that leads to the *normal form*:

$$\tfrac{\mathrm{d}}{\mathrm{d}t} \bar{A} = -i\omega_\mathrm{c} \bar{A} + \bar{g}_3 |\bar{A}|^2 \bar{A} + \ldots \qquad (4.11)$$

appropriate to the case at hand.

4.1.3.3 Slightly off the critical conditions

One last step remains to be performed. Up to now we have assumed that the control parameters had precisely the values that make the relevant modes marginal. This defines a *critical surface* in parameter space, $\mathbf{r} = \mathbf{r}_\mathrm{c}$. Close to this surface, the condition is no longer fulfilled but the real parts of the growth rates remain small.

A first-order Taylor expansion of these growth rates in a direction of the parameter space that is transverse (i.e. not tangent) to the critical surface allows one to get off the critical conditions. Let us simple denote the coordinate in that direction by r. For the pair of complex modes considered above, this comes to take:

$$s = \sigma \pm i\omega \quad \text{with} \quad \sigma \propto r, \quad \omega \approx \omega_c, \quad g_3 \approx \bar{g}_3,$$

and to *unfold* the critical dynamics (4.11) by correcting the growth rate, hence $-i\omega_c \mapsto s = \sigma - i\omega_c$. This corrected form[4] will be used below to account for the emergence of periodicity. A rewarding exercise is to derive it explicitly from Equation (2.42) governing the van der Pol oscillator, p. 61.

4.1.3.4 Universality and modelling

More generally, one may remark that the normal form and the perturbations that describe the departure from criticality depend in an essential way on resonance relations existing among the eigenvalues of the linearised stability problem at criticality. In the previous example we had:

$$s_1 = s_2 + 2s_1 \quad \text{for} \quad s_{1,2} = s_\pm = \pm i\omega_c,$$

which renders the term $A^* A^2$ resonant, which can be traced back to:

$$\exp(s_2 t) \exp(s_1 t)^2 \equiv \exp[(s_2 + 2s_1)t] = \exp(s_1 t)$$

with $s_2 = i\omega_c$ and $s_1 = -i\omega_c$, but the generalisation is immediate to any resonance and leads to the identification of nonlinear resonant terms.

This plain observation underlines the notion of *universality*:

All systems that bifurcate in the same way, have the same symmetries, and are controlled by the same number of parameters, have qualitatively similar dynamics.

To set this similarity on a quantitative footing requires complicated and tedious changes of variables, which can be avoided by having recourse to a *phenomenological modelling* of the considered system close to its critical conditions, that is:

1) Determine the linear normal form governing the driving modes, identify the corresponding resonance conditions and add the most general linear perturbations corresponding the departure from criticality, see §A.2.3.

2) Introduce all possible nonlinear resonant terms (different equivalent

---

[4] In line with Note 3, p. 132, it is assumed that $\sigma \ll \omega_c \sim \mathcal{O}(1)$ because admitting $\sigma \sim \omega_c$ is not compatible with the idea of a near-marginal situation.

forms are admissible) with appropriate phenomenological coefficients. In concrete applications these coefficients can be fitted from on-purpose well-designed experiments.

3) Take full advantage of available mathematical knowledge to account for the behaviour of the system at hand as its control parameters are varied.

4.2 Transition to Chaos

The low-dimensional dynamical systems framework being established, we now illustrate the theoretical approach to the growth of complexity until chaos is obtained at the end of an instability cascade. In turn, this cascade relies on the progressive increase of the dimension of the manifold on which the effective dynamics develops. By varying the control parameter, the number of potentially unstable modes is increased (see Figure 4.1) and, with it, the dimension of the effective dynamical system. Upon assuming continuous-time dynamics, one state variable is enough to account for transitions between time-independent regimes, two state variables are needed for periodic motion, and at least three for more complicated behaviour, regular (multi-periodic) or irregular (chaotic).

4.2.1 Time-independent and periodic regimes

4.2.1.1 Bifurcation between time-independent regimes

The simplest bifurcation corresponds to the transition from one time-independent regime to another, both represented by *fixed points* in state space. It can be described within an appropriate generalisation of the Landau model (1.19), p. 13, involving a single variable; see Exercise 4.4.4 for further relevant considerations.

Let us start with:

$$\tfrac{\mathrm{d}}{\mathrm{d}t} A = rA - A^3 + H, \tag{4.12}$$

describing the evolution of a system at a perturbed supercritical bifurcation. The dynamics described by (4.12) is typical of convection which, as we have seen, is a stationary instability that saturates beyond threshold ($r > 0$, $g > 0$). Here time t has been re-scaled with the natural evolution time τ_0 of the unstable mode.

Quantity H measures the intensity of some perturbation coupled to the instability mode. In convection this could be achieved with some device pro-

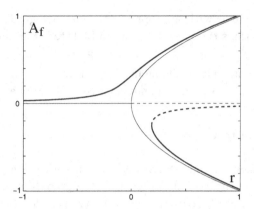

Fig. 4.4 *Imperfect fork bifurcation* accounted for by (4.12) with $g = 1$ for $H = 0.054$. Thick solid line (dashed line) corresponds to stable (unstable) fixed points as functions of r. The thin solid line is for the perfect bifurcation when $H = 0$ already displayed in Figure 1.4, p. 13. This corresponds to the *cusp catastrophe*, see below.

ducing a modulation of the background fluid density through non-uniform heating, which encourages the rising or sinking of the fluid already in the absence of temperature gradient.

The bifurcation diagram corresponding to (4.12), is displayed in Figure 4.4. The diagram corresponding to *perfect bifurcation*, with $H = 0$, features:

$$A_{\rm f} = \pm r^{1/2} \qquad (4.13)$$

and has already been displayed in Figure 1.4, p. 13. It is recalled here as a fine solid line. In the general case, the fixed points $A_{\rm f}$ of (4.12), obtained by solving it for A when $\frac{\rm d}{{\rm d}t}A = 0$, now depend on r and H. The introduction of a small field $H \neq 0$ induces a non-trivial response $A_{\rm f} \neq 0$ below the theoretical bifurcation point at $r = 0$. This response remains small as long as r is large and negative, so that

$$A_{\rm f} \approx (-1/r)H\,, \qquad (4.14)$$

but closer to the threshold, as $|r|$ decreases, the amplitude of the response increases up to a point where nonlinearity can no longer be neglected. For $r = 0$ this yields

$$A_{\rm f} = H^{1/3}\,. \qquad (4.15)$$

When r varies from $-\infty$ to $+\infty$, $A_{\rm f}$ follows a first branch of stable solutions, the thick line that continuously joins the axis $A = 0$ for $r \to -\infty$ to the arc of parabola $A_f = +r^{1/2}$ for $r \to +\infty$. Sufficiently above threshold ($r > 0$), a second branch appears, disconnected from the first, that joins the arc of parabola $A_f = -r^{1/2}$ in its stable part (continuous thick line) to the axis $A = 0$ for $r \to +\infty$ in its unstable part (dashed line). The singularity at $r = A = 0$ has disappeared and one speaks of an *imperfect bifurcation*.

This description is in fact in close correspondence with the Landau theory of 'second order' thermodynamic phase transitions (see §1.4.3, and especially p. 33). The mathematical counterpart of Landau theory is Thom's *catastrophe theory*, consult [Poston and Stewart (1978)]. Here also one is looking for extrema of a potential playing the role of the free energy in thermodynamics but, in full generality, one is no longer restricted to a one-dimensional state space. A gradient flow dynamics (§2.1.3) in the form:

$$\mathcal{G} \equiv \mathcal{G}(\{A_i; i = 1, d_c\}), \qquad \tfrac{d}{dt} A_i = -\partial \mathcal{G}/\partial A_i \qquad (4.16)$$

is assumed and the different cases arise from the *unfolding* of singular part of the potential $\mathcal{G}_{\text{sing}}$ by the most general perturbation $\mathcal{G}_{\text{pert}}$ in its normal form. The classification involves the number d_c of state variables and the number n_r of unfolding parameters, i.e. the dimension of the centre manifold and the number of control parameters. We list here the seven *elementary catastrophes* obtained for $d_c \leq 2$ and $n_r \leq 4$:

- with one state variable A (see Exercise 4.4.4):
  - the 'fold' with $\mathcal{G}_{\text{sing}} = A^3$ and $\mathcal{G}_{\text{pert}} = r_0 A$,
    which accounts for the *saddle-node* bifurcation *via* (4.16).
  - the 'cusp' with $\mathcal{G}_{\text{sing}} = A^4$ and $\mathcal{G}_{\text{pert}} = r_0 A + r_1 A^2$,
    giving back (4.12) with $r_0 = -H$ and $r_1 \equiv -r$.
  - the 'swallowtail' with $\mathcal{G}_{\text{sing}} = A^5$ and $\mathcal{G}_{\text{pert}} = r_0 A + r_1 A^2 + r_2 A^3$.
  - the 'butterfly' with $\mathcal{G}_{\text{sing}} = A^6$ and
    $\mathcal{G}_{\text{pert}} = r_0 A + r_1 A^2 + r_2 A^3 + r_3 A^4$.

- with two state variables A_1 and A_2:
  - the 'hyperbolic umbilic' with $\mathcal{G}_{\text{sing}} = A_1^3 + A_2^3$ and
    $\mathcal{G}_{\text{pert}} = r_0 A_1 + r_1 A_2 + r_2 A_1 A_2$.
  - the 'elliptic umbilic' with $\mathcal{G}_{\text{sing}} = A_1^3 - A_1 A_2^2$ and
    $\mathcal{G}_{\text{pert}} = r_0 A_1 + r_1 A_2 + r_2 (A_1^2 + A_2^2)$.
  - the 'parabolic umbilic' with $\mathcal{G}_{\text{sing}} = A_1^2 A_2 + A_2^4$ and
    $\mathcal{G}_{\text{pert}} = r_0 A_1 + r_1 A_2 + r_2 A_1^2 + r_3 A_2^2$.

Although the state space is two-dimensional, time-dependent asymptotic regimes are forbidden by the gradient flow property. We now raise this restriction and study the occurrence of oscillations.

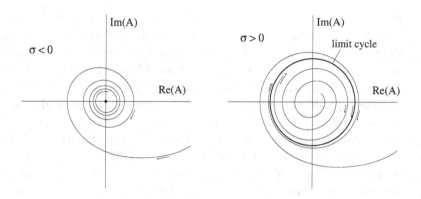

Fig. 4.5 Hopf bifurcation (birth of a limit cycle). The oscillations relax towards the origin below the threshold (left) and saturate at a finite amplitude above (right).

4.2.1.2 Emergence of periodicity

The emergence of oscillations, called the *Hopf* bifurcation, (or *Landau–Hopf, Hopf–Andronov, Poincaré–Andronov*) is best described through a complex representation of the relevant subspace where it develops. The generic model reads:

$$\tfrac{\mathrm{d}}{\mathrm{d}t} A = sA - g|A|^2 A, \qquad (4.17)$$

with $A \in \mathbb{C}$, $s = \sigma - i\omega$, and $g = g' + ig''$, see Figure 4.5.

Setting $A = \rho \exp(i\varphi)$, upon substitution and after simplification by $\exp(i\varphi)$ we get:

$$\tfrac{\mathrm{d}}{\mathrm{d}t}\rho + i\rho \tfrac{\mathrm{d}}{\mathrm{d}t}\varphi = (\sigma - i\omega)\rho - (g' + ig'')\rho^3$$

and, separating real and imaginary parts:

$$\tfrac{\mathrm{d}}{\mathrm{d}t}\rho = \sigma\rho - g'\rho^3, \qquad (4.18)$$
$$\tfrac{\mathrm{d}}{\mathrm{d}t}\varphi = -\omega - g''\rho^2. \qquad (4.19)$$

As already mentioned, the coefficients in (4.17) are functions of the control parameter r but, at leading order, we can assume ω and g constant, and introduce the dependence on r only where it is indispensable, i.e. to control the growth rate. Since it must be negative below threshold (damping) and positive above (amplification), we can simply take $\sigma \propto r$.

Solving (4.18) for ρ is similar to solving for A in the previous section, except that the symmetry $A \mapsto -A$ has to be enforced (no conjugate field

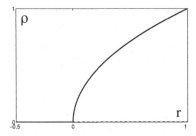 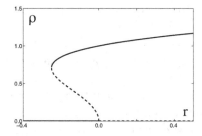

Fig. 4.6 The bifurcation is supercritical when $g' > 0$ (left). When $g' < 0$ (right), Equation (4.17) has to be completed by saturating higher order terms in the form $h_k |A|^{2k} A$ ($k > 1$) and the limit cycle that branches off at $r = 0$ is unstable (dashed line) while a large amplitude stable limit cycle also coexists with it (full line).

H) since it corresponds to the invariance through the change $\phi \mapsto \phi + \pi$. After elimination of the transient behaviour, we get:

$$\rho = \rho_* = (\sigma/g')^{1/2} \tag{4.20}$$

which, once inserted in (4.19), yields:

$$\varphi(t) = -\omega_* t + \varphi_0 \quad \text{with} \quad \omega_* = \omega + g'' r/g', \tag{4.21}$$

where φ_0 is a constant that depends on initial conditions. The so-obtained *limit cycle* is stable when the bifurcation is supercritical, which implies $g' > 0$, as illustrated in Figure 4.6. On the other hand, as understood from (4.21), coefficient g''/g' determines the change in angular frequency due to nonlinear couplings, i.e. the oscillation's nonlinear dispersion.

4.2.1.3 Dynamics in dimension two, general case

In fact one can show in full generality that the behaviour of a system with a two-dimensional state space which is isomorphic[5] to $\mathbb{R}^2$ cannot have regimes that are more complicated than periodic according to the *Poincaré–Bendixson theorem*. A qualitative idea of the reason why complex behaviour (chaos) is excluded for two-dimensional time-continuous systems stems from the observation that, at a regular point (i.e. $\mathcal{F}(\mathbf{X}) \neq 0$) the vector field defines only one tangent direction, so that two trajectories cannot cross at this point. In dimension 2, the consequence of this fact is particularly drastic: a trajectory is a line (dimension 1) that splits the space (dimension 2) in two disconnected parts. Once transported everywhere by the vector field,

---
[5] The surface of a torus is not isomorphic to $\mathbb{R}^2$ though tangent planes can be defined everywhere.

this topological constraint implies that trajectories corresponding to steady states can be either fixed points or closed curves, i.e. time-independent or time-periodic states. The rigourous mathematical proof of this result is not as easy as it seems; accessible references are [Hirsch and Smale (1974)] or [Lefschetz (1977)].

In contrast, many things become possible in dimensions strictly larger than two since trajectories have then enough room to wind in complicated ways without crossing by "escaping" in the supplementary dimensions. In fact, the same also holds for effective dynamical systems defined on two-dimensional manifolds with nontrivial topology such as the Mœbius band, which is non-orientable and cannot be applied on $\mathbb{R}^2$ without tear but can indeed be embedded in $\mathbb{R}^3$, hence the required supplementary dimension.

In what follows, we examine first the case of periodically forced two-dimensional systems, which are in fact three-dimensional systems with a particular structure. Stroboscopic analysis of trajectories is illustrated in Appendix B by means of numerical experiments. It is used to introduce *discrete-time* dynamical systems essential to the understanding of the subsequent transition steps toward chaos. Before doing this we examine first the respective roles of linear resonance and nonlinearity on the particularly simple case of the forced Duffing oscillator.

4.2.2 *Quasi-periodicity and resonance*

4.2.2.1 *Forced systems*

Up to now we have considered *autonomous systems* in which time t does not appear explicitly, so that the trajectory is independent of the instant chosen to specify the initial condition in state space, which is no longer the case of *forced systems* introduced on p. 41. It is easily seen that the effective dimension of a forced system formally written as:

$$\tfrac{d}{dt}\mathbf{X} = \mathcal{F}(\mathbf{X};t), \qquad (4.22)$$

is increased by one with respect to the dimension of the corresponding unforced system. In order to specify a trajectory completely, we have indeed to choose not only the initial condition $\mathbf{X}^{(0)}$ but, since the system is no longer time-translationally invariant, to tell also the time $t^{(0)}$ at which we start the system. The need for a supplementary piece of information already points to an increase of the effective dimension.

More formally, one can pass to an *extended state space* in which the system is autonomous by introducing an auxiliary state variable U trivially

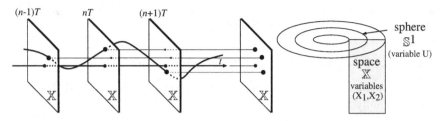

Fig. 4.7 Extended representation (left) and restricted representation (right) of the state space of a periodically forced system. The stroboscopic analysis corresponds to a series of "sections" in the extended state space, further identified modulo T in the restricted representation.

governed by $\frac{d}{dt}U \equiv 1$. Setting $\mathbf{Y} = \{\mathbf{X}, U\}$ we then obtain:

$$\tfrac{d}{dt}\mathbf{Y} = \mathcal{G}(\mathbf{Y}) \qquad \text{with} \qquad \mathcal{G} = \{\mathcal{F}, 1\} \qquad (4.23)$$

which is indeed autonomous. If the initial condition was $\mathbf{X} = \mathbf{X}^{(0)}$ at $t = t^{(0)}$ for (4.22), then for system (4.23), $U^{(0)} = t^{(0)}$ presents itself as the initial condition for U. The extended state space is thus the product of the original state space $\mathbb{X}$ by $\mathbb{R}$ which accounts for the time variable.

In practice, the most familiar case of non-autonomous system involves *periodic forcing* with period T (angular frequency $\omega = 2\pi/T$), i.e. $\mathcal{F}(\mathbf{X}; t + T) = \mathcal{F}(\mathbf{X}; t)$. From a mechanical viewpoint, one can understand the supplementary variable U as characterising the rotation of a wheel with a moment of inertia so large that its angular speed ω is independent of the state of the system to which it is coupled, acting on it like the connecting rod of a steam engine.

Stroboscopic analysis of the trajectories at the period of the forcing then comes to take pictures of the system at a series of times $t_k = kT$, $k \in \mathbb{N}$. Geometrically, this operation corresponds to a series of "sections" of the extended state space (Figure 4.7, left). During the time interval between two sections, the trajectory can be computed with as much precision as desired by integrating (4.22).

Owing to the periodicity of the forcing, we can represent the full trajectory by registering the state of the system in its (reduced) state space $\mathbb{X}$ while keeping track of the instant it passes through this state using label k. In this perspective the full state space of the system in restricted representation is the product of $\mathbb{X}$ by the periodized interval $[0, T]$; the periodized interval $[0; 1]$ is usually called the one-dimensional sphere $\mathbb{S}^1$, so that one can denote it as $\mathbb{X} \oplus T\mathbb{S}^1$ (Figure 4.7, right). The generalisation of this operation to the case of autonomous dynamical systems (Poincaré section) is a particularly efficient tool for the understanding of the emergence of chaos, as already mentioned in §1.3.3.

4.2.2.2 Steady state of a periodically forced oscillator

Let us consider a Duffing oscillator with natural period $\omega_0 = 1$, damped but externally forced:

$$\tfrac{\mathrm{d}^2}{\mathrm{d}t^2} X + 2\eta \tfrac{\mathrm{d}}{\mathrm{d}t} X + X + \varepsilon X^3 = f \cos(\omega t). \qquad (4.24)$$

We consider here only the case of a saturating nonlinearity $\varepsilon \geq 0$ (see also Exercise 2.5.7).

- *Linear preliminary.* When $\varepsilon = 0$, after the damping of transients, the steady state response is obtained by inserting $X = A_\mathrm{f} \cos(\omega t - \varphi)$ in (4.24), which yields:

$$[(1 - \omega^2) \cos(\omega t - \varphi) - 2\eta \sin(\omega t - \varphi)] A_\mathrm{f} = f \cos(\omega t).$$

Expanding $\cos(\omega t - \varphi)$ and $\sin(\omega t - \varphi)$, by identification of terms in $\cos(\omega t)$ and $\sin(\omega t)$, we get:

$$[(1 - \omega^2) \cos\varphi + 2\eta\omega \sin\varphi] A_\mathrm{f} = f,$$
$$-2\eta\omega \cos\varphi + (1 - \omega^2) \sin\varphi = 0.$$

We derive immediately from the second equation that

$$\tan\varphi = \frac{2\eta\omega}{1 - \omega^2}, \qquad (4.25)$$

and, upon elimination of $\cos\varphi$ and $\sin\varphi$

$$\left\{ (1 - \omega^2)^2 + 4\eta^2\omega^2 \right\} A_\mathrm{f}^2 = f^2, \qquad (4.26)$$

which leads to

$$A_\mathrm{f} = \frac{f}{\sqrt{(1 - \omega^2)^2 + 4\eta^2\omega^2}}. \qquad (4.27)$$

In Figure 4.8, this solution to the linear problem for $f = 1$, labelled $\varepsilon = 0$, is represented as a thick solid line for $\eta = 0.1$ and as a thin solid line in the limit $\eta = 0$.

- *Effect of the nonlinearity.* Let us assume that the forcing is sufficiently weak ($f \ll 1$) and that the response is not much distorted so that we can restrict ourselves to a first harmonic approximation, even close to the linear resonance. At steady state, we can still keep $X = A_\mathrm{f} \cos(\omega t - \varphi)$ which, once inserted in (4.24), adds to the terms already obtained the

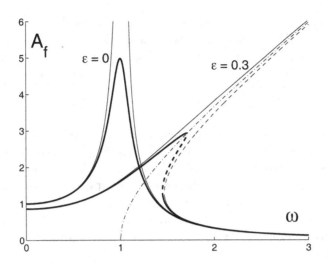

Fig. 4.8 Response A_f of the periodically forced Duffing oscillator upon sweeping the frequency around its resonance value. Thick and thin lines correspond to $\eta = 0.1$ and $\eta = 0$, respectively in the linear ($\varepsilon = 0$) and nonlinear ($\varepsilon = 0.3$) cases. Stable solutions are represented by solid lines and unstable ones by dashed lines.

quantity $\varepsilon A_f^3 \cos^3(\omega t - \varphi) = \varepsilon A_f^3 \left[\frac{3}{4} \cos(\omega t - \varphi) + \frac{1}{4} \cos(3(\omega t - \varphi)) \right]$. Only the first term of this sum is of interest to us now. Expanding it as before in $\cos(\omega t)$ and $\sin(\omega t)$, we immediately see that the solution can be obtained by replacing $(1 - \omega^2)$ by $\left[(1 - \omega^2) + \frac{3}{4}\varepsilon A_f^2 \right]$ everywhere.

Equation (4.26) now reads

$$\left\{ \left[(1 - \omega^2) + \tfrac{3}{4}\varepsilon A_f^2 \right]^2 + 4\eta^2 \omega^2 \right\} A_f^2 = f^2. \tag{4.28}$$

This equation is cubic in A_f^2 and can thus generically have one or three solutions, in contrast to the linear case ($\varepsilon = 0$) for which a unique solution is found.

Expanding (4.28) and considering the sign of its coefficients, one can easily show that there is always at least one positive root in A_f^2. In order to convince ourselves that it can also have other positive roots, we now examine the limiting case $\eta = 0$ which reads:

$$\left[(1 - \omega^2) + \tfrac{3}{4}\varepsilon A_f^2 \right]^2 A_f^2 = f^2.$$

In the absence of forcing ($f = 0$), in addition to the trivial solution $A_f = 0$, this equation has the double root:

$$A_f^2 = A_0^2 = \frac{4}{3\varepsilon}(\omega^2 - 1), \tag{4.29}$$

which is acceptable as soon as $\omega \geq 1$ (for $\varepsilon > 0$, subscript '0' in A_0 means zero forcing). This solution is perturbed by the introduction of a small forcing that suppresses the degeneracy. For ω sufficiently large, the pair of solutions that derives from it is given with a good approximation by

$$A_f^2 = A_0^2 \pm \tilde{A} \quad \text{with} \quad \tilde{A} = f\left[4(\omega^2 - 1)/3\varepsilon\right]^{-1/2},$$

where the correction $\tilde{A}$ remains small when compared to A_0^2, itself large in the considered limit $\omega \gg 1$. These two solution branches are traced for $f = 1$ and $\varepsilon = 0.3$ as thin lines in Figure 4.8. The solution for $f = 0$ derived from (4.29) is an arc of hyperbola that starts at $\omega = 1$ and serves as a common asymptote to these two branches (thin dash-dotted line). The thick line corresponds to the solution of the equation (4.28) including the viscous friction term with $\eta = 0.1$. As usual, unstable solutions are indicated by dashes.

In applications, one can fix f and change ω, or the reverse. In both cases one observes that there exist ranges of parameters over which multiple solutions exist and one can further show by methods adapted from those developed in Chapter 2 that the solution with intermediate amplitude is unstable and thus cannot be observed. By sweeping the control parameter (angular frequency or amplitude of the forcing) one may perform hysteresis cycles during which the oscillator jumps from a large amplitude to a small one or the reverse. One can also notice that when the nonlinearity is saturating ($\varepsilon > 0$), the curves bend in the direction of high frequencies, which is easily understood from the relation between the amplitude and the angular frequency for the free oscillator, Equation (2.48) p. 65 (increased average amplitude implies increased effective stiffness and thus shortening of the period).

• *Secondary resonance.* The modification of the response curve in the neighbourhood of the natural period of the oscillator, the so-called *primary* resonance is the most immediate effect of nonlinearity. The existence of *secondary* resonance at angular frequencies close to multiples or submultiples of this angular frequency is slightly less intuitive, though it still results from elementary trigonometric relations. The subharmonic resonance is most easily understood. As a matter of fact, if the forcing period $\omega = 1/3$ is apparently far from the resonance, the cubic nonlinearity induces some response at $\omega = 1$ since $\cos^3(t/3) = \frac{3}{4}\cos(t/3) + \frac{1}{4}\cos(t)$. In turn, this response can be seen as an external forcing of the fundamental mode, inducing its own resonant response. Considering a superposition

$X = \left[A\cos(t+\psi) + A_{\rm f}\cos(\omega t + \varphi)\right]$ and computing X^3 one finds terms in

$$\cos^3(t+\psi), \quad \cos^2(t+\psi)\cos(\omega t + \varphi),$$
$$\cos(t+\psi)\cos^2(\omega t + \varphi), \quad \cos^3(\omega t + \varphi),$$

which generate terms with angular frequencies ± 1, ± 3, $\pm 2 \pm \omega$, $\pm 1 \pm 2\omega$, $\pm \omega$ and $\pm 3\omega$, where the $\pm$ signs come from the fact that the resonance does not depend on the sign of the arguments of the sines and cosines. A resonance occurs every time a combined angular frequency is equal to the natural angular frequency (± 1).[6] In addition to $\omega \approx 1$ corresponding to the primary resonance, this rule adds $\omega \approx 1/3$ ($3\omega = 1$), $\omega \approx 3$ ($2 - \omega = -1$), and of course $\omega \approx 0$. However, the orders of magnitude of these different perturbations to the initial problem (the free, linear, friction-less oscillator) are different in each case, which imposes separate studies.

• *Simple or complex response?* As in Chapter 2, if one wants to go beyond the first harmonic approximation, one has to develop rigourous perturbation approaches by multiple scale methods. In the present context this would however lead us too far with little compensation as far as the insight into nonlinear phenomena is concerned. Asymptotic approaches to nonlinear oscillations can be found in [Nayfeh and Mook (1979)].

In practice, the relative simplicity of the response of the Duffing oscillator with a saturating nonlinearity ($\varepsilon > 0$) submitted to a harmonic forcing stems from the fact that its state space contains just one centre at the origin and no unstable elements. This would no longer be the case if, instead of $\frac{d^2}{dt^2}X + X + X^3 = 0$, we had taken $\frac{d^2}{dt^2}X - X + X^3 = 0$. The study of its phase portrait, similar to those in Exercises 2.5.6 and 2.5.8, shows that the centre at the origin is replaced by a saddle. Chaotic behaviour observed under external periodic forcing is precisely due to the instability inherent in this type of fixed point when the forcing increases beyond some threshold; see §4.3 below for an introduction as a foretaste of the detailed study presented in [Guckenheimer and Holmes (1983)]. The numerical study proposed in B.4.1 shows that the same result holds in the case of the periodically forced damped pendulum for the same reason: the presence of saddles in the phase portrait of the unperturbed system in Figure 2.9, p. 57. But before chaos, let us examine the case of self-sustained oscillations submitted to a periodic external forcing.

---

[6]Expanding the solution in terms of complex variables would lead to two conjugate resonance conditions, and at the end the same combinations, hence the ± 1.

4.2.3 Quasi-periodicity and locking

4.2.3.1 Case study: the periodically forced van der Pol oscillator

The phase portrait of the (unforced) van der Pol oscillator is less trivial than that of the standard damped Duffing oscillator since it displays an unstable focus at the origin and an attracting limit cycle surrounding it at some distance. We take it here in the form (2.60), p. 67 and add an external periodic forcing to obtain:

$$\tfrac{d^2}{dt^2}X - \epsilon(1-X^2)\tfrac{d}{dt}X + X = f\cos(\omega t). \tag{4.30}$$

Its different regimes will systematically be studied by means of stroboscopic analysis at the forcing period $T = 2\pi/\omega$.

In the absence of forcing ($f = 0$), the temporal evolution of some observable $\mathcal{W}(\mathbf{X})$ function of the state $\mathbf{X} \equiv \{X, \tfrac{d}{dt}X\}$ of the oscillator is a periodic function of time with period $T_0 = 2\pi/\omega_0$ ($\omega_0 \approx 1$ for $\varepsilon \ll 1$, see §2.3.2.2, p. 65).

When the forcing is weak ($f \ll 1$) and the imposed angular frequency is not commensurate with ω_0 ($\omega/\omega_0 = \alpha \notin \mathbb{Q}$), after transients have decayed, we expect $\mathcal{W}(\mathbf{X}(t)) = G(\omega_0 t + \varphi, \omega t + \psi)$ with $G(u+2\pi, v) \equiv G(u, v+2\pi) \equiv G(u,v)$, i.e. *two-periodic* behaviour.

In contrast, when ω and ω_0 are commensurate, i.e. $\omega/\omega_0 = \alpha \in \mathbb{Q}$, $\alpha = p/q$, integers p and q being relative primes, $\mathcal{W}(\mathbf{X})$ is a periodic function of time with period equal to the smallest common multiple of T and T_0, $\tilde{T} = pT = qT_0$.

• *Frequency locking.* When the intensity of the forcing increases, the oscillator can "feel better" if it leaves its own frequency and adopts that of the forcing. This is the *locking* phenomenon: over a full range of values of ω, around a condition such that $qT_0 = pT$, the oscillator adjusts its motion so that its effective period $\bar{T}_0$ still fulfils the resonance condition, $q\bar{T}_0 = \tilde{T} = pT$ so that the period of the forced system remains $\tilde{T}$.

Figure 4.9 displays the set of stroboscopic recordings X_{str} as a function of the forcing frequency ω (bifurcation diagram) in a given numerical simulation of (4.30), thus illustrating this spectacular persistence phenomenon of *locking windows* for $\omega \approx 1$ with $\varepsilon = 0.1$ and $f = 0.5$ (left), and for $\omega \approx 3$ with $\varepsilon = 0.1$ and $f = 1.0$ (right).

Let us consider first the window at $\omega \approx 1$. When the locking takes place, the system is periodic with period T so that, when sampled at period T, it takes one and a single value $X_{\text{str}}(\omega)$. On the contrary, as soon as ω gets out of the window, though the oscillator is still sensitive to the forcing,

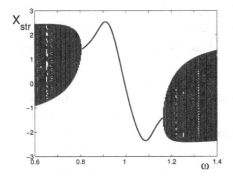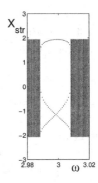

Fig. 4.9 Bifurcation diagram of the forced van der Pol oscillator. Left: $\omega \approx 1$, $\varepsilon = 0.1$, and $f = 0.5$. Right: $\omega \approx 3$, $\varepsilon = 0.1$, and a larger forcing, $f = 1.0$, to make its effects more visible; notice also the expanded scale for ω.

it recovers its independence and the signal becomes two-periodic. The stroboscopic analysis then produces a picture such as that for $\omega = 0.75$ in Figure 4.10 (left): the state of the oscillator (X and $\frac{d}{dt}X$) is registered at increasing multiples of the forcing period, $t_k = kT$, and it can be observed that the corresponding points arrange themselves along a closed curve that is continuously covered. Considering successive points shows that they jump from one position to the next with a slight shift, which explains the continuous coverage due to incommensurability of the two periods, while only a finite set of points would be reached in case of locking. The projection of this curve on the X axis produces the full vertical segments shown at

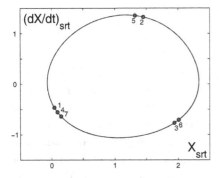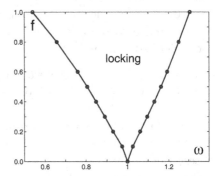

Fig. 4.10 Left: Stroboscopic analysis of a trajectory of the forced van der Pol oscillator for $\omega = 0.75$ and $f = 0.5$ (two-periodic unlocked regime). Right: Main resonance tongue $\omega \approx 1$: region of the (ω, f) parameter plane where locking is observed.

values of ω corresponding to unlocked behaviour on the bifurcation diagram of Figure 4.9. Notice that with $\omega = 0.75$ one has $4T_0 \simeq 3T$, i.e. close return every three samplings, which explains the proximity of points labelled 1, 4, and 7, etc.

The case of the window at $\omega \approx 3$ (Figure 4.9, right) is analogous but, inside it we now have three values of X_{srt} for each ω. This is easily understood from the fact that the trajectory is sampled at period T which is about 1/3 of the natural period of the oscillator, so that it is regularly sampled three times during one of its own turns. In the locked regime, it returns exactly at the same places (modulo 3) whereas outside the window, in the unlocked regime, return points shift slightly from one sampling to the next, which again gives continuous segments on the bifurcation diagram. Notice that, in spite of a much stronger forcing, the window is extremely narrow.

- *Emergence of complex behaviour.* The variation of the range in ω where the van der Pol oscillator is locked to the forcing, the *resonance tongue*, is displayed in Figure 4.10 (right) as a function of the intensity f of the forcing (on the vertical axis). For this resonance and sufficiently weak forcing, it widens linearly with f. Higher order resonances are narrower and widen more slowly. Frequency locking is a profoundly original manifestation of nonlinear effects. In the case of the van der Pol oscillator in a quasi-harmonic regime ($\varepsilon = 0.1$) it seems that one cannot observe phenomena more complicated than a decay of two-periodic behaviour into simply periodic behaviour, as long as one stays with reasonable values of the parameters. Things are different when the oscillator is more anharmonic (e.g. $\varepsilon = 1.0$). A wide variety of behaviour can be observed, with whole ranges of angular frequency where the response to forcing is chaotic, especially when the periodic driving is slow.

- *Mutual locking.* Up to now we have considered forcing in a strict sense, that is to say without feedback of the forced oscillator on the driving system. Besides, it is the presence of this immutable clock that makes the stroboscopic analysis easy to perform. Relaxing this condition, we can consider a system composed of two weakly coupled oscillators. Transposing the observations above, we may expect mutual locking in the neighbourhood of resonance conditions fulfilled by the angular frequencies of these oscillators, for example Huygens' twin pendulum clocks.[7] A system of two coupled oscillators is, as a whole, a four-dimensional system (2×2); see §5.4.1 for a

---
[7]See http://en.wikipedia.org/wiki/Christiaan_Huygens.

more general discussion. We now return to the case of autonomous systems in a dimension greater than two, to which we devote the end of this section.

4.2.3.2 Stability of a limit cycle in the general case

Let us consider self-sustained oscillations corresponding to a limit cycle in state space. Its stability must be studied not only on the effective surface containing the cycle, locally mapped onto the complex plane ($A \in \mathbb{C}$), which allows perturbations in the radial direction only, see Eq. (4.18), but in the full space in which the cycle is embedded, which permits escape in the supplementary dimensions. At this stage, it is useful to recall that we have arrived to an effective two-dimensional dynamics by adiabatic elimination of all stable modes, leaving us with a single relevant pair of centre modes. Upon increasing the stress applied to the system, we have to consider the possibility that among stable modes, some become "dangerous." Mathematically speaking, the situation is more complex than before when we had to deal with a fixed point, since now we have to perturb a well-established periodic dynamics.

Let us keep the geometrical approach introduced in §1.3.3 and sketched in Figure 4.7 illustrating the case of a periodic forcing, but now consider the intersections of the trajectories in a d-dimensional state space with a $(d-1)$-dimensional surface Σ, called the *Poincaré surface of section*, see Figure 4.11 (top-left). The oscillation that has settled can be viewed as a forcing for all other perturbations, and in the vicinity of the cycle, the state space has indeed this product structure of the sphere $\mathbb{S}^1$ times a relevant space in which *transverse* perturbations live, as extrapolated from Figure 4.7 (right). A correspondence between successive intersections of trajectories with Σ is thus established in the form of a map of Σ onto itself $\mathbf{M}' = \mathbf{\Phi}(\mathbf{M})$, called the *first return map* or the *Poincaré map*. This transition from *continuous-time* dynamics to *discrete-time* dynamics, from a d-dimensional differential system to a $(d-1)$-dimensional mapping, is therefore basically similar to the stroboscopic analysis used for periodically forced systems. The difference just comes from the fact that the time interval between two sections is no longer the external forcing period but varies slightly with the position in the neighbourhood of the cycle and tends to the period of the cycle as it is approached. This simply assumes that the differential dynamical system is smooth, so that the properties of the trajectories vary continuously, and that the surface of section is transverse (i.e. not tangent) to the cycle, so that the trajectories are correctly sampled.

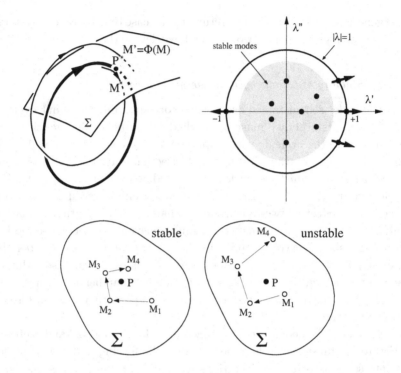

Fig. 4.11 Stability of a limit cycle from a geometrical viewpoint. Top-left: Poincaré section and first return (Poincaré) map $\Phi(\mathbf{M})$. Top-right: Spectrum $\{\lambda = \lambda' + i\lambda''\}$ of the linearised map $\mathbf{\Lambda}$ defined by $\lambda \delta \mathbf{M} = \mathbf{\Lambda} \delta \mathbf{M}$, where $\delta \mathbf{M}$ represents an infinitesimal departure from $\mathbf{P}$. Bottom: Stability/instability of the fixed point of the map.

Obviously, the cycle is associated with a fixed point $\mathbf{P}$ of the map Φ (i.e. $\mathbf{P} = \Phi(\mathbf{P})$) and will be stable is $\mathbf{P}$ is stable for Φ. In the neighbourhood of $\mathbf{P}$, the map Φ can be linearised into an operator represented by a square $(d-1) \times (d-1)$ matrix $\mathbf{\Lambda}$, and the stability properties of the cycle can be derived from its spectrum, as illustrated in Figure 4.11 (top-right).

In full generality, one expects non-degenerate complex eigenvalues and corresponding two-dimensional invariant subspaces that are easier to parameterize using complex variables. In such a subspace, the linearised dynamics then reads:

$$Z_{k+1} = \lambda Z_k, \tag{4.31}$$

$Z \in \mathbb{C}$ measuring the departure from the fixed point in the eigenspace of eigenvalue λ.

According to the definition, we have stability when the successive iterates of some initial condition $Z^{(0)}$ approach the origin as k increases and instability when they get away, Figure 4.11 (bottom). A given complex eigenmode is thus stable when $|\lambda| < 1$, marginal when $|\lambda| = 1$, and unstable when $|\lambda| > 1$. Writing λ as

$$\lambda = \rho \exp(2\pi i \alpha)$$

leads to a clear separation of the modulus ρ from the phase α of the eigenvalues.

Quantity α specifies the angular frequency of the mode that makes the cycle possibly unstable: $\alpha = \omega/\omega_0$. This can be irrational or rational like in the case of periodic forcing. $\lambda \in \mathbb{R}$ (and $= \pm 1$ when marginal) are special cases since (4.31) is then turned into a real iteration:

$$X_{k+1} = \lambda X_k, \qquad (4.32)$$

and we must distinguish the dynamics corresponding to $\lambda > 0$ ($\alpha = 0$, $\lambda \approx +1$) for which iterates tend to or depart from the origin in a monotonic fashion from that observed when $\lambda < 0$ ($\alpha = 1/2$, $\lambda = -1$) for which the iterates jump alternatively from one side of the origin to the other.

Eigenvalue λ can be complex or not, but the corresponding stability condition remains $|\lambda| < 1$ and a bifurcation takes place when at least one (pair of) eigenvalue(s) leaves the interior of the unit disc as the control parameters are varied. Different possibilities are indicated by arrows in Figure 4.11 (top-right).

Though the main purpose of the course is not a mathematical study of the different scenarios of transition to chaos in the perspective opened by Ruelle and Takens (see Figure 1.7, p. 17), let us mention that one can account for the simplest cases by completing (4.31) or (4.32) with appropriate terms. In the original Ruelle–Takens scenario, the destabilisation of the limit cycle still leads to a regular but two-periodic regime ($\lambda \in \mathbb{C}$). The iteration that accounts for it reads

$$Z_{k+1} = \lambda Z_k - \gamma |Z_k|^2 Z_k, \qquad \gamma = \gamma' + i\gamma''. \qquad (4.33)$$

After the damping of transients, it yields iterates evolving around a circle with radius $\sqrt{\rho - 1}$, where $\rho = |\lambda|$, Figure 4.12 (top). Trajectories in state space wind on an invariant *limit torus* which is the Cartesian product of two circles, one corresponding to the unstable limit cycle (longitude) and one for the new mode (meridian section by the Poincaré surface), as sketched in Figure 4.12 (bottom).

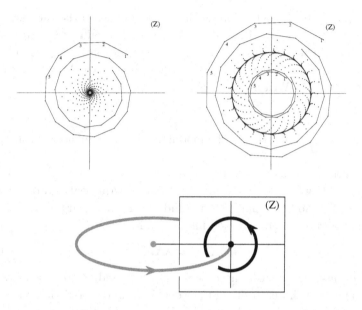

Fig. 4.12 Hopf bifurcation for a map (also called Neimark–Sacker bifurcation). Top: Iteration (4.33) for $|\lambda| < 1$ (left) and $|\lambda| > 1$ (right). Bottom: Perspective sketch of the skeleton of the *torus* over which the two-periodic dynamics develops asymptotically in time beyond bifurcation.

The picture obtained in this case is very similar to that for the forced van der Pol oscillator. The difference is just that the two angular frequencies in the system now have intrinsic origins, the first one, which plays the role of the forcing, stems from the original cycle and the second one from its most dangerous instability mode. They can easily be identified on the records of physical observables, e.g. the two-periodic regime observed in convection and illustrated in Figure 3.8, p. 106. As before, two-periodicity is observed only in the absence of resonant interaction between the two angular frequencies, that is to say as long as the ratio α remains irrational. Nonlinear couplings are therefore expected to be responsible for locking when α approaches a rational value p/q as the control parameters are varied. Locking is most conspicuous when the denominator q is a small integer.

4.2.3.3 *Resonance, weak and strong*

When $\alpha = p/q$ the two periods are resonant. Little is changed if q is large, i.e. $q \geq 5$. The resonance is *weak* and the system behaves roughly as in

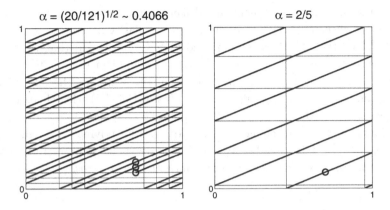

Fig. 4.13 Trajectories on the torus are represented by their phases: longitude $\theta = \omega_0 t + \theta^{(0)}$ (horizontal axis) and meridian coordinate $\varphi = \omega t + \varphi^{(0)}$ (vertical axis) defined modulo 2π. Up to a factor 2π, the torus is isomorphic to the square $\theta \in [0;1]$, $\varphi \in [0;1]$ with opposite sides identified. Left: for a two-periodic state, $\alpha \notin \mathbb{Q}$, the whole (θ, φ) square is covered by any given trajectory as $t \to \infty$, here $\alpha = 2\sqrt{5}/11 \approx 0.40656$. Right: when $\alpha = p/q$, the same state is reached after a time $qT_0 = pT$, here $\alpha = 2/5 = 0.4$, a weakly resonant value.

the non-resonant case: typically, the dynamics still takes place on the torus that has emerged from the bifurcation. Genuinely two-periodic unlocked regimes alternate with locked regimes inside narrow tongues as the control parameters are varied. In a quasi-periodic state the trajectory winds on the torus without closing (the torus is densely covered with it), whereas in case of locking the trajectory is a closed cycle wound on the torus, making p turns along the meridian and q turns in the longitudinal direction (Figure 4.13). Chaos may enter after a new instability mode sets in.

In contrast, when $q \leq 4$ the resonance is *strong* and has marked effects. The case $q = 4$ is less spectacular than when $q = 3$, a situation when the torus is transformed into a fractal manifold with wrinkles at all scales as nonlinearity increases, see Exercise 5.6.2 for a model. In the most extreme cases corresponding to $q = 2$ ($\alpha = 1/2$) and $q = 1$ ($\alpha = 0$), the torus is smashed into a band, either twisted (a Mœbius band, $q = 2$) or not ($q = 1$). The complex mapping (4.31) is then reduced to a real mapping (4.32) since for $p/q = 1/2$ one has $\lambda = \exp(i\pi) = -1$ and for $p/q = 1$, $\lambda = \exp(2i\pi) = 1$. Each of these two cases govern a specific scenario.

Two-periodic states as well as locked one-periodic states are regular asymptotic regimes. In order to observe a chaotic behaviour, it seems natural to think of an effective dimension still increased by one. This was precisely the context of the Ruelle and Takens approach, introducing *strange attractors* with properties accounting for irregularity in spite of determinism. More precisely, these authors showed (Note 5, p. 16) that when the system is smooth (C^∞), chaos is generic after supplementary bifurcations introducing two new angular frequencies in the system. The analyticity condition was later made milder (i.e. C^2 instead of C^∞) with chaos possibly occurring in three-frequency quasi-periodic systems.[8] Later, in §4.3.3, we shall consider two frequently observed scenarios relying on strong resonance, $\alpha = p/q$ with $q = 2$ or 1, but before, we spend some more time on the nature of that kind of aperiodic behaviour called *chaos*.

4.3 Beyond Quasi-Periodicity

Up to now, we have approached the emergence of complex dynamics by steps: time-independent, periodic, two-periodic, and multi-periodic, mostly following Landau's original idea (note 4, p. 16). Resting on a *superposition* of oscillatory modes this – in some sense too linear – scenario is not robust against nonlinear *interactions* as shown by Ruelle and Takens (note 5, p. 16), who insisted that multi-periodicity does account for the decorrelation of trajectories expected for turbulence and that sensitivity to initial conditions and small perturbations is essential in this respect.

Nonlinearity may have mild consequences, namely *saturation* as for the supercritical fork bifurcation (1.20), p. 13, or the supercritical Hopf bifurcation, (4.17), p. 138. But its effects can be much more severe, just by introducing the possibility of *multiple solutions*, some stable, others unstable. As soon as the dimension of the effective state space is strictly larger than two, escape in the third dimension permits *complexity*. Introduction of a third dimension has been performed first *via* periodic forcing in §4.2.2.1, and next as the result of the instability of a limit cycle against some new physical mechanism. In the latter case, the limit cycle could be viewed as an internal source of periodic forcing, and the definition of the Poincaré section technique and associated return map as a generalisation of stroboscopic analysis. In §4.2.2.2, a still simple situation resulted from periodic perturbation of a stable focus at the origin, with no other fixed point around (ordinary Duffing oscillator). A more complicated case was next examined where an unstable focus was the germ of a limit cycle and

---

[8]S. Newhouse, D. Ruelle, and F. Takens, "Occurrence of strange axiom A attractors near quasi-periodic flows on T^m, $m \leq 3$," Commun. Math. Phys **64** (1978) 35.

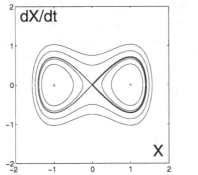 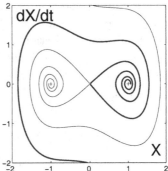

Fig. 4.14 Left: phase portrait of the modified Duffing oscillator with $\eta = 0$ (and $f = 0$). Right: the homoclinic loop is opened by the introduction of some friction, $\eta > 0$.

the periodic perturbation was applied, leading to the locking phenomenon, §4.2.3.1. Complexity inherent in chaos is best introduced by considering the periodic forcing of a saddle point as for the modified Duffing oscillator:

$$\tfrac{d^2}{dt^2}X + 2\eta \tfrac{d}{dt}X - X + \varepsilon X^3 = f\cos(\omega t), \qquad (4.34)$$

where the original '+' sign affecting the linear term on the l.h.s. in (2.44) has been replaced by a '−' sign. Without friction ($\eta = 0$) and forcing ($f = 0$), this equation accounts for the dynamics of a two-well system, with two symmetric stable elliptic points at $X = \pm 1$ and a saddle at $X = 0$ (see also Exercise 2.5.8, p. 79). Figure 4.14 (left) displays its phase portrait, in which one must notice the existence of a trajectory connecting the origin to itself, starting along its unstable direction and coming back along its stable direction. Such a trajectory is called *homoclinic*; in the same way a *heteroclinic* connection is a trajectory joining two different fixed points. Stretching due to instability and folding due to nonlinearity makes all the difference from the previously considered cases.

On general grounds, a stable (unstable) manifold $\mathcal{S}$ ($\mathcal{U}$) is the nonlinear extrapolation of the stable (unstable) subspace $\mathcal{E}_s$ ($\mathcal{E}_s$) corresponding to eigenvalues with negative (positive) real parts. Stable (unstable) manifolds associated to different limit sets cannot intersect because points belonging to the intersection would have to go to (come from) these different limit sets in the future (past) which is forbidden by determinism. In contrast stable and unstable manifolds may intersect since they relate to different time limits, future and past, respectively. Complexity arises mostly due to the possible existence of *homoclinic* and *heteroclinic points*, i.e. points belonging to the intersection of stable and unstable manifolds of a given limit set ("homo") or of different limit sets ("hetero"). This is discussed in the next subsection. Section 4.3.2 is then devoted to the description of basic properties of chaotic behaviour. We conclude the chapter (§4.3.3) by considering two frequently observed scenarios of transition to chaos arising in case of strong resonance with $q = 2$ and $q = 1$.

4.3.1 *Dynamics on stable and unstable manifolds*

Since detailed illustrations can be found in [Baker and Gollub (1996)] and mathematical considerations are developed in [Guckenheimer and Holmes (1983)], we just sketch the argument. The situation depicted in Figure 4.14 (left), with a homoclinic loop connecting the unstable fixed point to itself, is exceptional. The existence of this special trajectory is linked to the time-reversal symmetry (invariance under the change $t \mapsto -t$). Perturbations breaking this symmetry will open the loop. The trajectory issued from the fixed point along the unstable direction will miss it upon return, and the one that will reach it along its stable direction will have to come from elsewhere as shown in Figure 4.14 (right) where some friction has been added ($\eta > 0$).

But the unforced system is still two-dimensional. Let us now add a time periodic perturbation and perform a stroboscopic analysis of the system at the period of the forcing. Trajectories on the stable and unstable manifolds are now series of points sitting on curves which correspond to sections of these manifolds in the extended state space $\mathbb{X} \times \mathbb{R}$ by the successive planes $t = t_k = 2\pi k/\omega$, where ω is the frequency of the forcing in (4.34). At given friction coefficient η, two mutually exclusive situations can take place: either the stable and unstable manifold do not intersect, or they do, which depends on the strength of the forcing f. Intuitively, when friction is dominant and the forcing small, the manifolds do not intersect and the situation remains simple, namely oscillatory relaxation towards the trajectory arising from the former elliptic fixed points at $X = \pm 1$ now turned into stable spiral points. This is no longer the case if the intensity of the forcing increases because nonlinearity reshapes the manifolds far from the fixed point and makes them intersect. The intermediate case corresponds to homoclinic tangency that will take place at a specific threshold, the value of which can be obtained by a perturbation approach called Melnikov's method. Beyond threshold, the manifolds intersect transversally and the remarkable fact is that a single intersection implies the existence of an infinite, uncountable, number of intersections, which – as envisioned first by

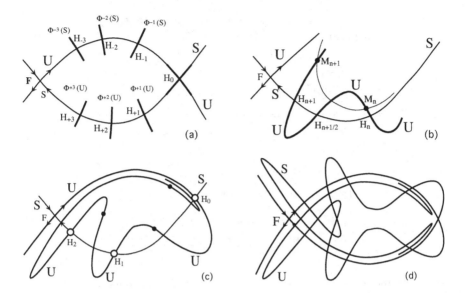

Fig. 4.15 Formation of a tangle from a homoclinic loop: (a) Assume the existence of a homoclinic point H_0 at the intersection of the unstable and stable manifolds $\mathcal{U}$ and $\mathcal{S}$ of fixed point F on the Poincaré section. The repeated application of the Poincaré map Φ carries a segment of the unstable manifold $\mathcal{U}$ into an series of segments intersecting $\mathcal{S}$ at points H_{+k} converging geometrically towards F along $\mathcal{S}$, whereas the inverse map Φ^{-1} generates the series H_{-k} along $\mathcal{U}$. (b) Continuity of $\mathcal{U}$ and $\mathcal{S}$ and preservation of the orientation imply the formation of lobes and an associated series of secondary points $H_{\pm m}$. (c) As they approach F the '+' lobes made by $\mathcal{U}$ pile up along $\mathcal{U}$ itself, while being compressed along $\mathcal{S}$. Their lengths increase geometrically at a rate given by the unstable eigenvalue (> 1) and at some stage, they cut $\mathcal{S}$ again making two new series of tertiary homoclinic points. (d) The same argument holds for lobes made by $\mathcal{S}$ under application of Φ^{-1}. The tangle is formed.

Poincaré (note 3, p. 16) – roots the complicated dynamics around the fixed point called *chaos*. The reason is sketched in the caption of Figure 4.15. The object created beyond tangency is called a *homoclinic tangle*; see, e.g. [Tabor (1989)].

4.3.2 *Nature of chaotic regimes*

Sufficiently strong nonlinearity forces the stable and unstable manifold to entangle. By construction, the set of homoclinic points is invariant under the action of Φ and its inverse Φ^{-1} but it has zero measure and cannot be observed as a trajectory in any experiment, since the probability that one of its points be an initial condition is strictly zero. However, its exis-

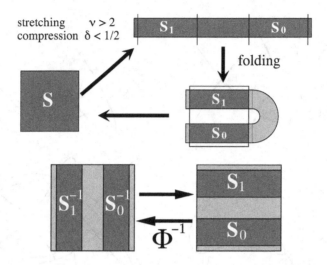

Fig. 4.16 Top: Construction steps of Smale's horseshoe map. Bottom: Definition and action of mapping Φ and its inverse Φ^{-1}.

tence controls the dynamics around the fixed point. Trajectories that come close to it are channelled by the complicated arrangement of its stable and unstable manifolds. They evolve chaotically for a while before leaving the neighbourhood of **F** to visit other regions of the state space. In order to better understand this property we need a model of homoclinic tangle. A simple one is Smale's horseshoe map. The map is an invertible transformation of the unit square $\{X, Y\} \in [0, 1] \times [0, 1]$ onto itself. It mimics the unstable dynamics along $\mathcal{U}$ by a stretching along the X coordinate and the stable dynamics along $\mathcal{S}$ by a compression in the Y direction. Nonlinearity is described as a folding of the stretched band and its replacement into the square, see Figure 4.16 (top).

Let us denote the map as Φ, the homoclinic set is then modelled as the set of points $\{\mathbf{H}\}$ invariant under Φ and Φ^{-1}. This set is the intersection of two sets constructed following the rules defining the ternary Cantor set: take a segment, remove the middle third, and iterate using the two segments left; here take the square, remove a central band, and iterate using the two bands left; see later §5.1.2 and especially Figure 5.3, p. 175, for an explicit map producing this result. One limit set is thus the result of iterating the square under Φ. It is the product of a ternary Cantor set in the direction of contraction (Y) by segment $[0, 1]$ in the direction of stretching (X). The second Cantor set, formed from Φ^{-1}, interchanges the roles of X and Y.

Points $(X, Y) \in \{\mathbf{H}\}$ can be labelled by the binary expansion of their coordinates, $X = \bullet \sigma_1 \sigma_2 \ldots$, $Y = \bullet \tau_1 \tau_2 \ldots$, with $\sigma_k = 0$ or 1 and $\tau_k = 0$ or 1 depending on the band it visited, S_0^{-1} or S_1^{-1}, S_0 or S_1, in the course of the iteration, respectively; see Figure 4.16 (bottom).

Next, considering the application of $\mathbf{\Phi}$, one can see that the ranks of all the digits of the expansion of X are decreased by 1 and the former leading digit, lost in this operation, becomes the leading digit of the expansion of Y with the rank of all its digits increased by 1. Application $\mathbf{\Phi}^{-1}$ exchange the roles of X and Y. Accordingly, points in the homoclinic set can be labelled as $\mathbf{H} = \{\ldots \tau_2 \tau_1 \bullet \sigma_1 \sigma_2 \ldots\}$ so that the action of $\mathbf{\Phi}$ ($\mathbf{\Phi}^{-1}$) comes a shift of the dot '$\bullet$' to the right (left). Upon coding, the transformation of points within the homoclinic set comes to a bidirectional *Bernoulli shift* (bidirectional because determinism implies that Poincaré map be invertible).

This representation and the *symbolic dynamics* which comes with it have the merit of pointing out several important properties of chaos:

– the existence of an infinite, uncountable, number of aperiodic orbits inside the homoclinic set: these are the points coded by aperiodic series of digits corresponding to irrational coordinates X and Y;

– the presence of an infinite, countable, number of periodic orbits: these are points with rational coordinates, which thus have periodic binary expansions;

– the instability of all orbits within the set since expansions that differ beyond rank k correspond to orbits differing by perturbations of $\mathcal{O}(2^{-k})$.

For a direct application see §5.3, p. 191, about chaotic advection. By construction, Smale's horseshoe map defines a set that has just the same topological structure as a homoclinic tangle. The stretching part of the transformation accounts for the sensitivity of trajectories to initial conditions and small perturbations advanced by Ruelle and Takens. More generally, a chaotic attractor may be defined as the closure[9] of an unstable manifold containing a homoclinic set.

4.3.3 Two classical scenarios of transition to chaos

Let us come back to frequently observed routes to chaos, both relying on the instability of a limit cycle in strong resonance resonance conditions, a local property in state space. Many other scenarios are possible, usually involving

---

[9]Mathematically, the closure of a set is the reunion of that set and all its accumulation points.

the global structure of state space. An example is *Shilnikov's spiral chaos* arising from a homoclinic trajectory at a saddle focus (an unstable fixed point with linear spectrum $\{\lambda, \sigma \pm i\omega\}$, with $\lambda > 0$ and $\sigma < 0$). Another example is the *attractor crisis*, when an attractor basin suddenly change its shape by colliding invariant manifolds of other limit sets.[10]

4.3.3.1 The period-doubling cascade

The *period-doubling cascade* observed for example in convection (Figure 3.6, p. 105) corresponds to the strong resonance $\alpha = 1/2$, whose normal form reads:

$$X_{k+1} = f_r(X_k) = -(1+r)X_k - X_k^3. \tag{4.35}$$

The corresponding scenario is illustrated here using the so-called *logistic map*[11]

$$X_{k+1} = 4rX_k(1 - X_k), \tag{4.36}$$

with $0 \leq r \leq 1$. The full bifurcation diagram in Figure 4.17 displays how the attractor (values of X reached as the iteration proceeds, after elimination of transients) changes as the control parameter r is varied.

For $r < 3/4$, a single point is obtained, corresponding to a periodic trajectory with period-one (period T_0 if one considers the underlying limit cycle). At $r = r_1 = 3/4$ a first period-doubling takes place: a period-two regime sets in with two values X_1 et X_2 alternatively visited as the system evolves. (Working out the nonlinear variable change that turns (4.36) into the normal form (4.35) close to the bifurcation point is an interesting exercise.)

At $r = r_2 = 0.86237\ldots$, this period-two attractor is destabilised against a period-four regime, ... The bifurcation cascade ends up at $r = r_\infty = 0.89248\ldots$, where the system becomes periodic with infinite period in the form 2^{2n}, $n \to \infty$. At this value, the so-called *edge of chaos*, the system is in fact aperiodic since it never returns exactly in the same state.

Beyond r_∞, continuous vertical segments, in fact densely covered with points, are observed on the bifurcation diagram for each value of r. They are the trace of chaotic attractors for which infinite sets of unequally distributed points are visited all along a trajectory.

---

[10] For a review of spiral chaos with references to original work, see L.P. Shilnikov and A. Shilnikov: *Shilnikov bifurcation*, Scholarpedia, 2(8):1891 (2007), and for crises E. Ott: *Crises*, Scholarpedia, 1(10):1700 (2006).

[11] For an introduction, see R.M. May, "Simple Mathematical models with very complicated dynamics," Nature **261** (1976) 459–467.

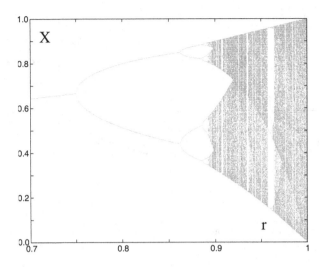

Fig. 4.17 Subharmonic cascade in (4.36).

In fact, the situation is also very complicated in the parameter space since the return to periodic attractors is clearly visible in some parameter windows, the corresponding periodic states themselves decaying to chaos through subharmonic cascades. Programming (4.36) or its variant (4.56), p. 169, is so easy that one should not miss exploring their bifurcation diagram numerically by oneself.

4.3.3.2 Type-I intermittency

Intermittency (type I, the one featured in Figure 3.9, p. 107, because several other types exist) takes place at a 1/1-resonance generically accounted for by the map:

$$X_{k+1} = r + X_k + X_k^2. \tag{4.37}$$

Before the transition, as long as $r < 0$, it has two fixed points solutions of:

$$X_\pm = r + X_\pm + X_\pm^2 \quad \Rightarrow \quad X_\pm = \pm\sqrt{-r}.$$

All trajectories starting with $X < X_{(+)}$ (and small enough) converge towards the fixed point at $X_{(-)}$. Other trajectories leave this limited region of state space, Figure 4.18. Beyond the transition, for $r > 0$, the fixed points have disappeared but the local structure of state space keeps a track of their presence as "ghosts": iterates go through the kind of tunnel opened

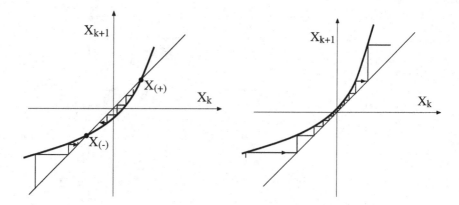

Fig. 4.18 Type-I intermittency according to (4.37). Left: $r < 0$, two fixed points, $X_{(-)}$ stable, $X_{(+)}$ stable. Right: $r > 0$, no fixed points but a narrow tunnel between the graph of the map and the line $X_{n+1} = X_n$.

between the graph of the map and the line $X_{k+1} = X_k$. They move very slowly since $X_{k+1} = X_k$ plus some tiny correction, approach the origin and then go away.

Now the *global* structure of state space may provide the opportunity for trajectories visiting remote regions of state space to come in the neighbourhood of the origin. When this is the case, for $r < 0$ the system coming close to the stable fixed point eventually converge to it. In contrast, fixed points no longer exist for $r > 0$: iterates sent in the vicinity of the origin, travel slowly through the tunnel, then get away and come back later.

The more or less uncontrolled evolution far from the previously existing fixed points correspond to chaotic bursts break the correlation between successive laminar intermissions, rendering their lengths unpredictable due randomness in the re-injection process.

While interpreting the experiments, one must remember that the map describes what happens after an appropriate section of the dynamics has been performed. At the bifurcation, twin limit cycles, one stable and the other unstable, collide and disappear. Laminar intermissions are made of slowly evolving regular periodic oscillations corresponding to the "ghost" limit cycles. They are interrupted by chaotic bursts during which the system explores remote regions in state space, until it is sent back in the laminar region, ready for a new intermission, hence the behaviour illustrated in Figure 3.9.

4.4 Exercises

4.4.1 Homogeneous instability in a confined context

Consider a spatially extended system governed by

$$\partial_t v = rv + \partial_{xx} v - v^3.$$

1) Linearised problem: determine the growth rate of Fourier modes $\propto \exp(ikx)$ with infinitesimal amplitude in a laterally unbounded system and obtain the marginal stability condition of the trivial solution $v \equiv 0$.

2) Adding boundary conditions $v(x) = 0$ at both ends of the interval $[0, \pi]$, check that one can take eigenmodes in the form $X_n = \sin(nx)$ and determine their growth rate as a function of n.

3) Find the condition on r such that only the first mode is excited, while others remain strongly stable. When this is the case, assuming $v(x,t) = A_1(t) X_1(x)$, determine the equation governing A_1 within a first-harmonic approximation. Show that the system experiences a supercritical fork bifurcation.

4) The cubic nonlinearity generates modes X_n, $n > 1$. Find the equation governing A_3, its order of magnitude at steady state when $r \ll 1$, and justify the approximation made.

4.4.2 Cellular instability in a confined context

Go back to model (3.39) and complete it by a nonlinear advection-like term to get:

$$\partial_t v + v \partial_x v = rv - (\partial_{xx} + 1)^2 v. \qquad (4.38)$$

Then consider a confined system with boundary conditions $v = \partial_{xx} v = 0$ at $x = 0$ and $x = \ell$. Check that modes in the form $\sin(k_n x)$ are appropriate and insert expansion:

$$v(x,t) = \sum_{n=1}^{\infty} A_n(t) \sin(k_n x), \qquad k_n = n\pi/\ell, \qquad (4.39)$$

in (4.38). Further separate the different harmonics and obtain the set of ordinary differential equations governing the amplitudes A_n. Write down explicitly the equations for A_1, A_2 and A_3 when $\ell = \pi$ for $r \ll 1$. Show that in this limit $A_2, A_3, \ldots$ are enslaved to A_1. Propose a coherent truncation of the system above some order $N \geq 2$ fixed in advance.

When truncated at order 2 the system reads:

$$\tfrac{d}{dt} A_1 = rA_1 + \tfrac{1}{2} A_1 A_2, \qquad (4.40)$$

$$\tfrac{d}{dt} A_2 = -9 A_2 - \tfrac{1}{2} A_1^2. \qquad (4.41)$$

Sketch its phase portrait in the reduced state space (A_1, A_2). Determine the effective dynamics of A_1 obtained by adiabatic elimination of A_2 and the nature of the bifurcation as r increases from negative to positive values.

4.4.3 Stability and attractors of a two-dimensional system

Consider the system:

$$\tfrac{d}{dt} X = -aX + Y + XY, \qquad (4.42)$$

$$\tfrac{d}{dt} Y = -bY - X^2, \qquad (4.43)$$

where a and b are constants. Apart from the name of variables, this system is close to (4.40-4.41) since just a non-diagonal linear term has been added and coefficient a is no longer assumed to be small.

By computing the eigenvalues and eigenvectors of the system when linearised around the origin $X = Y = 0$, show that it remains locally stable (stable with respect to infinitesimal perturbations) as long as $a > 0$ and $b > 0$. Sketch the phase portrait of the linearised system for $a < b$.

Consider now the system with its nonlinear terms and determine the fixed points as functions of $\Delta = 1 - 4ab$. This quantity will serve as a control parameter in the following. Show that a saddle–node bifurcation takes place when Δ changes its sign. Find the eigenvalues and eigenvectors of the vector field linearised around each non-trivial fixed point.

Sketch the phase portrait of the system for $\Delta < 0$, $\Delta = 0$, and $\Delta > 0$ (take e.g. $b = 1$ and values of a that lead to simple numerical applications, in particular $a = 1/4, 3/16$). Observe that the stable manifold of the unstable fixed point that extrapolates its stable eigendirection is the boundary between the attraction basins of the two stable fixed points when they exist. Notice that nonlinearity profoundly modifies the aspect of the phase portrait expected from the linear analysis close to the origin. Numerical simulation of (4.42, 4.43) is recommended.[12]

---

[12] For a discussion of the physical relevance of this exercise, to be reconsidered in chapter 7, see: O. Dauchot et P.M., "Local versus global concepts in hydrodynamic stability theory," J. Phys. II France **7** (1997) 371–389.

4.4.4 Landau model of bifurcation for one real mode

Let A be a real variable governed by a first-order differential equation:

$$\tfrac{\mathrm{d}}{\mathrm{d}t} A = \mathcal{F}(A). \tag{4.44}$$

In Chapter 2, §2.1.3 it was shown that this equation can be written in gradient form $\tfrac{\mathrm{d}}{\mathrm{d}t} A = -\partial \mathcal{G}/\partial A$, so that $\mathcal{G}$ decreases as A evolves. Consider now the neighbourhood of a time-independent base state A_0 such that $\mathcal{G}$ is stationary, $F(A_0) = -\partial_A \mathcal{G}(A_0) \equiv 0$.

1) Rewrite the equation governing the dynamics of a perturbation A' defined by $A = A_0 + A'$ as a Taylor expansion

$$\tfrac{\mathrm{d}}{\mathrm{d}t} A' = \sum_n a_n (A')^n, \tag{4.45}$$

express the linear stability condition of state A_0 in terms of the derivatives of $\mathcal{F}$ at A_0 and interpret this condition in terms of $\mathcal{G}$

We now assume that $A_0 = 0$ and that the linear growth-rate σ of A is simply given by $\sigma = r$, which define $r = 0$ as the bifurcation threshold. Equation (4.45) then reads

$$\tfrac{\mathrm{d}}{\mathrm{d}t} A = rA + \sum_{n \geq 2} a_n A^n. \tag{4.46}$$

Assuming that the system can be truncated beyond some order N, using either $\mathcal{F}$ or $\mathcal{G}$, show that the dynamics described by (4.46) is meaningful in the sense that A remains finite for all times, provided that N is an odd integer and a_N is negative. (When N is even, the model is only locally valid and should be completed by terms of higher degree to acquire a more global validity). Find the conditions fulfilled by coefficients a_n in (4.46) when the system is invariant through the symmetry $A \mapsto -A$.

2) When $N = 2$, consider the systems:

$$\tfrac{\mathrm{d}}{\mathrm{d}t} A = r_0 - A^2 \tag{4.47}$$

and

$$\tfrac{\mathrm{d}}{\mathrm{d}t} A = r_1 A - A^2. \tag{4.48}$$

Draw the graphs giving their fixed points A_* (bifurcation diagrams) against the control parameters r_0 or r_1, and indicate their stability properties, using solid (dashed) lines for stable (unstable) fixed points. Justify the terms *saddle–node* and *trans-critical* for these two bifurcations.

3) Same exercise when $N = 3$ with systems:

$$\tfrac{\mathrm{d}}{\mathrm{d}t} A = rA - A^3 - H, \tag{4.49}$$

producing the imperfect bifurcation (Figure 4.4, p. 136) and
$$\tfrac{d}{dt}A = rA - A^2 - A^3. \tag{4.50}$$
accounting for a perfect bifurcation perturbed by a breaking of the symmetry $A \mapsto -A$. Observe that, in both cases, elements of the case $N = 2$ are recovered locally in the plane (r, A_*).

4) Consider the case $N = 5$ with symmetry $A \mapsto -A$:
$$\tfrac{d}{dt}A = rA + aA^3 - A^5, \tag{4.51}$$
coefficient a being positive or negative.

a) Determine the position of the fixed points at given (r, a).

b) When a is positive, show that the bifurcation at $r = 0$ is subcritical and that the system experiences hysteresis cycles as r is varied. Draw the bifurcation diagram and observe the presence of situations already studied for smaller maximum degree N. Determine the shape of potential $\mathcal{G}$ corresponding to the different cases.

c) Consider briefly (4.51) with $a < 0$ and identify the different nonlinear regimes according to which of the nonlinear terms controls the position of the fixed point.

Remark. When replaced in the context of catastrophe theory, p. 137, (4.47) is said to *unfold* the quadratic nonlinearity. It is more general than (4.48) which additionally supposes the persistence of the fixed point at $A = 0$ [incidentally, this is also the case of (4.46)]. On more general grounds, one can always find the translation $A \mapsto A + \alpha$ that suppresses the coefficient a_{N-1}, and re-scale A so that $a_N = \pm 1$. The generic unfolding of $\tfrac{d}{dt}A = \pm A^N$ then reads:
$$\tfrac{d}{dt}A = r_0 + r_1 A + \cdots + r_{N-2} A^{N-2} \pm A^N,$$
where the *unfolding parameters* r_n are perturbations, i.e. must remain small when compared to the absolute value of the coefficient of the highest degree term (± 1). This system may have no fixed points, as seen above for $N = 2$ with (4.47) when $r_0 < 0$.

4.4.5 Variation on the theme of Exercise 4.4.4

Consider the system
$$\tfrac{d}{dt}A = \mathcal{F}_r(A) = -A/r + 2A^2 - A^3, \quad r > 0.$$

1) Determine explicitly the potential $\mathcal{G}_r(A)$ from which it derives, up to an arbitrary constant that will be taken as zero.

2) Find the fixed points as functions of the control parameter r. Show that a pair of fixed points appear through a saddle–node bifurcation at some value of r to be determined. Study the stability of the different fixed points and compute the value of the corresponding potentials.

3) Find the asymptotic behaviour of the non-trivial fixed points as $r \to +\infty$ and draw the bifurcation diagram.

4) In the Landau theory of phase transitions $\mathcal{G}$ would be a good model of free energy for a first-order transition, e.g. the liquid-gas transition. Compute the value r_M of parameter r corresponding to the Maxwell plateau at which the free energies of the two competing stable states are equal. Draw the graph of $\mathcal{G}_r$ for $r < r_M$, $r = r_M$, and $r > r_M$.

4.4.6 Excitable system

Consider a dynamical system in the form

$$\tfrac{d}{dt} X = F_Y(X) = X - X^3 - Y, \tag{4.52}$$

where Y is, for the moment, a parameter.

Determine the fixed points of (4.52), their number and nature (stable or unstable) as Y is varied. Determine the potential from which this equation derives (see Exercise 4.4.4, above).

Suppose now that Y is a dynamical variable of its own, governed by

$$\tfrac{d}{dt} Y = \epsilon(X - r), \tag{4.53}$$

where $0 < \epsilon$ is a constant and r the control parameter.

Determine the character of the fixed point of system (4.52, 4.53) as a function of r, assuming that $1 - 3r^2 \neq 0$ and $\epsilon \ll |1 - 3r^2|$.

Interpret the phase portrait of the system in the two cases $|r| < 1/\sqrt{3}$ and $|r| > 1/\sqrt{3}$. In the first case, explain the shape of the relaxation oscillations observed after damping of the transient. [Answer: Figure 4.19.]

In the second case, the system is said to describe an excitable medium such that, for initial conditions $X > r$, the relaxation of the system towards its fixed point occurs after a large pulse. A reaction–diffusion system (see p. 27) with a reaction part in the form (4.52, 4.53) and r large enough will develop fronts separating excited regions from relaxed one.

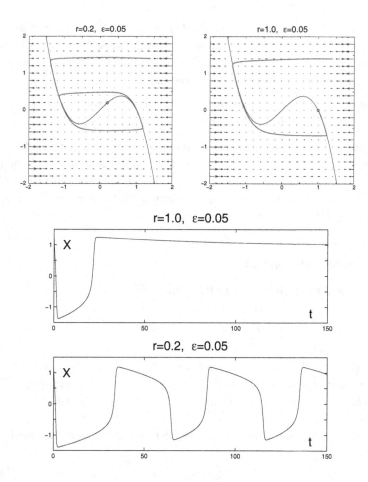

Fig. 4.19 Excitable system. Top: phase portraits with variable X (Y) along the horizontal (vertical) axis. Bottom: time series of variable X

4.4.7 Brusselator

The system obtained in Exercise 1.5.2, p. 36, reads

$$\frac{d}{dt}X = A - (B+1)X + X^2Y, \qquad (4.54)$$
$$\frac{d}{dt}Y = BX - X^2Y. \qquad (4.55)$$

Find its fixed points as functions of parameters A and B and discuss their linear stability properties. Show that the reaction can bifurcate from a time-independent state towards an oscillatory regime beyond some critical value B_c to be determined as a function of A.

Compute the eigenvalues at lowest order in $\varepsilon = B - B_c$ Adapt the first harmonic approximation of Chapter 2, §2.3.1.2, p. 62 to derive the amplitude equation governing the system beyond threshold.

4.4.8 Locking

In a permanent regime, an oscillator can be described using a phase variable θ defined modulo 2π and governed by $\frac{d}{dt}\theta = \omega$ where ω is the angular frequency of the oscillator. Consider two such oscillators weakly but nonlinearly coupled governed by:

$$\frac{d}{dt}\theta_1 = \omega_1 - f(\theta_2 - \theta_1),$$
$$\frac{d}{dt}\theta_2 = \omega_2 - f(\theta_1 - \theta_2).$$

Justify the fact that function f has to be periodic with period 2π. In the following, take $f(\varphi) = \frac{1}{2}K\sin(\varphi)$ and write down the system for $\bar{\theta} = \frac{1}{2}(\theta_1 + \theta_2)$ and $\varphi = \frac{1}{2}(\theta_1 - \theta_2)$ (define $\delta\omega = \frac{1}{2}(\omega_1 - \omega_2)$).

Determine the condition on K ensuring the existence of fixed points to the equation for φ. Discuss the nature of the asymptotic regime when this condition is (is not) fulfilled. Qualitatively describe the phase intermittency regime that takes place when the existence condition just misses being fulfilled.

4.4.9 Logistic map (variants)

Consider the map
$$X_{k+1} = f_a(X_k) = a - X_k^2 \tag{4.56}$$
where $a > 0$ is the control parameter. Draw the graph of f_a, find its fixed points and study their stability properties.

In the following, consider the dynamics of a trajectory starting in the neighbourhood of $X_*^{(+)} > 0$ for a close to a_0 where the fixed point bifurcates. Setting $a = a_0 + \epsilon$, rewrite the nonlinear map for $X' = X - X_*^{(+)}$ in terms of $\epsilon \ll 1$ at first order in ϵ.

Check that the map for X' has no fixed point in the neighbourhood of $X' = 0$ and that the system bifurcates towards a period-2 cycle by studying the iterated map $X'_{n+2} = g_\epsilon(X'_n)$ as ϵ crosses zero. By expanding to third order in X'_n, conveniently simplified thanks to the assumption $\epsilon \ll 1$, show that the bifurcated cycle is stable for ϵ small enough.

Derive the change of variables that put (4.56) in the equivalent forms $Y_{k+1} = rY_k(1 - Y_k)$ and $Z_{k+1} = 1 - bZ_k^2$.

4.4.10 Delayed logistic map

Consider the map

$$X_{k+1} = rX_k (1 - X_{k-1}) . \qquad (4.57)$$

Two initial conditions are necessary to start this recurrence which is thus a second-order discrete-time dynamical system. Its non-delayed version is one of the equivalent variants of the logistic map used in the previous exercise.
1) Convert (4.57) into a first-order system by setting $Y_k = X_{k-1}$, compute the Jacobian of the map and determine its inverse when it is possible.
2) Find the fixed points of the map $X_{k+1} = X_k = Y_{k+1} = Y_k$ and study the stability of the non-trivial fixed point $X_* \neq 0$ as a function of $r \in [1,4]$. In order to do this, find the eigenvalues of the linearised system and their nature – real or complex – as a function of r. Show that the system displays a Hopf bifurcation in the sense of (4.33) for some value of r to be determined.

4.4.11 Arnold's cat

Consider the discrete-time two-dimensional dynamical system $(X,Y) \mapsto (U,V) = \boldsymbol{S}(X,Y)$ defined on the torus $\mathbb{T}^2$, the unit square (modulo 1), with opposite sides identified, called "Arnold's cat map":

$$U = 2X + Y$$
$$V = X + Y$$

This system is a classical example of chaotic iteration. Determine the inverse map $(U,V) \mapsto (X,Y) = \boldsymbol{T}(U,V)$ and check that $\boldsymbol{S} \circ \boldsymbol{T} = \boldsymbol{I}$ where $\boldsymbol{I}$ is the identity map. Compute the determinant of its Jacobian matrix and shows that the system is conservative (Chapter 2, Exercise 2.5.1, p. 73). Illustrate this property by considering the transform of the square $(0,0), (0,1/3), (1/3,0), (1/3,1/3)$. Show that all trajectories are unstable. Compute the second iterate $\boldsymbol{S}^2 \equiv \boldsymbol{S} \circ \boldsymbol{S}$.

Chapter 5

Characterising and Using Chaos

5.1 Quantifying Chaos

As conceived by Ruelle, *chaos* is a dynamical regime characterised by a specific *sensitivity* of trajectories to *initial conditions* and small perturbations. In state space, it is accounted for by *strange attractors* that are robust attracting limit sets on which small departures between two trajectories are indefinitely amplified (the general meaning of instability) as a result of a stretch-and-fold process analogous to the the kneading of dough (baker map, see Figure 5.3, p. 175).

The main interest of these concepts is to reconcile *determinism* and *stochasticity* in dissipative dynamical systems, which was initially supposed to shed some light on the problem of the 'nature of turbulence' as introduced in Chapter 1, p. 16.

In this section we thus focus on a quantitative estimate of the two main facets of chaotic dynamics: the *longitudinal* instability of trajectories measured by *Lyapunov exponents* and the *transverse* foliated structure of state space characterised by *fractal dimensions*.

5.1.1 *Instability of trajectories and Lyapunov exponents*

The instability of trajectories on a strange attractor is illustrated here by means of an expanding iteration called the *dyadic map* (Fig. 5.1, top-left). It reads:

$$X_{k+1} = 2X_k \quad (\text{mod } 1). \tag{5.1}$$

It turns out to be convenient to visualise the trajectory $\{X_k, k = 0, 1, \ldots\}$ as a walk in the complex plane $Z_{k+1} = Z_k + \exp(2\pi i X_k)$. Here Z has no dynamical significance in contrast to X which is governed by the map. This

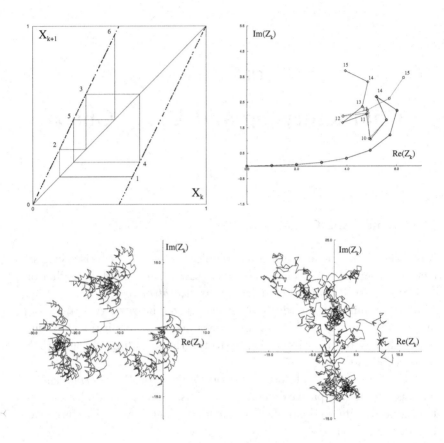

Fig. 5.1 Interpretation of chaos. Top left: Dyadic map. Top right: Divergence of neighbouring trajectories. Bottom left: Dyadic walk. Bottom right: Random walk.

simple representation helps us to clearly illustrate the divergence of trajectories starting at neighbouring points (Fig. 5.1, top-right) and the long-term evolution (Fig. 5.1, bottom-left) that leads one to think of the random walk that would be obtained by drawing the successive X_k uniformly at random over the unit interval (Fig. 5.1, bottom-right).

The divergence rate of trajectories is a good measure of chaotic behaviour. Let us show how the analysis proceeds in a simple example and consider a map f of a single real variable X:

$$X_{k+1} = f(X_k),$$

more general than (5.1). To begin with, a given trajectory $\{X_k; k = 0, 1, 2 \ldots\}$ starting at X_0 is taken as a reference and we consider a neighbouring trajectory $\{\tilde{X}_k; k = 0, 1, 2 \ldots\}$ starting at $\tilde{X}_0 = X_0 + \delta X_0$. Denoting $f' = \mathrm{d}f/\mathrm{d}X$ we have:

$$\tilde{X}_1 = X_1 + \delta X_1 = f(X_0 + \delta X_0) = f(X_0) + f'(X_0)\,\delta X_0$$
$$\Rightarrow \quad \delta X_1 = f'(X_0)\,\delta X_0$$

and using the chain rule:

$$\delta X_k = \left(\prod_{m=0}^{k-1} f'(X_m)\right) \delta X_0\,.$$

Assuming a geometrical growth/decay as $|\delta X_k| \sim \gamma^k |\delta X_0| \equiv \exp(k\lambda)|\delta X_0|$, which defines both γ and $\lambda = \log\gamma$, we get:

$$\gamma = \left(\left|\frac{\delta X_k}{\delta X_0}\right|\right)^{1/k} = \left(\prod_{m=0}^{k-1} |f'(X_m)|\right)^{1/k}.$$

At this stage, γ still depends on k. To get rid of this dependence we take the limit $k \to \infty$ which leads to the definition of the *Lyapunov exponent*:

$$\lambda = \log(\gamma) = \lim_{k\to\infty} \frac{1}{k} \sum_{m=0}^{k-1} \log\left(|f'(X_m)|\right). \tag{5.2}$$

In other words, the Lyapunov exponent is thus the time average of the local divergence rate $\log(|f'|)$. It presents itself as a measure of *long term unpredictability* since trajectories diverge in the mean when $\lambda > 0$ (the dyadic map (5.1) gives $\lambda = \log 2$). This property can be taken as defining chaotic dynamics.

The extension to maps with several variables leads to the definition of the *Lyapunov spectrum*, which rests on the analysis of the asymptotic behaviour of the product of Jacobian matrices obtained from the chain rule. This generalises the eigen-spectrum of the stability matrix at a fixed point to the case of arbitrary trajectories. Technically, the matter is difficult but, in the limited context of this book, it is sufficient to know that this can be done and that, by successive generalisations, one can determine Lyapunov spectra for differential systems (from time-τ maps, see p. 42) and for continuous media governed by partial differential equations (after spectral approximation or discretization leading to finite-order ordinary differential systems, see Appendix B).

The Lyapunov spectrum can be ordered by decreasing values of the individual exponents. The system is then declared to be chaotic when the largest exponent is positive. The empirical determination of Lyapunov exponents from time records will be examined later, see p. 189.

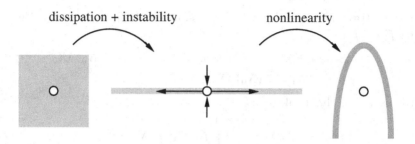

Fig. 5.2 Expansion in unstable direction, contraction due to dissipation, folding by nonlinearity, all in combination, produce *horseshoes* typical of chaotic attractors.

Let us stress the fact that the whole approach in terms of dynamical systems is a progressive extension of linear instability concepts from time-independent regimes (fixed points of continuous-time systems) to periodic regimes (limit cycles then seen as fixed points of discrete-time systems) and finally to irregular aperiodic regimes.

5.1.2 Fractal features

Up to now we have been interested in expansion properties *along* the attractor. However, the systems of interest are supposed to be dissipative, which implies overall contraction of volumes in state space. As we have seen, chaotic behaviour is marked by instability, which means expansion in some directions. This expansion has to be more than compensated for by stronger contraction in other directions. Some folding must then take place in order to maintain trajectories in a bounded region of the state space, as sketched in Figure 5.2. (The case of regular regimes would be much less anisotropic with, at most, neutral directions on average.)

What happens in the *transverse* directions can be concretely illustrated using a celebrated simple two-dimensional map called the *baker map* (Fig. 5.3, top-left). This mapping of the unit square $[0,1] \times [0,1]$ expands it onto itself by a factor $\kappa = 2$ in the direction of the first coordinate and by a factor $\kappa' = \alpha/\kappa$ in the other direction ($\kappa' < 1 \Rightarrow$ contraction).

When $\alpha = 1$, that is $\kappa' = 1/2$, the baker's map is conservative (areas are preserved). In order to obtain a dissipative system, contraction must be larger than expansion, i.e. $\alpha < 1$, so that $\kappa\kappa' = \alpha < 1$. The strange attractor obtained in that case displays a characteristic *fractal* transverse structure (Fig. 5.3, top-right). This fractal structure is a triadic *Cantor*

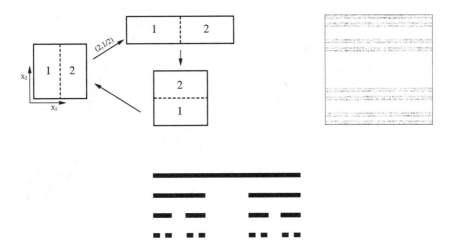

Fig. 5.3 Top-left: Conservative *baker map* (1st step: expand along X_1 ($\kappa = 2$) and contract along X_2 ($\kappa' = 1/2$); 2nd step: cut the right half of rectangle and place it back in square). Top-right: Strange attractor for the dissipative map with $\kappa' = 1/3$ (expansion rate $\kappa\kappa' = 2/3 < 1$). Bottom: Few steps of the construction rule of the triadic Cantor set, to be read from top to bottom. (Notice that, to build the dissipative baker attractor, we used a map that translates part (2) above part (1) and puts it upside down so that the fractal structure follows from the conventional Cantor middle-third rule.)

set classically obtained by removing the open middle third of a segment (i.e. $]1/3, 2/3[$ out of $[0, 1]$, yielding $[0, 1/3] \cup [2/3, 1]$) and repeating the operation indefinitely on the two segments left apart in that operation as shown in Figure 5.3 (bottom), yielding a self-similar set that is invariant upon magnification by a factor of three.

In practice, nonlinearity usually folds the trajectories as seen with the Hénon map (Note 7, p. 17):

$$X_{k+1} = 1 - aX_k^2 + bY_k, \qquad (5.3)$$
$$Y_{k+1} = X_k. \qquad (5.4)$$

The attractor corresponding to $a = 1.4$ and $b = 0.3$ is presented in Figure 5.4. The folding originates from the nonlinearity in (5.3), which is of 'logistic' type (see Exercise 4.4.9). Coefficient b in (5.4) guarantees the dissipative character of the map provided that $|b| < 1$, as shown by performing Exercise 2.5.1 with this map. The fractal character of the attractor is particularly obvious from the magnification of the boxed region displayed in the right part of the figure.

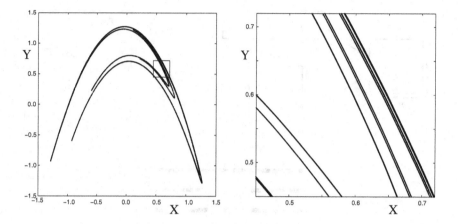

Fig. 5.4 Hénon attractor obtained by iterating map (5.3, 5.4) for $a = 1.4$ and $b = 0.3$ from some initial condition after elimination of the transient: Full attractor already illustrated in Fig. 1.8 (left) and zoom on the region in the box.

The *fractal dimension* gives a good idea of the way the attractor occupies the state space. In a d-dimensional space, it is obtained by covering the considered set, here the attractor, by elementary volume elements of size ε^d, next counting the number $\mathcal{N}(\varepsilon)$ of such elements necessary to cover it, and finally study how $\mathcal{N}(\varepsilon)$ grows as the linear size ε of the volume elements tends to zero. The fractal dimension d_f is thus defined as:

$$d_\mathrm{f} = \lim_{\varepsilon \to 0} \frac{\log\left(\mathcal{N}(\varepsilon)\right)}{\log(1/\varepsilon)}. \tag{5.5}$$

For an ordinary, connected, continuous set, this definition yields the usual *topological dimension* (0 for a point, 1 for a line, 2 for an ordinary surface, etc.). Let us see how it works for an indefinitely fragmented object like the triadic Cantor set by considering it at step k of the construction process. It can be covered by segments of length $\varepsilon = 1/3^k$, and 2^k such segments are needed at that step, hence:

$$d_\mathrm{f}(k) = \frac{\log(2^k)}{\log\left(1/(1/3)^k\right)} = \frac{\log(2^k)}{\log(3^k)} = \frac{\log(2)}{\log(3)} \approx 0.63092975,$$

somewhere between the dimension of a countable set of isolated points (0) and that of a continuous segment (1).

Here the evaluation does not depend on k because the set is strictly self-similar but more generally the limit $\varepsilon \to 0$ must really be taken. The computation of the fractal dimensions of other classical self-similar sets from a direct application of (5.5) is the subject of exercise 5.6.1.

Turning to the dissipative baker map, let us focus on what happens in the transverse direction X_2. We take for granted that, owing to the expanding character of the map along X_1, the attractor is continuous in that direction, which just adds 1 to the dimension found for the transverse part. Following the same idea as for the triadic Cantor set, we can observe that, after one iteration, the length of a segment along X_2 is multiplied by $\kappa' < 1$, which suggests us to take $\varepsilon = (\kappa')^k$ after k iterations. Each application of the map brings about κ reduced copies of the full set, so that we have $\mathcal{N}(\epsilon) = \kappa^n$. From the formula, we get $d_f = \log(\kappa)/\log(1/\kappa') = -\log \kappa / \log \kappa'$ along the contracting direction and thus $1 - \log \kappa / \log \kappa'$ for the full attractor displayed in Figure 5.3 (middle part). With $\kappa = 2$ and $\kappa' = \alpha/\kappa$ with $\alpha = 2/3$, this yields $d_f \approx 1.63092975$.

As an exercise one can look at the aspect of dissipative generalised baker maps with definitions more complex than that illustrated in Figure 5.3, e.g. with $\kappa > 2$ and different values of α for different pieces, and then try to determine the dimension of the corresponding (transverse) Cantor sets, which might be less simple than it seems.

5.2 Empirical Approach to Chaotic Systems

Let us consider a chaotic regime observed in a given experiment, e.g. in convection. The need of an empirical approach becomes obvious when it appears impossible from a practical point of view to get an *ab intio* understanding of its nature. Most often, even something as qualitative as the type of transition scenario that develops under specific conditions cannot be predicted. For example, in Chapter 3, §3.2.3, we have seen that not only the physical properties of the fluid matter, but the experimental configuration and, in case of attractor coexistence, the history of the experiment also plays a role in the transition (though the system, when engaged in a given scenario, follows all its steps at a quantitative level).

Sufficiently far from the threshold of the primary instability, a multiplicity of different possible permanent regimes can be reached by following specific experimental procedures. The effective state space is thus already very complicated and poorly understood, even when confinement effects select a small number of driving modes. It is then advisable to search for a representation of the dynamics in some *reconstruction space* obtained from the experimental records, a space in which the evolution can be described, the amount of chaos can be measured and, hopefully, techniques of control can be developed.

The output of experiments, either in the lab or using a computer, generally presents itself in the form of *time series* of some observable, i.e. a series of numerical values[1] taken by some function $W = \mathcal{W}(\mathbf{X})$ of the system's state $\mathbf{X} = (X_1, \ldots, X_d)$ in its state space $\mathbb{X}$ with dimension d. We thus assume a regularly sampled time series in the form:

$$\{W_k, k = 0, 1, \ldots\}, \quad \text{with} \quad W_k = \mathcal{W}(\mathbf{X}(t_k)), \quad t_k = k\tau,$$

where τ is the inverse of the sampling frequency.

Just having a look at the plotted time series is a step that should never be skipped since this gives valuable information on the recurrent or intermittent character of the signal. But clearly more objective analyses are required, especially when the system has reached some sort of attractor, i.e. a permanent regime characterised by statistically stationary data, and for which finite-length series are typical of the dynamics, provided that they are not too short (i.e. correspond to several "turns" around the attractor).

Given a time series of length K, $\{W_0, W_1, \ldots, W_{K-1}\}$, one should first compute its average:

$$\overline{W} = \frac{1}{K} \sum_{k=0}^{K-1} W_k,$$

and its variance:[2]

$$\sigma_W^2 = \frac{1}{K} \sum_{k=0}^{K-1} (W - \overline{W})^2,$$

and then to re-scale it by making the changes:

$$\frac{(W_k - \overline{W})}{\sigma_W} \mapsto W_k, \quad k = 0, 1, \ldots, K-1.$$

For simplicity, we assume in the following that this pre-processing has been performed, i.e. that we work with a signal supposed to be stationary, with zero mean and unit variance.

---

[1] For systems evolving in space, the output may also be pictures. This opens the vast field of image processing which basically extends to two-dimensional quasi-continuous arrays the problem of treating one-dimensional discrete scalar series considered here.

[2] The *unbiased* variance should be defined by dividing by $K - 1$ instead of K but we are always interested in K large so that this makes no difference.

5. Characterising and Using Chaos

In a second instance, one usually considers the auto-correlation of the signal defined as:

$$C(\Delta t) = \lim_{T \to \infty} \frac{1}{T} \int_0^T W(t) W(t + \Delta t) \, dt. \tag{5.6}$$

In the case of a finite series of discrete records one has to replace this mathematical relation by a practical definition:

$$C(\kappa) = \frac{1}{K - \kappa} \sum_{k=0}^{K-\kappa-1} W_k W_{k+\kappa}, \tag{5.7}$$

upon assuming $\Delta t = \kappa \tau$, which suggests one to keep $\kappa \ll K$ in order to keep the theoretical contents of the correlation function to a reasonable approximation.

The auto-correlation is the simplest device to identify periodicities in the signal. It is usually not computed from its definition but rather using Fourier transforms. The next section thus begins with a brief reminder about them and continues with an introduction to the Hilbert transform, a useful tool to perform the demodulation of periodic signals with superimposed slow amplitude and phase variations.

5.2.1 Standard analysis using Fourier transforms

On general grounds the direct Fourier transform is defined as:

$$\tilde{f}(\omega) = \frac{1}{2\pi} \int_{-\infty}^{+\infty} f(t) \exp(-i\omega t) \, dt \tag{5.8}$$

and its inverse as:

$$f(t) = \int_{-\infty}^{+\infty} \tilde{f}(\omega) \exp(i\omega t) \, d\omega. \tag{5.9}$$

The Fourier transform of the auto-correlation function then reads:

$$S(\omega) = \frac{1}{2\pi} \int_{-\infty}^{\infty} C(\tau) \exp(-i\omega \tau) \, d\tau$$

and a straightforward computation shows that

$$S(\omega) = |\tilde{W}(\omega)|^2, \tag{5.10}$$

where $\tilde{W}(\omega)$ is the Fourier transform of $W(t)$. $S(\omega)$ is called the *Fourier spectrum* of the signal.[3] In practice, (5.8, 5.9, 5.10) are usually computed

---
[3] This result is called the *Wiener–Kintchine theorem*. The derivation uses a few tricks, among which the fact that the Fourier transform of a constant $f \equiv 1$ is a Dirac distribution $\delta_D(\omega)$.

using Fast Fourier Transforms and care has to be taken about how they are implemented and how the number K of data points is dealt with in the normalisation of the output.

Direct inspection of the signal sometimes suggests that a periodic process modulated in amplitude and/or phase is at work. An elegant way to perform the *demodulation* consists in constructing a complex signal $Z(t)$, whose real part is the primitive signal $W(t)$ and whose imaginary part is constructed so as to make a $\pi/2$ phase angle with it. This is easily understood from the consideration of a strictly periodic signal: let $W(t) = A\cos(\omega t)$, the signal shifted by $\pi/2$ is then $A\cos(\omega t - \pi/2) = A\sin(\omega t)$, so that $Z(t) = A\exp(i\varphi(t))$, with amplitude $A = |Z|$ and phase φ; $d\varphi/dt = \omega$ is then the angular frequency.

In the general case, the instantaneous amplitude of the modulated signal is given by the modulus of Z and its instantaneous period is derived from the argument of Z by differentiation with respect to time, provided that how to obtain the signal at $\pi/2$ is known. This can be done directly from the Fourier transform of $W(t)$, by copying the case of the periodic signal taken as an example above, frequency by frequency:

Starting from

$$W(t) = \int_{-\infty}^{+\infty} \tilde{W}(\omega)\exp(i\omega t)\mathrm{d}\omega, \tag{5.11}$$

and setting:[4]

$$\tilde{W}(\omega) = \tfrac{1}{2}[A(\omega) + iB(\omega)] = \tilde{W}(-\omega)^*, \tag{5.12}$$

one can observe that $A(\omega)$ and $B(\omega)$ are even and odd functions of ω, respectively. In full generality we have $B(0) = 0$, and also $A(0) = 0$, since $W(t)$ is assumed to have zero mean. Expression (5.11) can then be rewritten as:

$$W(t) = \int_0^{+\infty} [A(\omega)\cos(\omega t) - B(\omega)\sin(\omega t)]\,\mathrm{d}\omega.$$

Now, let us construct the signal shifted by $\pi/2$ as above, Fourier component after Fourier component. Denoting this signal W', we get:

$$W'(t) = \int_0^{+\infty} [A(\omega)\cos(\omega t - \pi/2) - B(\omega)\sin(\omega t - \pi/2)]\,\mathrm{d}\omega,$$

which more simply reads:

$$W'(t) = \int_0^{+\infty} [B(\omega)\cos(\omega t) + A(\omega)\sin(\omega t)]\,\mathrm{d}\omega.$$

---

[4]The complex conjugation property arises from the fact that the observable $\mathcal{W}$ is supposed to be a real function of **X**.

By definition of $Z = W + iW'$, we obtain:

$$Z(t) = \int_0^{+\infty} \exp(i\omega t)[A(\omega) + iB(\omega)]\,\mathrm{d}\omega, \qquad (5.13)$$

which involves a sum restricted to the positive angular frequencies (the so-called 'analytical signal'). Comparing (5.13) and (5.11, 5.12), we get the Fourier transform $\tilde{Z}(\omega)$ by setting to zero all components of the Fourier transform of W corresponding to negative angular frequencies and by doubling all the others. The analytic signal $Z(t)$ itself is then recovered by computing the inverse transform of $\tilde{Z}$, the instantaneous amplitude is given by $|Z|$ and the instantaneous frequency as $\frac{\mathrm{d}}{\mathrm{d}t}\phi$, where ϕ is the argument (phase) of Z. Mathematically W and W' are Hilbert transform of each other[5] hence the expression 'Hilbert transform demodulation' (the full procedure is implemented by hilbert.m in the MATLAB software).

5.2.2 Reconstruction using delays

Let us now take a point of view more in line with the theory of dynamical systems. A difficulty arises immediately from the fact that no specific assumption can be made, except that a deterministic framework is relevant, so that one can just write formally $\frac{\mathrm{d}}{\mathrm{d}t}\mathbf{X} = \mathcal{F}(\mathbf{X})$ for unspecified states living in some space $\mathbb{X}$. In particular, the dimension d_{eff} and the nature of the manifold supporting the dynamics,[6] and the explicit relation between states and the observable $W = \mathcal{W}(\mathbf{X})$ are not known beforehand.

Practically all reconstruction techniques derive from the *method of delays* mathematically formalised by Takens in 1981.[7] This method is numerically more robust than a previous approach based on the evaluation of successive time derivatives of the experimental signal in terms of finite differences,[8] which has the drawback of amplifying the noise.

Let us illustrate the reconstruction approach using a discrete-time system for simplicity. At the beginning we assume that

$$\mathbf{X}_{k+1} = \mathcal{F}(\mathbf{X}_k)$$

---

[5] Formally, $\tilde{W}'(\omega) = -(1/\pi)\,\mathcal{P}\int_{-\infty}^{\infty}[\tilde{W}(\omega')/(\omega - \omega')]\mathrm{d}\omega'$, where $\mathcal{P}$ denotes the Cauchy principal part of the integral.

[6] The dimension of the physical system is usually infinite; think of a continuous medium.

[7] F. Takens: "Detecting strange attractors in turbulence," Lect. Notes Math. **898** (1981) 366–381.

[8] N.H. Packard *et al.*: "Geometry from a time series," Phys. Rev. Lett. **45** (1980) 712–716.

and that the time series of some scalar observable $\mathcal{W}$ is available (the method is, at least conceptually, easy to extend to the case of several observables). Reconstructing the dynamics means determining an empirical relation between the $\mathbf{X}_k$ in their state space only from the knowledge of the W_k, $k = 0, 1, \ldots$

A single measure $W_0 = \mathcal{W}(\mathbf{X}_0)$ is not sufficient to determine the state $\mathbf{X}_0$ since we surely need more than one coordinate to define it. But we assume that the next value W_1 corresponds to a point $\mathbf{X}_1$ that evolves from $\mathbf{X}_0$ under some mapping $\mathcal{F}$, unknown but supposed to exist. The second measurement thus adds a piece of information about the coordinates of $\mathbf{X}_0$ through $W_1 = \mathcal{W}(\mathbf{X}_1) = \mathcal{W}(\mathcal{F}(\mathbf{X}_0))$. The third one, $W_2 = \mathcal{W}(\mathbf{X}_2) = \mathcal{W}(\mathcal{F}(X_1)) = \mathcal{W}(\mathcal{F}(\mathcal{F}(\mathbf{X}_0)))$ adds another piece, etc. In principle, a sufficiently long series of d_{test} successive measurements, $\{W_0, \ldots, W_{d_{\text{test}}-1}\}$, should serve us to specify $\mathbf{X}_0$. In the same way, $\{W_1, \ldots, W_{d_{\text{test}}}\}$ would do so for $\mathbf{X}_1$, etc. Eventually a whole trajectory would then be reconstructed from the series of vectors in $\mathbb{R}^{d_{\text{test}}}$:

$$\mathbf{V}_k = [W_k; \ldots; W_{k+d_{\text{test}}-1}] \, .$$

The concrete implementation of the method is by increasing the dimension of the reconstruction space until a consistent quantitative assessment of the observations can be given. In practice, the problem can be reformulated in terms of the *reliability* of the reconstruction. The space of vectors $\mathbf{V}_k$ has to be in correspondence with the region of state space visited by the system when the permanent regime is reached. Whereas we can accept redundancy, i.e. a dimension d_{test} that is too large, we must not lose useful dynamical information. In mathematical terms the representation of the system must be *injective*, so that different states are reconstructed differently:

$$\mathbf{X}_k \neq \mathbf{X}_{k'} \quad \Rightarrow \quad \mathbf{V}_k \neq \mathbf{V}_{k'} \, . \tag{5.14}$$

What was defined as a tentative number of component d_{test} is thus more mathematically understood as the dimension of the space in which the effective state space can be *embedded* by means of some injective map. To stick with this abstract point of view more closely, we now define d_{test} as the *embedding dimension*, denote it d_{e}, and thus specify states in the embedding space as:

$$\mathbf{V}_k = [W_k; \ldots; W_{k+d_{\text{e}}-1}] \, , \tag{5.15}$$

Takens' method of delays is sketched in Figure 5.5. His theorem states that the $\mathbf{V}_k$ defined above, where the observable W is defined by a differentiable

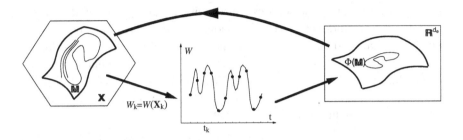

Fig. 5.5 Illustration of Takens' theorem. $\mathbb{M}$ is the manifold in the space $\mathbb{X}$ over which the effective dynamics takes place. Coordinates of points representing the system in $\mathbb{R}^{d_e}$ are obtained from the series of measurements W_k through (5.15).

functional on state space $\mathcal{W}$, achieve a reliable reconstruction provided that the dimension d_e is large enough: $d_e \geq 2d_{\text{eff}} + 1$, where d_{eff} is the topological dimension of the manifold supporting the effective dynamics.

This theorem can be understood as an extension to the present context of Witney's theorem that states that a compact d-dimensional differentiable manifold $\mathbb{M}$ can be embedded in the Euclidean space $\mathbb{R}^{2d+1}$. As an illustration of the latter, one can consider a loop (dimension 1) that would be projected on a plane as the figure eight. In order to resolve its structure, and in particular to check that the intersection is a fake, one must be able to look at the loop from another side, i.e. to stand in a $2 \times 1 + 1 = 3$ dimensional space.

In fact the reconstruction proposed by Takens is more general than the one introduced in (5.15) since the $\mathbf{V}_k$ can be any series of d_e measurements, $[W_k; W_{k+\kappa_1}; ...; W_{k+\kappa_{d_e-1}}]$, and nothing forbids it to take irregularly distributed intervals κ_q. It is however natural to take κ_q as the successive multiples of some basic κ, i.e. $\kappa_q = q\kappa$, $q = 1, 2,..., d_e - 1$. When the signal is obtained from the time sampling of a continuous-time system with period τ this corresponds to a sub-sampling at period $\kappa\tau$.

The mathematical viewpoint developed so far is apparently strong. However our enthusiasm must be somewhat tempered and a pragmatic perspective has to be taken, for it is not clear that the physical system of interest fulfils the theoretical conditions underlying the theorem, and first of all that the d_{eff}-dimensional manifold over which the dynamics takes place is sufficiently smooth. As it is usually the case for chaotic nonlinear systems, the attractor has a fractal dimension $d_f \geq d_{\text{eff}}$ and, heuristically, one can replace d_{eff} by d_f in the inequality for d_e. This still does not lead to any concrete estimate, in part because measurements are always polluted by noise that comes and hinders the reconstruction. Strategies have thus been developed to get around these difficulties and determine more or less optimally the two basic ingredients of any reconstruction: the base delay κ and the embedding dimension d_e. This is what we now briefly discuss, inviting the interested reader to consult [Abarbanel (1996); Kanz and Schreiber (1997); Weigend and Gershenfeld (1993)] for details.

5.2.3 *Sampling frequency and embedding dimension*

For more specificity we use a synthetic signal obtained by numerical simulation of a noisy limit cycle governed in the complex plane $Z = X + iY$ by a *Langevin equation*

$$\tfrac{\mathrm{d}}{\mathrm{d}t} Z = (1+i)Z - (1-i)|Z|^2 Z + \zeta(t) ,$$

where $\zeta(t)$ is a Gaussian noise. We assume that the trajectories are computed by a second order Runge–Kutta scheme (B.9–B.10) with time step $\delta t = 0.01$. A noise with amplitude 0.05 is added to X and Y at each time step (for the generation of Gaussian noise see [Press et al. (1986)]). The trajectory used for this example is displayed in Figure 5.6 (top). The period of the (deterministic) signal is $T = \pi$. The signal $W(t)$ that we take is just the X component of the trajectory and we mimic the sampling process by recording its value every $\tau = 3\delta t$.

The first problem is that of the sampling time, and thus that of κ. An exaggeratedly high frequency is not an advantage since a huge volume of data is produced, the main part of which is redundant owing to the noise that blurs the information. It is then believed that the delay between two successive records must be sufficient to bring novel information. A practical rule is to take the delay corresponding to the first zero of the auto-correlation as the effective sampling time $\kappa\tau$. For the present (deterministic) sinusoidal signal, $W(t) = \sin(2t)$, the auto-correlation defined through (5.6) is $C(\Delta t) = \cos(2\Delta t)$ and this rule prescribes $\Delta t = \pi/4$, i.e. a quarter of a period: two measurements out of phase by such a shift are indeed fully discriminated since one is maximum when the other is zero.

The discrete estimate of the auto-correlation function (5.7) is displayed in the left part of Figure 5.7 where one can observe that the noise adds a slight damping to the behaviour expected for a periodic signal. Applying the rule leads to $\kappa \simeq 25$, which indeed corresponds to a quarter of a period once it is recalled that $\tau = 3\delta t = 0.03$. Two-dimensional reconstructions

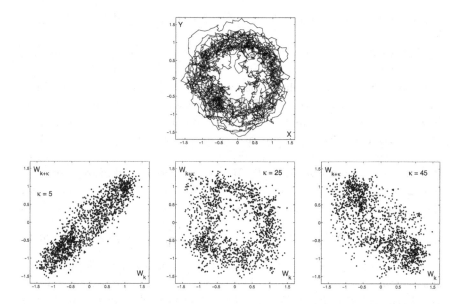

Fig. 5.6 Top: Original noisy periodic signal in its own state space. Bottom: Reconstruction with sampling time $\tau = 0.03$ and delays $\kappa = 5$, 25 and 45 of observable $W(t) \equiv X(t)$.

in the plane $(W_k, W_{k+\kappa})$ for $\kappa = 5$, 25 and 45 are displayed in Figure 5.6 (bottom) which shows that the representation with $\kappa = 25$ is the most similar to the original signal, whereas for $\kappa = 5$ and $\kappa = 45$, corresponding to measurements too close in time or practically in phase opposition, the

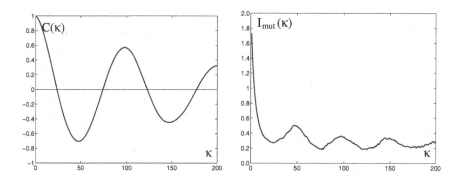

Fig. 5.7 Auto-correlation $C(\kappa)$ and mutual information $I_{\text{mut}}(\kappa)$ of signal W in Figure 5.6 as a function of the delay κ.

reconstructed signals align along directions $W_{k+\kappa} = +W_k$ and $W_{k+\kappa} = -W_k$, respectively.

The study of the auto-correlation function does not always lead to a satisfactory choice of κ. Let us now mention a more general criterion which rests on a similar philosophy. This criterion bears on the *mutual information* contained in two records shifted by some time amount when compared to that contained in a single record.[9] Again, the signal's stationary character is assumed. First the probability distribution $\mathcal{P}(W)$ of $W_k = W$ is obtained by an appropriate normalisation of the histogram of the values of W_k. Next, the joint probability $\mathcal{P}_\kappa(W', W'')$ is determined in the same way but for couples $(W_k = W'; W_{k+\kappa} = W'')$, with $\kappa = 1, 2, \ldots$. From these, the *mutual information* is defined as:

$$I_{\mathrm{mut}}(\kappa) = \sum_{W',W''} \mathcal{P}_\kappa(W', W'') \log\left(\frac{\mathcal{P}_\kappa(W', W'')}{\mathcal{P}(W')\mathcal{P}(W'')}\right). \qquad (5.16)$$

This quantity is a measure of the redundancy in the signal: when κ is small, points in state space are highly correlated so that learning about $W_{k+\kappa}$ when W_k is known does not bring much novel information; in contrast when $\kappa \gg 1$, the points become de-correlated and $\mathcal{P}_\kappa(W_k, W_{k+\kappa})$ is essentially equal to the product $\mathcal{P}(W_k)\mathcal{P}(W_{k+\kappa})$, so that the mutual information is nearly zero, the information gained is the same as from independent drawings W_k and $W_{k+\kappa}$ using $\mathcal{P}(W)$. As can be seen from the right part of Figure 5.7 that displays the graph of $I_{\mathrm{mut}}(\kappa)$ obtained by applying (5.16) to our signal, the mutual information does not decrease monotonically but in general presents a first minimum at some intermediate κ that defines an optimal value κ_{opt} corresponding to a minimum of the redundancy, before ultimately decaying owing to statistical de-correlation: taking simultaneously W_k and $W_{k+\kappa_{\mathrm{opt}}}$ should thus give the best information about the dynamical evolution. Here, without surprise κ_{opt} is the same as that given by the previous rule, but more generally this is not the case.

It should be noted that, when no clear minimum of $I_{\mathrm{mut}}(\kappa)$ is obtained, this can mean either the presence of a very large noise, or that the observable has been under-sampled, or that too many degrees of freedom are involved, all cases where methods of the theory of low dimensional deterministic dynamical systems are of little help.

Once the parameter κ has been determined optimally, one can re-sample the time series with the new step and discard redundant information contained in the intermediate values that are too closely correlated to the

---
[9] A.M. Fraser, H.L. Swinney, "Independent coordinates for strange attractors from mutual information," Phys. Rev. A **33** (1986) 1134–1140.

retained ones. In fact, the κ parallel time series obtained by changing the phase of the reconstruction modulo κ are not independent but yield κ equivalent sets for one and the same trajectory. Comparing these reconstructed trajectories may serve to appreciate the amount of noise quantitatively.

Renumbering the time series, we can assume $\kappa = 1$ and focus on our second problem: finding the most appropriate embedding dimension d_e.

Keeping in mind that we want to obtain an injective representation, we reverse the implication (5.14), now stated:

$$\mathbf{V}_k = \mathbf{V}_{k'} \quad \Rightarrow \quad \mathbf{X}_k = \mathbf{X}_{k'},$$

which suggests to analyse the reliability of the identification of states (r.h.s.) from comparisons of different reconstructed trajectories (l.h.s) in spaces with different dimensions: neighbours in some space are "true neighbours" if they remain so for all reconstruction spaces. This property can be checked by increasing d_e, i.e. enlarging the width of the window swept on the data.

The *method of false neighbours*[10] is an efficient strategy to decide when to stop adding coordinates. Consider a trial dimension d', i.e. $[W_k; W_{k+1}; \ldots; W_{k+d'-1}]$ and the $(d'+1)$-dimensional reconstruction obtained by adding a component $W_{k+d'}$. Choose a distance in reconstruction space and a criterion to decide which is neighbour and which is not (depending on the noise amplitude). Next determine the number of false neighbours, i.e. the number of pairs of points that were neighbours in d' dimensions and are no longer neighbours in $d'+1$ dimension, as illustrated in Figure 5.8. Then increase d' up to the point when the fraction of false neighbours decreases significantly and choose that value as optimal embedding dimension.

The Euclidean distance derived from the L^2 norm is usually not a convenient choice for neighbourhood evaluations, since it requires a lot of computations. In contrast, the distance derived from the L^∞-norm, i.e.

$$\text{dist}\left(\mathbf{V}^{(1)}, \mathbf{V}^{(2)}\right) = \sup_k \left|V_k^{(1)} - V_k^{(2)}\right|.$$

only requires comparisons and is more economical.

Choosing the embedding dimension as given by the false-neighbour method should appropriately unfold the structure of the attractor.[11] The

---
[10] M.B. Kennel et al.: "Determining embedding dimension for phase-space reconstruction using a geometrical construction," Phys. Rev. A **45** (1992) 3403–3411.
[11] Improvements over the basic method are discussed by L. Cao, "Practical method for determining the minimum embedding dimension of a scalar time series," Physica D **10** (1997) 43–50.

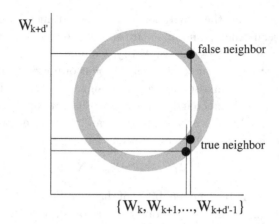

Fig. 5.8 Comparison of the reconstruction in a d'-dimensional space featured by the horizontal axis (coordinates $\{W_k; \ldots; W_{k+d'-1}\}$, and a reconstruction in the $(d'+1)$-dimensional space obtained by adding coordinate $W_{k+d'}$.

final presentation of the result can be improved by changing from the canonical basis in the reconstruction space to a basis correlated to the data in the least-square sense.[12] The method known as *proper orthogonal decomposition* or *singular value decomposition* is implemented in numerical programs such as MATLAB and is also used in the field of pattern recognition, which may be of interest in the analysis of spatiotemporal chaos to be introduced in Chapter 6.

5.2.4 *Application*

From a d_e-dimensional reconstruction one can next extract quantitative information on the system, and in particular the amount of chaos present, using quantities such as Lyapunov exponents or fractal dimensions.

In contrast with the theoretical approach, where the expression of the dynamical system is known, here we have just a (very long) time series of some observable at given control parameter. The attractor is then first reconstructed by the method of delays from this time series, Figure 5.9(a). We next assume that the permanent regime is reached and that the system explores its attractor repeatedly and satisfactorily in a statistical sense. If

---

[12](a) The idea was introduced by D.S. Broomhead and G.P. King, "Extracting qualitative dynamics from experimental data," Physica D **20** (1986) 217–236. (b) For a concrete implementation, see R. Vautard *et al.*: "Singular spectrum analysis: a toolkit for short, noisy chaotic signals," Physica D **58** (1992) 95–126.

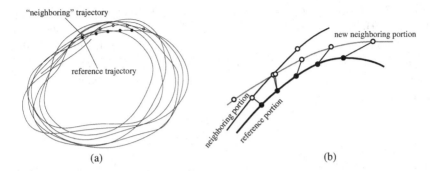

Fig. 5.9 Empirical determination of the largest Lyapunov exponent of a system in its reconstructed state space. (a) Reconstructed attractor. (b) Re-scaling of the distance between the reference portion of trajectory and its neighbour.

this is the case, when the system comes in a given region of state space, it never comes exactly at the same place but in some neighbourhood. Taking a portion of the trajectory as a reference, the vector field in its neighbourhood can be estimated from the set of trajectory pieces followed in close approaches to that reference portion, e.g. through a least-square adjustment of the coefficients of the local Jacobian matrix. This is in general a hard matter. A quantity that can, however, be determined more easily is the largest Lyapunov exponent. The procedure used in early attempts[13] reflects particularly clearly its nature as an average trajectory divergence rate: since Lyapunov exponents are quantities relative to the tangent evolution, the distance between pieces of trajectories serving to compute the divergence rate must remain small enough; when following the reference trajectory, one is therefore obliged to look for pieces of trajectories as close as possible to the current point, in the direction of fastest divergence, as suggested in Figure 5.9(b).

A quantity that can serve to characterise the fractal properties of strange attractors is the *correlation dimension* introduced by Grassberger and Procaccia.[14] It is extracted from the distribution of distances between pairs of points on the reconstructed attractor $\Delta(\mathbf{V}_i, \mathbf{V}_j)$, obtained by computing

$$C(R) = \lim_{N \to \infty} \frac{1}{N^2} \sum_{\{\mathbf{X}_i, \mathbf{X}_j\}} \Upsilon\left(R - \Delta(\mathbf{V}_i, \mathbf{V}_j)\right), \qquad (5.17)$$

---

[13] A. Wolf, J.B. Swift, H.L. Swinney, J.A. Vastano, "Determining Lyapunov exponents from a time series," Physica D **16** (1985) 285–317.

[14] P. Grassberger, I. Procaccia, "Measuring the strangeness of strange attractors," Physica D **9** (1983) 189–208.

where $\Upsilon = 0$ for $u < 0$, and $\Upsilon(u) = 1$ for $u > 0$ (Heaviside distribution). Provided that the statistic is sufficient, this quantity measures the number of points in volume elements of radius R around each point, roughly speaking its mass, as it varies with R. For a compact d-dimensional object, one would have $C(R) \sim R^d$ as $R \to 0$. The correlation dimension is thus defined as

$$\nu = \lim_{R \to 0} \frac{\log\big(C(R)\big)}{\log(R)} . \tag{5.18}$$

The way it is defined makes exponent ν similar to d_f. As such, it gives an easily determined measure of the fractal character of the attractor from the statistics of a long time record in the embedding space.

This approach has been developed many times. The example given here is the first application to a concrete convection experiment in confined geometry, by Malraison et al.[15] It is seen in Figure 5.10 (left) that the correlation integral (5.17) plotted in log–log scale as a function of R indeed display a linear part at small R from which an exponent ν can be derived. At the time of the experiment, the determination of the optimal embedding dimension was made by just increasing d_e progressively.

If the embedding dimension is chosen too small, one observes that ν is close to d_e which is easily understood from the fact that the reconstruction does not contain enough information to attest the deterministic character of the dynamics. Points fill the **V**-space homogeneously like a random signal would do. On the contrary, if upon increasing d_e, it happens that ν saturates at some finite value, as here, it is a good indication that some deterministic dynamics is at work producing low dimensional chaos, see Figure 5.10 (right). If, unfortunately, ν continues to increase with d_e, either confinement effects are too weak and the effective signal is not low dimensional, or the noise level is too high.

To conclude this section, let us remark that a reliable quantitative determination of the amount of chaos requires a lot of data but that useful information can anyway be obtained from the reconstruction technique, e.g. to attempt chaos control, see §5.4. Software packages are available for an automatic treatment of experimental data.[16]

---

[15]B. Malraison et al.: "Dimension d'attracteurs étranges: une détermination expérimentale en régime chaotique de deux systèmes convectifs," C.R. Acad. Sc. Paris **297** Série II (1983) 209–214.

[16]For example, in Note 12(b), or by R. Hegger et al.: "Practical implementation of nonlinear time series methods: the TISEAN package," Chaos **9** (1999) 414. (http://www.npipks-dresden.mpg.de/~tisean).

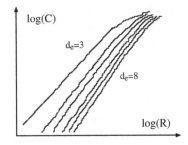 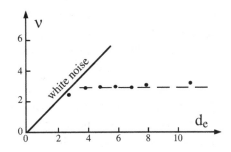

Fig. 5.10 Determination of the correlation dimension in a convection experiment. Left: Distribution of distances in state space for several embedding dimensions. Right: Variation of the slope at small R as d_e is increased. Exponent ν is seen to saturate at 2.8 which is the correlation dimension of the considered attractor. When d_e is too small, trajectories fill the reconstruction space homogeneously, as if the signal was a white noise, hence the corresponding line $\nu = d_e$. After Malraison et al., Note 15.

5.3 Mixing by Stirring

5.3.1 *Mixing: diffusion or/and advection?*

Achieving homogeneity in heterogeneous media is an important problem in engineering applications. Attached to test particles immersed in a fluid substrate, the property which has to be made as uniform as possible is called a *tracer*. The tracer is *passive* if its distribution does not affect the motion of the fluid. In other words the problem is purely kinematic. In contrast, when gravitational effects cannot be neglected, the temperature or the concentration of a solute cannot be considered as passive tracers since their fluctuations change the density so that supplementary dynamical effects are expected from buoyancy, i.e., convection. In the following, we consider only passive tracers and further assume that the test particles strictly follow the fluid. This assumption fails if, though otherwise passive, the tracer is lighter (gas bubbles) or heavier (solid grains) than the fluid itself, and a flow at finite Reynolds number is considered so that differential inertia has to be taken into account (though any modification of the flow by the tracer can still be neglected).

Homogeneity can be obtained through molecular diffusion according to the Fick equation

$$\partial_t c = \mathcal{D}\boldsymbol{\nabla}^2 c, \qquad (5.19)$$

where c is the concentration and $\mathcal{D}$ the molecular diffusivity. It thus takes a time of order $\tau_\mathrm{m} = \ell^2/\mathcal{D}$ ('m' for 'molecular') to wipe out the lack of homogeneity at scale ℓ, hence the smaller ℓ, the shorter τ_c. Equation (5.19) describes diffusion in a fluid at rest. If the fluid is stirred, the presence of the velocity field **v** transforms it into an *advection-diffusion* equation:

$$(\partial_t + \mathbf{v} \cdot \boldsymbol{\nabla}) c = \mathcal{D} \boldsymbol{\nabla}^2 c. \quad (5.20)$$

When the velocity field moves the tracer particles so as to increase the concentration gradient by introducing small scales of motion, diffusion is enhanced and the diffusion time is shortened. *Turbulence* can generate such small scale fluctuations by a cascading process from the large scale L where fluid motion is maintained, with typical velocity gradient U/L, down to the small scale where energy is consumed by viscous friction (kinematic viscosity ν), as sketched in §1.3.5, p. 21 and further analysed in chapter 8. It is indeed common practice to turn the spoon vigourously in one's cup to get milk particles uniformly distributed in tea. It seems easy to stir milk in tea because this aqueous mixture has low viscosity and because, with little effort, we input a comparatively large energy in the fluid. Agitation would be obtained at much higher cost with a large viscosity fluid.

An estimate of the agitation in the fluid with typical velocity gradient u/ℓ at some scale ℓ can be obtained from the value of the corresponding Reynolds number $R_\ell = u\ell/\nu$, comparing the viscous time $\tau_v = \ell^2/\nu$ to the advection time $\tau_a = \ell/u$ over that scale (p. 9). When R_ℓ is small the velocity field u is smooth, otherwise it may vary wildly from point to point. A similar comparison can be made for mixing *via* molecular diffusion, which leads to the definition of the *Peclet number* $P_\ell = \tau_m/\tau_a = u\ell/\mathcal{D}$. A tracer distribution that is smooth at scale ℓ is thus expected at small P_ℓ.

The quantity $S = P_\ell/R_\ell = \nu/\mathcal{D}$, called the *Schmidt number*, is the ratio of two molecular properties and does not depend on ℓ. Since, in condensed fluids,[17] molecular diffusivity $\mathcal{D}$ is usually small when compared to kinematic viscosity ν, one has $S \gg 1$.

In the turbulent situation alluded to above, the mixing of tracers is a by-product of turbulence. Let ℓ_K be the typical scale of the velocity fluctuations at the end of the turbulent cascade (§8.1.3, p. 318) where the effective Reynolds number is $R_{\ell_K} \sim 1$, the corresponding Peclet number is $P_{\ell_K} = SR_{\ell_K} \sim S$, which is not necessarily small. Notwithstanding our intuition, turbulent stirring may not be good at mixing. So, we must first specify what we really understand by 'mixing' and next consider whether

---

[17] In gases, thermal agitation mixes momentum, energy and molecules at similar rates so that the Prandtl number $P = \nu/\kappa$ (p. 93), the Lewis number $L = \mathcal{D}/\kappa$ (p. 120), and the Schmidt number $S = \nu/\mathcal{D}$ are all of order unity.

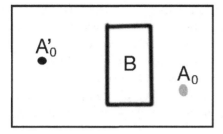 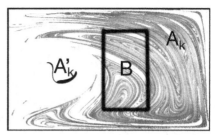

Fig. 5.11 Illustration of mixing, adapted from Ottino (note 18): Here a cavity with sliding walls is filled with glycerine. At time t_0, two blobs of fluorescent tracer are placed in the system at A_0 and A'_0. Then the walls are set into motion periodically according to some protocol. At time t_k, after k periods (here $k = 10$) the blob initially at A_0, is now stretched and folded into A_k and occupies a large part of domain B. In contrast, the blob initially at A'_0 has been little deformed so that A'_k has no chance to visit domain B. Mixing may thus not be uniform, a situation that has do be dealt with and avoided as much as possible. The experiment considered here will be re-examined in §5.3.2.1.

it can be achieved in non-turbulent instances.[18]

In connection with the formal definition of *mixing* found in dynamical systems theory,[19] loosely speaking we shall say, upon modelling the evolution of the fluid during a time interval T as a mapping $\mathbf{\Phi}$ over the physical space[20] and considering a test region $\mathcal{B}$ and a blob of volume $\mathcal{A}_0$ at time t_0, if originally $\mathcal{A}_0$ does not intersect $\mathcal{B}$ but there exists a time k such that for all $k' > k$, $\mathcal{A}_{k'}$ intersects $\mathcal{B}$, then the flow is *mixing*. Such a situation is illustrated in Figure 5.11.

Let us therefore consider a system in the limit when the Schmidt number goes to infinity ($\mathcal{D} \to 0$). Property c of the test particle is then not diffused but just transported with it by the velocity field: $\frac{d}{dt}c \equiv (\partial_t + \mathbf{v} \cdot \boldsymbol{\nabla})c = 0$. See p. 7 where the *Lagrangian* point of view that focuses on the fluid particle *displacement* was opposed to the more familiar *Eulerian* point of view that focuses on its *velocity* and is linked to it through the definition of the *material derivative*. In practice, even a smooth laminar (Eulerian) velo-

---

[18] A popular introduction is: J. Ottino "The mixing of fluids," Scientific American **260** (1989) 56–67. Expanded, more technical versions are: J.M. Ottino, "Mixing, chaotic advection, and turbulence," Annu. Rev. Fluid Mech. **22** (1990) 207–253, and the book [Ottino (1989)].

[19] For a brief introduction, consult: J.L. Lebowitz, O. Penrose, "Modern ergodic theory," Physics Today, February 1973, pp. 23–29.

[20] In line with our previous approach in terms of dynamical systems, points in $\mathcal{A}_0$ evolve into points in $\mathcal{A}_1 = \mathbf{\Phi}(\mathcal{A}_0)$, and next into points in $\mathcal{A}_2 = \mathbf{\Phi}(\mathcal{A}_1) = \mathbf{\Phi}^2(\mathcal{A}_0)$, ...

city field can produce good mixing provided that, considered as a vector field, it generates (Lagrangian) chaotic trajectories in *physical space* which turns out to be the state space of the problem. This is so because folding and stretching inherent in chaos as described earlier, and especially in §4.3, p. 154ff, make the fluid particle randomly visit whole regions of the physical domain where the flows develops. The archetypal example of dough kneading, with stretching and folding supposed to disperse salt uniformly, has been illustrated in Fig. 5.3, p. 175.

From the expression of the velocity field **v** as the material derivative of the position **X** of fluid particles in an incompressible flow we have:

$$\tfrac{\mathrm{d}}{\mathrm{d}t}\mathbf{X} = \mathbf{v}(\mathbf{X},t) \qquad \text{with} \qquad \boldsymbol{\nabla}\cdot\mathbf{v} = 0\,. \tag{5.21}$$

Since we assume that the tracer follows the fluid, Equation (5.21) also governs the displacement of test particles. We know from previous studies that space-filling chaotic trajectories can appear in a continuous-time dynamical system provided that its dimension is at least 3, which is the case here since (5.21) with prescribed field **v** defines a three-dimensional dynamical system. Such a situation is called *chaotic advection* or *Lagrangian chaos*.

A celebrated (but somehow academic) example of time-independent three-dimensional flow with chaotic streamlines is the ABC flow:

$$u = A\sin Z + C\cos Y\,, \quad v = B\sin X + A\cos Z\,, \quad w = C\sin Y + B\cos X\,,$$

which is an exact solution of the incompressible Euler equations[21] in a $2\pi\times2\pi\times2\pi$ domain with periodic boundary conditions. Chaotic trajectories are present in certain ranges of the parameters A, B, and C which, by the way, are the initials of Arnold, Beltrami, and Childress.[22]

If geometrical constraints impose an essentially two-dimensional flow pattern, chaos is possible only when the flow is time-dependent, most often periodically forced, as discussed in the previous chapter. In the following we give examples of some relevant cases, beginning with time-periodic two-dimensional closed or open flows where time-periodicity allows one to perform standard stroboscopic analysis, and next considering so called *static mixers* where a steady, three-dimensional, open flow is achieved in a pipe made of a series of sections with some well-chosen shape, and where effective time periodicity is produced by the periodic repetition of identical up-stream/down-stream flow conditions within a section. A brief catalogue of interesting cases is given in Ottino's articles mentioned in Note 18.

---

[21] It is also an exact equally academic solution of the Navier–Stokes equations in the presence of a spatially periodic forcing just compensating viscous dissipation.

[22] For a detailed approach with references, see: T. Dombre *et al.*: "Chaotic streamlines in the ABC flows," J. Fluid Mech. **167** (1986) 353–391.

5.3.2 Time-periodic two-dimensional configurations

In two dimensions,[23] a time-independent, divergence-free, velocity field cannot generate chaotic trajectories since it derives from a stream function such that

$$\tfrac{\mathrm{d}}{\mathrm{d}t}X \equiv v_x = -\partial_y \Psi, \qquad \tfrac{\mathrm{d}}{\mathrm{d}t}Y \equiv v_y = \partial_x \Psi. \tag{5.22}$$

This is an area-preserving system and orbits fulfil $\tfrac{\mathrm{d}}{\mathrm{d}t}\Psi = 0$: they are level curves of $\Psi(x,y)$ which plays the role of a Hamiltonian in analytical mechanics (§2.1.2, p. 45). The dimension of the state space, here the physical space, is not large enough to permit the divergence of neighbouring trajectories characteristic of chaos as implied by the Poincaré–Bendixson theorem (§4.2.1.3). The situation changes if the flow field is made time-dependent and more specifically time-periodic since the evolution has to be appreciated in the enlarged state space (the product of the state space and the time axis) and trajectories in that space sampled at the period of the forcing, which produces return maps after identification of points modulo the period (see Figure 4.7, p. 141).

5.3.2.1 Boxes with moving walls

Two examples are illustrated in Figure 5.12 which, in the left part, displays the case of the flow in a rectangular box with moving walls already used to illustrate the concept of mixing in Figure 5.11 and, on the right, the flow between eccentric cylinders called the journal-bearing flow.[24]

Let us consider *creeping flow* conditions when the Reynolds number is sufficiently small that inertial effects can be neglected and the velocity field obtained within the Stokes approximation that neglects the advection term $\mathbf{v} \cdot \boldsymbol{\nabla} \mathbf{v}$ in the Navier–Stokes equations. From the no-slip boundary condition, by entrainment, the motion of the walls generates a nontrivial velocity field inside the box. As long as the wall speed is kept time-independent, cellular flow patterns are obtained as illustrated in Figure 5.13 and Figure 5.14. Trajectories remain regular, with elliptic and hyperbolic points

---

[23] Circumstances under which the flow is essentially two dimensional are easily obtained, either with a very short system in the third direction (Hele-Shaw configuration) or on the contrary, very long. Otherwise, three-dimensional features of the flow cannot be neglected.

[24] The moving wall apparatus has been studied in detail by: C.W. Leong, J.M. Ottino: "Experiments on mixing due to chaotic advection in a cavity," J. Fluid Mech. **209** (1989) 463–499. In the context of chaotic advection, the journal bearing flow has been introduced by J. Chaiken *et al.*: "Experimental study of Lagrangian turbulence in Stokes flow," Proc. R. Soc. Lond. A **408** (1986) 165–174.

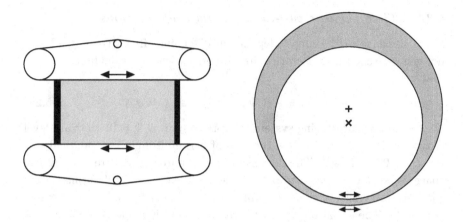

Fig. 5.12 Two examples of mixing systems with moving walls. Left: Rectangular box closed with left/right walls fixed and top/bottom circulating bands. Right: Journal-bearing flow. The fluid is located in the region in grey. The walls are moving, as suggested by the arrows. As long as that motion is constant in time, only simple regular trajectories are generated, but when the forcing becomes time-dependent, e.g. periodic, chaotic trajectories are obtained in certain regions of the system.

as key features (§2.2.2, p. 51). In contrast, when the walls are moved in a time-dependent fashion according to some protocol, e.g. stop-and-go and/or alternating, entrainment becomes irregular and can wrap trajectories in a complicated way. For example, mixing in Figure 5.11 is obtained with the top wall moving from left to right during some time, next the bottom wall from right to left at the same speed during the same length of time, and this time sequence repeated at will, showing that the simple flow

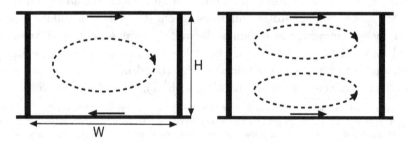

Fig. 5.13 Sketch of the simple flow patterns obtained in a flat box ($H < W$) with walls sliding at constant speed. A single cell is obtained when the two walls go in opposite directions. Two counter-rotating cells are produced when the walls move in the same direction (right).

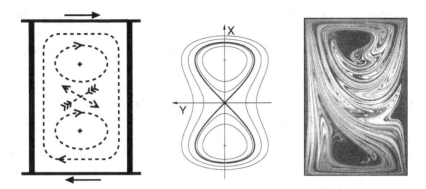

Fig. 5.14 A two-cell pattern is obtained in a tall box ($H > W$) with walls sliding in opposite directions (left). The figure-eight set of flow lines is topologically equivalent to the phase portrait of the Duffing system with negative linear stiffness (middle). Chaotic advection obtained upon periodic forcing (after Leong and Ottino, note 24).

sketched in Figure 5.13 (left) can induce chaotic advection in a large part of the box once made time-dependent.

The flow configuration is a function of the wall velocity and the aspect ratio H/W of the box. In a tall box ($H/W > 1$), when the top and bottom walls of the box are moving in opposite directions, a two-cell flow pattern is produced has sketched in Fig. 5.14 (left). This pattern can be modelled by the phase portrait of a two-well oscillator system $\mathcal{V}(X) = -\frac{1}{2}X^2 + \frac{1}{4}X^4$ yielding

$$\frac{\mathrm{d}^2}{\mathrm{d}t^2}X - X + X^3 = 0 \quad \text{or} \quad \frac{\mathrm{d}}{\mathrm{d}t}X = Y, \quad \frac{\mathrm{d}}{\mathrm{d}t}Y = X - X^3, \quad (5.23)$$

as illustrated in the middle panel. The flow is next made time-periodic according to various protocols, e.g. the one producing the image on the right panel: move the top wall to the right during 20 sec, after a 5 sec pause move the bottom wall to the left during 20 sec and repeat the cycle after another second 5 sec pause; the snapshot shown in the right panel of Figure 5.14 was taken after 4 full periods. Mathematically, the complex behaviour of trajectories observed is linked to the presence of the saddle point in the middle of the cell and the formation of a homoclinic tangle from the opening of homoclinic loops under periodic forcing as explained in §4.3.1. This feature is generally present in time-dependent two-dimensional flows producing chaotic mixing, as understood from considering Figure 4.15, p. 157, especially its panel (c) which illustrates so-called *lobe dynamics*, i.e. the transformation of phase space domains limited by the stable ($\mathcal{S}$) and unstable ($\mathcal{U}$) manifolds of the saddle point under time periodic perturbation.

Remembering that points in phase space in that figure correspond to fluid particles in physical space, one sees that, after a few periods, points originally in a lobe to the right of S are reintroduced in a lobe located on its left, which ends in chaotic motion upon iterating the transformation a large number of times. It should be noted that, in dynamical systems theory, one is mostly interested in asymptotic properties at the '$t \to \infty$' limit. In the present context, short term properties (number of iterations kept small) are of more concern since efficient mixing is searched for.

Another mixing system with moving walls is the journal-bearing flow sketched in Figure 5.12 (right). The creeping flow between the eccentric cylinders, which depends on the ratio of cylinders' diameter and the eccentricity, is the superposition of the flow induced by the outer cylinder rotating at angular speed ω_{out} assuming the inner one at rest, and the flow induced by the inner cylinder with angular speed ω_{in} assuming the other at rest. As long as ω_{out} and ω_{in} are independent of time, the trajectories of the test particles are non-chaotic as before but chaotic advection arises as soon as some time-dependent rotation protocol is adopted, e.g. periodic with period T and α the fraction of period during which the outer cylinder is set into motion:[25] $\omega_{\text{out}} = \omega$, $\omega_{\text{in}} = 0$ for $t \in [0, \alpha T[$ (mod T) and $\omega_{\text{out}} = 0$, $\omega_{\text{in}} = -\omega$ for $t \in [\alpha T, T[$ (mod T). Other examples are introduced in Exercise 5.6.4.

5.3.2.2 Stirring with rods

Moving a rod, or a set of rods, in a fluid is the most natural way of stirring. This is what the whisks of an eggbeater do in a mixing bowl (Fig. 5.15, left). In a highly viscous fluid, the rods drag the fluid in a way that wraps the trajectories of the test particles in a possibly complex way inducing chaotic advection, depending on the geometry, especially the shape of the container, the number of rods and the stirring protocol (Fig. 5.15, right). The mixing effect is best understood in the Lagrangian perspective by considering the effects of the flow produced by the movement of the rods on a material line/surface. The points belonging to this set follow trajectories generated by the dynamical system (5.21) and as such cannot intersect, which induce topological constraints of utmost importance and a lot of important results

---

[25] For illustrations, see note 18 and the cover of the corresponding issue of Scientific American; see also P.D. Swanson and J.M. Ottino, "A comparative computational and experimental study of chaotic mixing of viscous fluids," J. Fluid Mech. **213** (1990) 227–249.

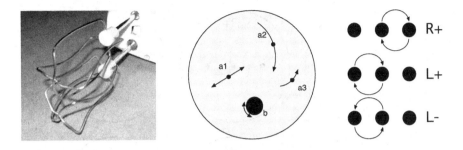

Fig. 5.15 Left: The intertwined motion of the whisks efficiently mixes cooking ingredients. Middle: Their effects can be modelled as moving rods translated in a container according to some pattern here seen from above, e.g. rods a1, a2, a3; if, as rod b, they have a sizeable diameter, they can also stay at fixed positions but be rotated around their axis according to some protocol, achieving a flow that can then be modelled by a set of vortices as in the second part of Exercise 5.6.4. Right: example of translation protocol for three rods exchanging their positions. R+ correspond to a clockwise permutation of the two right-most rods, L+ and L− to clockwise and anti-clockwise permutations of the left-most rods.

derive from topological considerations.[26] The time history of the motion is in fact less important than these considerations which explain why the mixing efficiencies of the two protocols of rod permutation described in Figure 5.15 are so different. When the rods are always permuted clockwise (R+L+), the material line joining the wall of the container to a rod is wrapped always in the same direction and can be unwrapped so that the topology of the line remains simple. In contrast, when the permutation of the pair on the right is clockwise and the other permutation anti-clockwise (R+L−), the line is wrapped in a complex way due to topological obstructions implied by the crossing prohibition (Fig. 5.16). Another view on the problem is obtained by following the position of the rods in an extended state space product of the discussed representing the container and the time axis in much the same way as in Figure 4.7, p. 141, explaining the stroboscopic analysis. Doing this one obtains braids – here with three strands – that are either just twisted (R+L+) or entangled (R+L−).

Other protocols can be considered for example a single rod can be dragged through the fluid along a figure-eight route, the clockwise and anti-clockwise parts of the route again producing mixing for the same reason. In a similar but more complicated way, the whisks of the eggbeater in Figure 5.15 (left) can be modelled by two sets of four rods rotating in

---
[26]See P.L. Boyland, et al.: "Topological fluid mechanics of stirring," J. Fluid Mech. **403** (2000) 277–304.

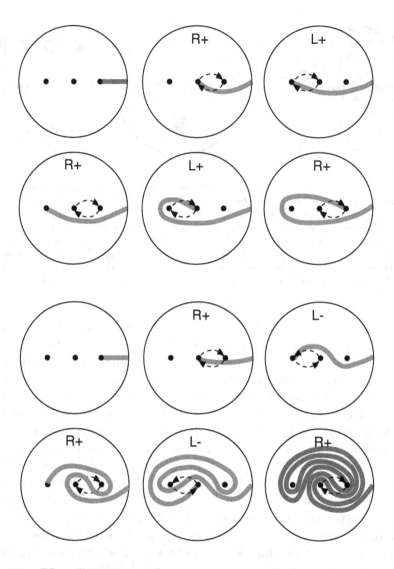

Fig. 5.16 Effect of mixing protocols on a segment of material line initially joining the wall of the container to the right-most rod. Protocol R+L+R+L+... (top) is less efficient than protocol R+L−R+L−... (bottom).

opposite directions.

Stirring with rods is not limited to closed systems but can also be used in open flow configurations which are important in industrial applications since they allow continuous operation, e.g., food processing, glass or paper

Fig. 5.17 Mixing in a free-surface channel by two stirrers moved along circles (dotted) by rotating arms (1: anti-clockwise, 2: clockwise). In the image, the viscous fluid (cane sugar syrup) flows from bottom to top. The mixing pattern, visualised by the stretching of a small India ink blob, extends over the channel's width thanks to the fact that the rods scrape the lateral walls against the stream and bring back the fluid in the centre of the channel ahead of the region where the circular paths of the rods overlap. The mixing is then most effective. Notice the existence of the separation points at the lateral walls (a and b) and the wrapping of the filaments issued from the blob deposited near a. In contrast, when the arms rotate in the opposite direction (1: clockwise, 2: anti-clockwise), the rods still drag the fluid to the centre of the channel but downstream the mixing region, which is less efficient (Courtesy E. Gouillart, Saint-Gobain-Recherche, France.)

making.[27] In that case, the fresh material enters a mixing region of limited extent and mixed material is collected at the exit. An example is given in Figure 5.17.

5.3.3   Time-dependent two-dimensional open flows

Essentially two-dimensional flows occur frequently in the natural environment. In particular the fluid envelopes of the Earth have vertical scales of the order of the kilometre or less whereas dynamics in the ocean and the atmosphere develops over much larger scales (called synoptic) ranging from hundreds of kilometres in the ocean to thousands of kilometres in the atmosphere. Within layers or sub-layers of these envelopes, motions are mostly horizontal and can be modelled as two-dimensional time-dependent eddies. In this geophysical context, the Lagrangian perspective taken in the study of mixing is particularly relevant to problems such as pollutant dispersal or

---

[27] E. Gouillart *et al.*: "Open-flow mixing: Experimental evidence of strange eigenmodes," Phys. Fluids **21** (2009) 023603.

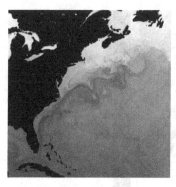
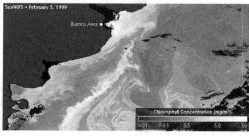

Fig. 5.18 Grey level pictures from satellite imaging (© NASA). Left: Sea Surface Temperature in the NW Atlantic: Darker grey corresponds to tropical waters and lighter grey to polar waters. Currents are easy to identify. The Gulf Steam is an unstable jet generating gyres (long lived vortices) and smaller scale eddies forming hot/cold filaments within cold/hot waters (from NASA-NOAA-N project). Right: Chlorophyll distribution is a tracer of phytoplankton activity. Filamentary patterns are conspicuous in coastal waters off Argentina along a high productivity band associated with the convergence of the Brazil warm current and the Malvinas cold current (from NASA SeaWIFS project).

marine biological productivity. Figure 5.18 illustrates two typical oceanic situations.[28]

As a model of open flow that induces mixing, let us consider the system introduced by Rom-Kedar et al.[29] made of two counter-rotating vortices of equal strength perturbed by an oscillatory extensional velocity field.[30] In the absence of perturbations the flow lines are as sketched in Figure 5.19 (left). One vortex being dragged by the velocity field of its companion, the pair travels to the right at some velocity which is next subtracted by changing to a moving frame. The flow lines are easily guessed from the nature of the flow around the vortices and the symmetry of the problem which forces

---

[28] For a review of the oceanic case, consult: K.V. Koshel, S.V. Prants, "Chaotic advection in the ocean," Physics–Uspekhi **49** (2006) 1151–1178 or S. Wiggings, "The dynamical systems approach to Lagrangian transport in oceanic flows," Annu. Rev. Fluid Mech. **37** (2005) 295–328. An interesting illustration in the atmospheric context is T.-Y. Koh, B. Legras, "Hyperbolic lines and the stratospheric polar vortex," Chaos **12** (2002) 382–394.

[29] V. Rom-Kedar et al.: "An analytical study of transport, mixing and chaos in an unsteady vortical flow," J. Fluid Mech. **214** (1990) 347–394.

[30] Except for the presence of two vortices and not one, the situation bears some similarity with that in the first part of Exercise 5.6.4. Here, the stream-function reads: $\Psi = -\Gamma/4\pi \log\{[(x-x_v)^2+(y-y_v)^2]/[(x-x_v)^2+(y+y_v)^2]\} - V_v y + \epsilon xy \sin(\omega t)$, where $(x_v, \pm y_v)$ are the coordinates of the vortices, V_v their average velocity, ϵ the intensity of the perturbation and ω its frequency. The components of the velocity field are $v_x = \partial_y \Psi$ and $v_y = -\partial_x \Psi$.

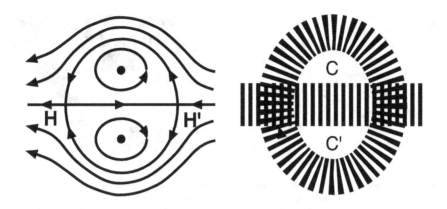

Fig. 5.19 Flow lines around two counter-rotating vortices. Left: unperturbed case; H and H' are two stagnation points in the moving frame. Right: the effect of a periodic perturbation is to blur the neighbourhood of the heteroclinic trajectories; fluid particles remain confined to the core regions C and C', or else experience mixing for an unpredictable lapse of time if they happen to visit the hatched region, or feel the presence of vortices only little when they travel far enough in the outside free flow region.

the existence of two saddle points at H and H', which are stagnation points for the flow (in the moving frame). These points are thus *heteroclinic* in the sense of §4.3 and the special trajectories going to and from these points are easily interpreted as their stable and unstable manifolds. Upon periodic perturbations these manifolds are forced to wiggle and produce a *tangle* as explained in §4.3.1. The *stochastic layer* so generated is featured as hatched thick lines in Figure 5.19 (right).

In the absence of forcing (Figure 5.19, left), fluid particles can follow two sorts of trajectories, either outside the separatrices joining H to H', which corresponds to free flow, or trapped close to one of the two vortices inside the separatrices. Weak periodic forcing transforms the vicinity of the formerly smooth separatrices into a messy region as sketched in Figure 5.19 (right). We now have three kinds of trajectories. As before, those in the external region (free flow) sweep the vicinity of the vortices, those close to the vortices remain trapped in the so-called core regions C and C'. But those that were close enough to the stable manifold of H far on the right upstream, now enter the tangle and follow the complicated evolution implied by a *lobe dynamics* analogous to that alluded to previously for mixing in the box with sliding walls (§5.3.2.1). This chaotic evolution is however transient and sooner or later the fluid particles get back into the smooth region far downstream. The *residence time*, the time spent in the mixing region,

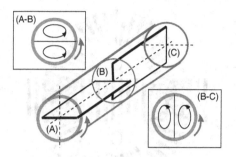

Fig. 5.20 Left: Partitioned-pipe mixer: Lagrangian chaos results from a combination of axial flow and the recirculation imposed by the cylinder rotation. Right: Kenics KM Series static mixer. Excellent mixing is obtained by alternating left-handed and right-handed partitioning helical blades.

depends sensitively on the position at the entrance. Residence times turn out to be distributed according to decreasing exponentials.[31]

5.3.4 Time-independent three-dimensional open flows

Of great engineering interest are open mixers in which material to be mixed is forced trough a pipe at the inlet and mixture retrieved at the outlet. Examples are displayed in Figure 5.20. On the left a system called the partitioned-pipe mixer is made of a series of fixed blades at right angles, two of which are shown. These blades are placed in a pipe which is rotating about its axis. At low flow rates, i.e. in the laminar regime, the trajectories in a portion of the pipe, e.g. (A-B), are helical lines traced on cylindrical stream-surfaces as a result of a superposition of the longitudinal transport and the azimuthal flow imposed by the rotation of the pipe which generates the cellular patterns, one for each of the half-cylinders, sketched in the inset. The position of a particle in section (B) can in principle be computed from its position in section (A), taken as an initial value problem of the Stokes equations. A similar computation can be performed in portion (B-C). A mapping from section (A) onto section (C) is obtained which closely corresponds to a Poincaré map, when it is recognised that the position along the axis plays the role of time. Indeed, if one neglects developing flow in the very first portions, the flow is spatially periodic and axial average transport

---

[31]This can be understood by working out Exercise 5.6.3 about *transient chaos*. As discussed in §4.3.2, the tangle involves a horseshoe map, the unstable part of which can be modelled by the tent map considered in that exercise.

make the particle trajectories time periodic. When going from one pipe portion to the next, particles jump from one kind of stream-surfaces to the other. Lagrangian chaos is present is a large part of the flow. Once again, efficiency of mixing has to be appreciated as a short term property of the iterated mapping, and not asymptotically in time since the partitioned-pipe arrangement has a finite length.

The set-up considered above requires a rotation of the cylinder containing the blades in order to generate the torsion of flow lines necessary to complicate the plain Poiseuille flow that would take place otherwise. This torsion can be obtained in a completely static configuration by using helical blades as shown in Figure 5.20 (right).

Another system producing chaotic advection is a twisted pipe made of a series of bends,[32] which can have some importance for heat-and-mass transfer applications. In a curved pipe, longitudinal vortices of opposite circulation develop due to inertia. This recirculation plays a role similar to the one induced by the rotation of the cylinder in the presence of the partitioning plates in the case discussed above. They generate critical streamlines with stable and unstable directions ending in homoclinic tangles upon periodic perturbation as in the previous cases. The interest of the system is here that a numerical approach can be implemented more easily, thanks to analytical formulas available to describe the flow through one bend in the small-curvature approximation. Mixing efficiency can be tuned by changing the curvatures and diameters the pipes and the pitch of the overall helical arrangement of the bends.

5.4 Control of Chaos

5.4.1 *Synchronisation of dynamical systems*

Let us consider a time-continuous system written in the form

$$\tfrac{\mathrm{d}}{\mathrm{d}t}\mathbf{X} = \mathcal{F}(\mathbf{X}, t), \tag{5.24}$$

in which the origin of explicit time dependence is not specified for the moment. The concept of stability of a given trajectory $\mathbf{X} = \mathbf{X}(t)$ issued from some initial condition $\mathbf{X} = \mathbf{X}_0$ at $t = t_0$ follows from the study of the evolution of infinitesimal perturbations governed by

$$\tfrac{\mathrm{d}}{\mathrm{d}t}\delta\mathbf{X} = \partial_{\mathbf{X}}\mathcal{F}(t)\,\delta\mathbf{X} \tag{5.25}$$

---

[32] S.W. Jones *et al.*: "Chaotic advection by laminar flow in a twisted pipe," J. Fluid Mech. **209** (1989) 335–357.

where $\partial_{\mathbf{X}}\mathcal{F}(t)$ is a short hand notation for the Jacobian matrix of $\mathcal{F}$ computed all along that trajectory.

For fixed points or limit cycles, that problem has time-independent or time-periodic coefficients. Here the Jacobian matrix has an arbitrary time dependence so that the situation is the same as for the search of the Lyapunov exponents of a given trajectory governed by an autonomous dynamical system (§5.1.1). The trajectory is then linearly stable if the largest Lyapunov exponent extracted from (5.25) is strictly negative.

5.4.1.1 *One-way coupled systems*

In what follows, we restrict ourselves to the consideration of a system forced by the time evolving state $\mathbf{X}_\mathrm{e}$ of some autonomous external system:

$$\tfrac{\mathrm{d}}{\mathrm{d}t}\mathbf{X}_\mathrm{e} = \mathcal{F}_\mathrm{e}(\mathbf{X}_\mathrm{e}). \tag{5.26}$$

Accordingly, we rewrite (5.24) as

$$\tfrac{\mathrm{d}}{\mathrm{d}t}\mathbf{X} = \mathcal{F}(\mathbf{X}, \mathbf{X}_\mathrm{e}). \tag{5.27}$$

In this expression, $\mathbf{X}_\mathrm{e}$ is a function of time. It represents a particular realisation of the forcing, $\mathbf{X}_\mathrm{e}(t)$. For a periodic forcing, the external system is an oscillator and this specific realisation can be parameterized by the phase of the oscillator, the value it takes at $t = t_0$ in the absolute time reference frame. If the forcing is multi-periodic, e.g. the compound of several independent oscillators, several phase variables have to be specified, one for each frequency. If the forcing results of the coupling with a chaotic system, the trajectory followed by the system depends sensitively on the full initial condition, its own one and that of the forcing. After a while, corresponding to the elimination of a transient, the forcing is a stochastic process with stable statistical characteristics typical of the corresponding strange attractor. The sensitivity of $\mathbf{X}_\mathrm{e}(t)$ to its initial condition forbids the realisation of two experiments in strictly identical conditions.

The linear stability of the forced system is now derived from

$$\tfrac{\mathrm{d}}{\mathrm{d}t}\delta\mathbf{X} = \partial_{\mathbf{X}}\mathcal{F}(\mathbf{X}, \mathbf{X}_\mathrm{e})\,\delta\mathbf{X},$$

where partial derivatives in the expression of the Jacobian matrix involve only the components of $\mathbf{X}$. It is specific to the trajectory $\mathbf{X}(t)$ followed by the forced system but is also conditioned by the trajectory of the forcing system $\mathbf{X}_\mathrm{e}(t)$. In this case, the concept of stability is best understood by considering two twin forced systems coupled to the same realisation of the forcing system:

$$\tfrac{\mathrm{d}}{\mathrm{d}t}\mathbf{X}_1 = \mathcal{F}(\mathbf{X}_1, \mathbf{X}_\mathrm{e}),$$
$$\tfrac{\mathrm{d}}{\mathrm{d}t}\mathbf{X}_2 = \mathcal{F}(\mathbf{X}_2, \mathbf{X}_\mathrm{e}),$$

and studying the evolution of the distance $\|\mathbf{X}_1 - \mathbf{X}_2\|$ between the two trajectories $\mathbf{X}_1$ and $\mathbf{X}_2 = \mathbf{X}_1 + \delta\mathbf{X}$, while keeping one and the same $\mathbf{X}_\mathrm{e}$. Adapting the calculations described in §5.1.1, one thus decide for stability from a whole spectrum of *conditional Lyapunov exponents*, where the term "conditional" is there to stress that one focus, not on the full autonomous system $(\mathbf{X}, \mathbf{X}_\mathrm{e})$ but only on the forced subsystem $\mathbf{X}$, thus not on that part of instability that stems from the sensitivity to initial conditions of the forcing system when it is chaotic. The trajectory of the forced system will thus be *stable* if the largest conditional Lyapunov exponent is *negative*. Accordingly the response of the system will be unique (the distance between trajectories starting from two slightly different initial conditions decays to zero) and the forcing is successful. Otherwise, when the largest conditional Lyapunov exponent is positive, the response of the forced system is sensitive to perturbations.

5.4.1.2 *Synchronisation of identical systems*

In the case considered above, one cannot not expect that, when its dynamics is stable, the forced system strictly follows its master $(\mathbf{X}(t) \equiv \mathbf{X}_\mathrm{e}(t))$ since the forcing system may have an arbitrary structure unrelated to that of the forced system. The result is just that $\mathbf{X}(t)$ is uniquely determined once the specific realisation of the forcing $\mathbf{X}_\mathrm{e}(t)$ is specified. Now we demand more, i.e. that the slave can follow the master strictly, but we must restrict to systems that permit such full synchronisation, whatever the nature of the dynamical regime, which is usually called *identical synchronisation*.

A first way to build systems displaying this property is to take two identical systems, define one as the master

$$\tfrac{\mathrm{d}}{\mathrm{d}t}\mathbf{X}_\mathrm{e} = \mathcal{F}(\mathbf{X}_\mathrm{e}), \tag{5.28}$$

while the other, the slave $\mathbf{X}$, is governed by

$$\tfrac{\mathrm{d}}{\mathrm{d}t}\mathbf{X} = \mathcal{F}(\mathbf{X}) + \mathcal{F}_\mathrm{c}(\mathbf{X}, \mathbf{X}_\mathrm{e}), \tag{5.29}$$

with a coupling term $\mathcal{F}_\mathrm{c}(\mathbf{X}, \mathbf{X}_\mathrm{e})$ such that $\mathcal{F}_\mathrm{c}(\mathbf{X}, \mathbf{X}) \equiv 0$, e.g. a simple linear feed-back $\mathcal{F}_\mathrm{c}(\mathbf{X}, \mathbf{X}_\mathrm{e}) = K(\mathbf{X}_\mathrm{e} - \mathbf{X})$.

Another strategy consists in starting again from two systems with identical initial structure $\mathcal{F}$, take one as the master (5.28) and modify the second one:

$$\tfrac{\mathrm{d}}{\mathrm{d}t}\mathbf{X} = \boldsymbol{\mathcal{F}}_\mathrm{m}(\mathbf{X}, \mathbf{X}_\mathrm{e}) \qquad (5.30)$$

by substituting some of the components of $\mathbf{X}$ by their equivalent taken from $\mathbf{X}_\mathrm{e}$ in the expression of $\boldsymbol{\mathcal{F}}$, i.e. $X_i(t) \mapsto X_{\mathrm{e},i}(t)$ for some i, either everywhere in $\boldsymbol{\mathcal{F}}$ or in some of its components $\mathcal{F}_j$ only, thus achieving *total* or *partial* substitution, respectively. By construction, this coupling is obviously compatible with identical synchronisation.

In the following we shall not specify which kind of coupling we consider but just assume that $\boldsymbol{\mathcal{F}}_\mathrm{m}(\mathbf{X}, \mathbf{X}) \equiv \boldsymbol{\mathcal{F}}(\mathbf{X})$. The largest conditional Lyapunov exponent relative to trajectory $(\mathbf{X}, \mathbf{X}_\mathrm{e})$ is obtained from

$$\tfrac{\mathrm{d}}{\mathrm{d}t}\delta\mathbf{X} = \partial_\mathbf{X}\boldsymbol{\mathcal{F}}_\mathrm{m}(\mathbf{X}, \mathbf{X}_\mathrm{e})\,\delta\mathbf{X}, \qquad (5.31)$$

which explicitly depends on the trajectory $\mathbf{X}$ considered:

• *identical synchronisation*: we have $\mathbf{X}(t) \equiv \mathbf{X}_\mathrm{e}(t)$ by assumption. When $\lambda_\mathrm{max}^\mathrm{cond}$ is *negative* this regime is stable and can be observed in practice, provided that the initial condition has been properly chosen, i.e. sufficiently close to the initial state of $\mathbf{X}_\mathrm{e}$. This is called *strong synchronisation*. On the contrary, if it is *positive*, perturbations are amplified and $\mathbf{X}$ cannot stay synchronised with $\mathbf{X}_\mathrm{e}$.

• *arbitrary trajectory*: again two cases, either $\lambda_\mathrm{max}^\mathrm{cond}$ is negative or positive. When $\lambda_\mathrm{max}^\mathrm{cond} > 0$, the trajectory is unstable and nothing more can be said. In the opposite case, $\lambda_\mathrm{max}^\mathrm{cond} < 0$, perturbations to trajectory $\mathbf{X}(t)$ relax to zero. Following the previous interpretation of stability, the distance between two twin systems started approximately in the same way continue to behave so: they are synchronised between themselves but not to the driving. This is called *weak synchronisation*.

These definitions are thus best understood from the consideration of systems in the form

$$\tfrac{\mathrm{d}}{\mathrm{d}t}\mathbf{X}_\mathrm{e} = \boldsymbol{\mathcal{F}}(\mathbf{X}_\mathrm{e}), \qquad (5.32)$$

$$\tfrac{\mathrm{d}}{\mathrm{d}t}\mathbf{X}_1 = \boldsymbol{\mathcal{F}}_\mathrm{m}(\mathbf{X}_1, \mathbf{X}_\mathrm{e}), \qquad (5.33)$$

$$\tfrac{\mathrm{d}}{\mathrm{d}t}\mathbf{X}_2 = \boldsymbol{\mathcal{F}}_\mathrm{m}(\mathbf{X}_2, \mathbf{X}_\mathrm{e}), \qquad (5.34)$$

by wondering which situation prevails after transients have been eliminated: either $\mathbf{X}_\mathrm{e} = \mathbf{X}_i$, $i = 1, 2$ (strong synchronisation) or $\mathbf{X}_\mathrm{e} \neq \mathbf{X}_i$ and $\mathbf{X}_1 = \mathbf{X}_2$ (weak synchronisation) or $\mathbf{X}_1 \neq \mathbf{X}_2$ (no synchronisation).

5.4.1.3 Illustrating case study

Concepts introduced above do not depend on whether time is continuous or discrete. Here, let us consider a discrete time system:

$$X_e^{k+1} = \mathcal{F}(X_e^k), \tag{5.35}$$

$$X^{k+1} = (1-g)\mathcal{F}(X^k) + g\mathcal{F}(X_e^k), \tag{5.36}$$

with the coupling constant $g \in [0,1]$ and $\mathcal{F}$ is the logistic map

$$\mathcal{F}(X) = 4X(1-X). \tag{5.37}$$

Computed using (5.2), the corresponding Lyapunov exponent is $\lambda = \log 2$. By construction, identical synchronisation is a possible regime. The conditional Lyapunov exponent is obtained from $\delta X^k = (1-g)\mathcal{F}'(X^{k-1})\delta X^{k-1}$. When $X^k = X_e^k$ for all k, the computation is the same as for the forcing trajectory apart from the factor $(1-g)$. One then obtains $\lambda^{\text{cond}} = \log(1-g) + \lambda$, so that strong synchronisation is thus achieved as long as $\lambda^{\text{cond}} < 0$, i.e. $2(1-g) < 1$, hence $g > 1/2$. When $g < 1/2$ strong synchronisation is no longer possible and one must perform the numerical computation for a trajectory X^k obtained by simulating system (5.35–5.36) starting from an initial condition (X_e^0, X^0) with $X^0 \neq X_e^0$. The result is displayed in Figure 5.21: As g is decreased below $g_{\text{is}} = 1/2$, the conditional Lyapunov exponent remains negative down to $g_{\text{gs}} \approx 0.33$ (threshold for generalised or weak synchronisation), value below which the coupling intensity is too weak to maintain some sort of organisation in the system.

Figure 5.22 illustrates typical situations in state space for twin systems coupled to the forcing one as in (5.32–5.34), and different values of g after elimination of a long transient. For $g = 0.10$, much smaller than the threshold of weak synchronisation, all three systems evolve chaotically and nearly independent of each others: traces of the coupling can just be found in the non-uniformity of the distributions of points. The tendency to weak synchronisation can be seen in the lower panel of the middle column where the probability to find $X_1^k \approx X_2^k$ increases rapidly as g_{gs} is approached from below ($g = 0.31$). When weak synchronisation takes place one gets $X_1^k = X_2^k$ for all k, here for $g = 0.42$. The last case, not represented, corresponds to strong synchronisation for $g > g_{\text{is}} = 1/2$ and $X_e^k = X_1^k = X_2^k$ for all k. Notice first that, since the logistic map (5.37) is chaotic, in all cases the state variables fluctuate wildly, whether synchronised, strongly, weakly or not at all, and that nothing dramatic occurs in the (X_e, X_1) planes (top line) as the threshold for generalised synchronisation is crossed.

210 *Instabilities, Chaos and Turbulence*

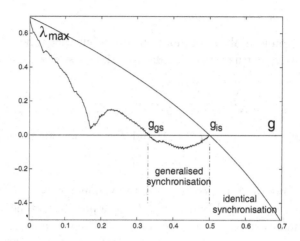

Fig. 5.21 Conditional Lyapunov exponent for the coupled logistic maps (5.35–5.36).

5.4.1.4 *Mutual coupling*

Up to now, we have considered one-way coupling, i.e. no feedback of the slave on its master, which is in line with control ideas. A more general case

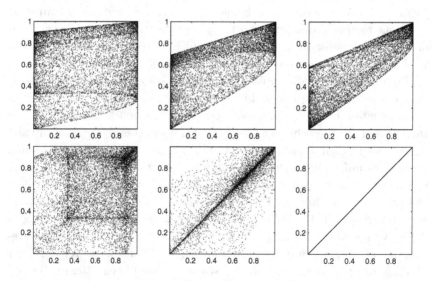

Fig. 5.22 Different projections of trajectories in state space $\mathbb{X}_e \times \mathbb{X}_1 \times \mathbb{X}_2$ for different values of the coupling constant g. From left to right: $g = 0.10$ (no synch.), $g = 0.31$ (no synch.), and $g = 0.42$ (weak synch.). Top line: Projections in the plane $\mathbb{X}_e \times \mathbb{X}_1$, Bottom line: Projections in the plane $\mathbb{X}_1 \times \mathbb{X}_2$.

is that of *two-way* or *mutual* coupling. All parts of the system are then reset to an equal footing, even if the couplings have not the same strengths and/or the systems not identical. This field is more a subject of general dynamical systems theory than of control properly speaking, and when the number of subsystems becomes large, it becomes a subject of statistical physics and network theory. Here we just give a few indications on some simple cases. Let us then consider an assembly of N coupled subsystems. The dimension of subsystem j being d_j, the dimension of the system is $d = \sum_{j=1}^{N} d_j$. The system

$$\tfrac{d}{dt}\mathbf{X}_j = \mathcal{F}_j(\mathbf{X}_j, \{\mathbf{X}_{j'}\}), \qquad j, j' = 1, \ldots N,$$

will admit $\mathbf{X}_1 = \mathbf{X}^{(j)}$, $j = 2, \ldots N$, as a solution if

$$\mathcal{F}^{(j)}(\mathbf{X}, \{\mathbf{Y} = \mathbf{X}\}) \equiv \mathcal{F}(\mathbf{X}).$$

A simple example is given by identical subsystems of dimension d defined as

$$\tfrac{d}{dt}\mathbf{X}_j = \mathcal{F}(\mathbf{X}_j) + C\sum_{j' \neq j}(\mathbf{X}_{j'} - \mathbf{X}_j), \tag{5.38}$$

where g is a coupling constant. This defines (5.38) as a *globally coupled system*. Another example is a ring of diffusively coupled systems:

$$\tfrac{d}{dt}\mathbf{X}_j = \mathcal{F}(\mathbf{X}_j) + g(\mathbf{X}_{j+1} - 2\mathbf{X}_j + \mathbf{X}_{j-1}), j = 1, \ldots N,$$

$j \pm 1$ being understood modulo N. By construction, identical synchronisation is then a possible solution and a dynamics governed by $\mathcal{F}$ develops on a manifold of dimension d called the *synchronisation manifold*.[33] The stability of the synchronised state is then analysed with respect to fluctuations in the $(N-1)d$ transverse dimensions. When stable, the corresponding attractor belongs to this manifold.

5.4.2 Delay systems and chaos control using delays

The basic idea is to try to stabilise a periodic regime with period T using a control signal that is a function of the state of the system at time $t - T$. Before presenting the method in more detail, let us examine the general properties of delayed systems.

---

[33]The manifold is defined by $N-1$ conditions $\mathbf{X}_1 = \mathbf{X}_j$, $j = 2, \ldots N$, each corresponding to d equations. In the state space of dimension Nd, $(N-1)d$ relations indeed define a manifold of dimension d.

5.4.2.1 Delay systems

Up to now we have considered systems, the evolution of which only depends on their states at the time t of interest. We now assume delayed interaction, that is to say a dynamics depending also on some other time in the past $t - T$:
$$\tfrac{\mathrm{d}}{\mathrm{d}t}\mathbf{X}(t) = \mathcal{F}\big(\mathbf{X}(t), \mathbf{X}(t - T)\big). \tag{5.39}$$
This simple assumption in fact makes the problem a much more difficult one. Let us indeed consider the simplest possible integration scheme, an explicit first-order Euler algorithm (Appendix B):
$$\mathbf{X}(t_{k+1}) = \mathbf{X}(t_k) + \delta t\, \mathcal{F}\big(\mathbf{X}(t_k), \mathbf{X}(t_{k-n})\big), \tag{5.40}$$
where $t_k = t_0 + k\, \delta t$ and $T = n\, \delta t$. In order to start the iteration, we have to specify an initial condition made of $n+1$ entries $\mathbf{X}(t_0), \mathbf{X}(t_{-1}), \ldots, \mathbf{X}(t_{-n})$. In addition to $\mathbf{X}(t_0)$, $\mathbf{X}(t_{-n})$ is needed to compute $\mathbf{X}(t_1)$, $\mathbf{X}(t_{-n+1})$ to compute $\mathbf{X}(t_2)$, and so on. Since n goes to infinity as δt goes to zero, given T, the problem is therefore infinite-dimensional. In practice, the number of effective degrees of freedom is however not infinite since the presence of the differential operator implies a temporal coherence with effects analogous to the spatial coherence introduced by space differentiation inherent in the description of dynamics in continuous media (§1.2, §6.4).

Let us consider the case of a scalar X function of time with (5.39) taken in the form
$$\tfrac{\mathrm{d}}{\mathrm{d}t}X(t) = -X(t) + \mathcal{F}\big(X(t - T)\big), \tag{5.41}$$
where the first term on the r.h.s. accounts for a linear relaxation of X towards its equilibrium value $X = 0$ with time constant $\tau = 1$. A typical delay forcing is the one proposed by Mackey and Glass[34] displayed in Figure 5.23.

The linear stability analysis of the steady states of (5.41), roots of $X_\mathrm{f} = \mathcal{F}(X_\mathrm{f})$, are obtained as usual by inserting $X = X_\mathrm{f} + \delta X$ in (5.41), and expanding the equation to first order in δX, which yields
$$\tfrac{\mathrm{d}}{\mathrm{d}t}\delta X = -\delta X + g\, \delta X(t - T) \tag{5.42}$$
with $g = \partial_X \mathcal{F}(X_\mathrm{f})$. Assuming $\delta X(t) = \delta X(0) \exp(st)$, one obtains that s must be a solution to
$$s = -1 + g\, \exp(-sT). \tag{5.43}$$

---
[34] M.C. Mackey, L. Glass: "Oscillation and chaos in physiological control systems," Science 197 (1977) 287–289. For a more general introduction with references, consult L. Glass, M. Mackey: "Mackey-Glass equation," Scholarpedia (2009).

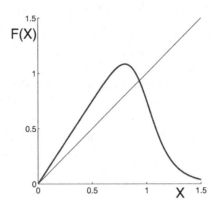

Fig. 5.23 The Mackey–Glass forcing term has been introduced to model some blood pathology: $\mathcal{F}(X) = aX/(1+X^\alpha)$, here with $a = 1.5$ and $\alpha = 10$.

Setting $s = s' + is''$, (5.43) reads

$$s' = -1 + g\exp(-s'T)\cos(s''T), \qquad (5.44)$$
$$s'' = -g\exp(-s'T)\sin(s''T). \qquad (5.45)$$

The reader is invited to show that for functions such as those depicted in Fig. 5.23 the nontrivial fixed point can be unstable against oscillatory modes ($s'' \neq 0$) when $g < -1$. The bifurcation points ($s' = 0$) are given by

$$1 = g\cos(s''T) \quad \text{et} \quad s'' = -g\sin(s''T),$$

which yields

$$s''T = \pm\arccos(1/g) + 2k\pi \quad \text{with} \quad s'' = \mp g\sin(\arccos(1/g)).$$

Assuming that the control parameter is the delay T (> 0), one gets a series of threshold values:

$$T_k = \frac{\arccos(1/g) + 2k\pi}{|g|\sin(\arccos(1/g))}. \qquad (5.46)$$

so that, upon increasing T, the first instability is obtained for $k = 0$. An approach paralleling the one developed in the previous chapter can be developed and a chaotic response is expected when the delay is large enough since this corresponds to an increasing number of linearly unstable degrees of freedom, typically for $T > T_3$ with four active modes in the system.

5.4.2.2 Control by delayed feedback (Pyragas' method)

As illustrated in § 4.3.2, chaotic regimes are characterised by the presence of unstable periodic regimes with different periods, in fact a countable infinity of such regimes. It is of course not possible to force a given system to follow an arbitrary chosen orbit at low cost but it is relatively easy to force it in order to stabilise one of its own unstable periodic orbits, if a dynamics sufficiently close to that orbit is of interest. In its simplest form, the method proposed by Pyragas[35] consists in introducing a delayed feedback with a delay time equal to the period of the system one is interested in.

Let us consider the autonomous system

$$\tfrac{\mathrm{d}}{\mathrm{d}t}\mathbf{X} = \mathcal{F}(\mathbf{X})\,, \tag{5.47}$$

an observable $\mathcal{W}(\mathbf{X})$, the time series of which, $W(t) = \mathcal{W}(\mathbf{X}(t))$, is recorded for control purposes and an unstable T-periodic orbit of interest $\mathbf{X}^{(0)}(t)$ ($\mathbf{X}^{(0)}(t+T) \equiv \mathbf{X}^{(0)}(t)$).

The control signal $f(t)$ is computed from the value of the observable $\mathcal{W}$ at time $t-T$. It has to cancel as long as the system follows the desired orbit and modify it in an appropriate way as soon as the trajectory drifts away. A simple choice is thus to take a correction proportional to $(W(t)-W(t-T))$. Formally, this correction is a functional defined on the state space:

$$f(t;T,K) = K[\mathcal{W}(\mathbf{X}(t)) - \mathcal{W}(\mathbf{X}(t-T))] \tag{5.48}$$

where K is a gain parameter. This choice is just a particularly simple form of a more general expression

$$f(t) = \int_0^\infty \phi(t')[\mathcal{W}(\mathbf{X}(t-t')) - \mathcal{W}(\mathbf{X}(t-T-t'))]\mathrm{d}t',$$

where ϕ is a kernel that averages the correction to be applied on a neighbourhood of time $t - T$, $\phi(t) = \delta(t)$ giving back the simple choice above. In fact an arbitrarily complex feedback can be chosen provided that $f \equiv 0$ when $\mathbf{X}(t) \equiv \mathbf{X}^{(0)}(t)$.

The action of the control signal usually amounts to a "force" exerted on the system that can be expressed as a modification of some component of the vector field $\mathcal{F}$. Our aim is here to develop a control strategy that is as independent as possible of the concrete system considered, So we formally write the modified system in the form

$$\tfrac{\mathrm{d}}{\mathrm{d}t}\mathbf{X} = \mathcal{F}_\mathrm{m}(\mathbf{X}(t), f(t;K,T))\,, \tag{5.49}$$

---

[35] K. Pyragas: "Continuous control of chaos by self-controlling feedback," Phys. Lett. A **170** (1992) 421–428.

where $f(t; K, T)$ is the control signal (5.48). Subscript 'm' being sufficient to recall that we deal with the modified system, f, K, and T will be omitted in the following.

By construction we have:

$$\tfrac{d}{dt}\mathbf{X}^{(0)} = \mathcal{F}_m(\mathbf{X}^{(0)}) \equiv \mathcal{F}(\mathbf{X}^{(0)})$$

since $\mathbf{X}^{(0)}$ is periodic with period T and $f \equiv 0$ when evaluated with $\mathbf{X}^{(0)}$. Assuming that the control is effective and that the controlled orbit stays close to the target orbit, one should have $\mathbf{X} - \mathbf{X}^{(0)}$ small and thus $|f|$ small, which permits the use of standard tools of linear stability analysis of periodic orbits so that the approach sketched in the previous section is recovered.[36] An important case is when the target orbit is unstable against period doubling, for which it can be shown that the gain K in (5.48) has to be large enough to achieve control ($K > K_{\min}$) but not too large ($K < K_{\max}$) so that the trajectory does not escape due to an oscillatory instability in the control scheme which, in addition, will work only when the instability rate of the target orbit is moderate enough. In order to control periodic orbits that are too much unstable, a possibility is to include delayed feedback terms at multiples of the period T of the target orbit, e.g. $f(t; T, K_1, \ldots K_M) = \sum_1^M K_m[\mathcal{W}(\mathbf{X}(t)) - \mathcal{W}(\mathbf{X}(t - mT))]$. In practice, this control method has many tricky limitations but it remains popular because it is usually easy to implement and does not require much information on the system to be controlled, especially not much pre-treatment of the data. For more technical details see Just's contribution in [Schuster (1999)].

5.4.3   The OGY method

We now consider the chaos control method introduced by Ott, Grebogi, and Yorke;[37] see the review by Grebogi and Lai in [Schuster (1999)]. This method basically assumes prior detailed knowledge of the system, either directly through its analytical definition or else from a good empirical reconstruction (§5.2). This makes it somewhat more difficult to implement than the previous method. On the other hand, control is obtained by monitoring the value of its control parameter, which seems a natural approach. Below, we first illustrate the principle of the method on an abstract simple case and next sketch the general problem.

[36] W. Just et al.: "Mechanism of time-delayed feedback control," Phys. Rev. Lett. **78** (1997) 203–206.

[37] T. Shinbrot et al.: "Using small perturbations to control chaos," Nature **363** (1993) 411–417.

5.4.3.1 *Principle of the method*

Let us consider a one-dimensional, discrete-time system (a map of the real line onto itself):
$$X_{k+1} = \mathcal{F}(X_k, r),$$
in the neighbourhood of an unstable fixed point X_* corresponding to a nominal value r_* of its control parameter r, that is $X_* = \mathcal{F}(X_*, r_*)$.[38] The aim of the control is to keep X_k as close as possible to X_*. So, let us assume that a trajectory closely approaches X_* at some time k, $X_k \simeq X_*(r_*)$. At the next step, if r is kept equal to r_*, X_{k+1} remains close to X_* but slightly further since the fixed point is unstable. The idea is to modify r away from r_* to make the iterate closer to X_*. Departures from (r_*, X_*) are supposed sufficiently small to permit Taylor expansions truncated just beyond first order. So iteration $k+1$ will be performed using a control parameter $r = r_* + \delta r$. This, in turn, changes X_* to $X_* + \delta X_*$. The control aims at finding δr that achieves $X_{k+1} = X_*$. Using $r = r_* + \delta r$, iteration $k+1$ moves the fixed point of interest to a new position:

$$X_*+\delta X_* = \mathcal{F}(X_*+\delta X_*, r_*+\delta r) = \mathcal{F}(X_*, r_*) + \delta X_* \partial_X \mathcal{F}(X_*, r_*) + \delta r\, \partial_r \mathcal{F}(X_*, r_*).$$

Using the fact that $X_* = \mathcal{F}(X_*, r_*)$ by assumption, we get:
$$\delta X_* = \frac{\partial_r \mathcal{F}(X_*, r_*)}{1 - \partial_X \mathcal{F}(X_*, r_*)} \delta r,$$
which can be more compactly denoted as $\delta X_* = (\partial_r X_*)\delta r$. Performing the iteration yields
$$X_{k+1} = X_* + \delta X_* + \partial_X \mathcal{F}(X_* + \delta X_*, r_* + \delta r)[X_k - (X_* + \delta X_*)]$$
which is required to be equal to the nominal value X_*. At lowest order, one obtains
$$X_{k+1} = X_* + (\partial_r X_*)\delta r + \partial_X \mathcal{F}_*[X_k - X_* - (\partial_r X_*)\delta r], \tag{5.50}$$
where $\partial_X \mathcal{F}_*$ and $\partial_r X_*$, short-hand notations for $\partial_X \mathcal{F}(X_*, r_*)$ and $\partial_r X_*(r_*)$ respectively, are the two parameters describing the effects of perturbations in state space and in control parameter space in the neighbourhood of the fixed point X_* of interest.

Imposing $X_{k+1} - X_* = 0$ immediately gives the control rule
$$\delta r = \frac{(X_k - X_*)\partial_X \mathcal{F}_*}{[\partial_X \mathcal{F}_* - 1]\partial_r X_*}. \tag{5.51}$$

---

[38] The same strategy is easily transposed to the case of a periodic point of order n, which is nothing but a fixed point of $\mathcal{F}^n$ (application of $\mathcal{F}$ repeated n times).

Since higher order terms have been neglected, iteration $k+1$ with correction term δr_{k+1} given by (5.51) above does not bring X right on X_* so that one can continue. Since the fixed point is unstable against infinitesimal disturbances, departures from it are unavoidable so that one has to continue indefinitely.

The control strategy is successful provided that the trajectory has visited the fixed point sufficiently closely. Since by assumption one deals with a chaotic system that wanders on its attractor in an ergodic fashion, sooner or later it will enter the neighbourhood of the fixed point and a criterion can be found to decide when algorithm (5.51) can be applied. In the presence of strong external noise, the system may however leave that neighbourhood. Application of the rule is then suspended until the trajectory comes back. The width of the neighbourhood where the algorithm is applied is a matter of optimisation: when it is too narrow, the time during which the system is not controlled may be large; when its is too wide, higher order terms neglected upon linearisation spoil the control.

5.4.3.2 Control of high-dimensional maps

The nontrivial features of the method arise as soon as one considers maps involving several state variables. To begin with, let us take a two-dimensional map

$$\mathbf{X}_{k+1} = \mathcal{F}(\mathbf{X}_k, r), \quad \mathbf{X} \in \mathbb{R}^2, \quad r \in \mathbb{R},$$

and study the stabilisation of trajectories around some unstable fixed point $\mathbf{X}_*$ for a nominal value r_* of the control parameter. The equation corresponding to (5.50) now reads

$$\mathbf{X}_{k+1} = \mathbf{X}_* + \partial_r \mathbf{X}_* \delta r + \partial_\mathbf{X} \mathcal{F}_*(\mathbf{X}_k - \mathbf{X}_* - \partial_r \mathbf{X}_* \delta r), \quad (5.52)$$

in which $\partial_\mathbf{X} \mathcal{F}_* \equiv \partial_\mathbf{X} \mathcal{F}(\mathbf{X}_*, r_*)$ is the Jacobian matrix of $\mathcal{F}$ computed at the nominal fixed point, and $\partial_r \mathbf{X}_* \equiv \partial_r \mathbf{X}_*(r_*)$ is a vector computed for $r = r_*$ that accounts for the displacement of $\mathbf{X}_*$ due to the control parameter change.

It is not possible to achieve $\mathbf{X}_{k+1} = \mathbf{X}_*$ with a single free parameter δr_k since this condition implies two equations, one for each component of $\mathbf{X}$. Let us denote the two eigenvalues of $\partial_\mathbf{X} \mathcal{F}_*$ as λ_u and λ_s. The unstable nature of the fixed point implies $|\lambda_\mathrm{u}| > 1$ and $|\lambda_\mathrm{s} < 1|$ and two manifolds, unstable and stable, that are one-dimensional each. The idea put forward by Ott et al., illustrated in Figure 5.24 consists in imposing that the iterate be projected on the stable manifold of the fixed point, whereas the

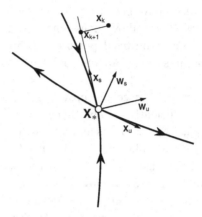

Fig. 5.24 Neighbourhood of an unstable fixed point and principle of the OGY control method.

convergence towards the fixed point is ensured by the contracting character of the dynamics along the stable direction. Requiring only one condition to be matched thanks to an adjustment of r, control is thus achieved along the unstable direction only.

In fact, the algebra is developed by just asking that the iterate $\mathbf{X}_{k+1}$ computed with $r + \delta r_k$ reach the stable linear subspace at $\mathbf{X}_*$ generated by $\mathbf{X}_s$ with eigenvalue λ_s, and not the stable manifold. In addition to $\mathbf{X}_u$ with eigenvalue λ_u, let us also define $\mathbf{W}_u$ orthogonal to $\mathbf{X}_s$, different from $\mathbf{X}_u$ since the Jacobian matrix $\partial_{\mathbf{X}}\mathcal{F}_*$ is presumably not self-ajdoint.[39] Writing that the projection of $\mathbf{X}_{k+1} - \mathbf{X}_*$ along $\mathbf{W}_u$ is zero leads to:

$$0 = \mathbf{W}_u \cdot [\partial_r \mathbf{X}_* \delta r + \partial_{\mathbf{X}}\mathcal{F}_*(\mathbf{X}_k - \mathbf{X}_* - \partial_r \mathbf{X}_* \delta r)]$$

that is,

$$\delta r = \frac{\mathbf{W}_u \cdot [\partial_{\mathbf{X}}\mathcal{F}_*(\mathbf{X}_k - \mathbf{X}_*)]}{\mathbf{W}_u \cdot [(\partial_X \mathcal{F}_* - \mathcal{I})\partial_r \mathbf{X}_*]}, \qquad (5.53)$$

where $\mathcal{I}$ is the identity operator.

In order to recover the form in which relation (5.53) is generally presented, we have to think in terms of row-vectors and column-vectors, or in terms of left-eigenvectors (the $\mathbf{W}_{u,s}$ transposed) and right-eigenvectors (the $\mathbf{X}_{u,s}$). The Jacobian operator can then be written as $\partial_{\mathbf{X}}\mathcal{F}(\mathbf{X}_*, r_*) = \lambda_u \mathbf{X}_u \otimes \mathbf{W}_u + \lambda_s \mathbf{X}_s \otimes \mathbf{W}_s$. Using the orthogonality properties of the $\mathbf{X}_{s,u}$ and

---

[39] Here, $\mathbf{W}_u$ is the eigenvector associated with eigenvalue λ_u of the adjoint to the Jacobian operator, i.e. its transposed $\partial_{\mathbf{X}}\mathcal{F}_*^t$ since we deal with a real vector space. In the same way, $\mathbf{W}_s$ associated with λ_s is orthogonal to $\mathbf{X}_s$. For a reminder, see §A.3, p. 387.

$\mathbf{W}_{s,u}$ to compute scalar products, one finds $\mathbf{W}_u(\lambda_u\mathbf{X}_u\otimes\mathbf{W}_u+\lambda_s\mathbf{X}_s\otimes\mathbf{W}_s) = \lambda_u\mathbf{W}_u$, so that one can write (5.53) in the form:

$$\delta r_k = \frac{\lambda_u \mathbf{W}_u(\mathbf{X}_k - \mathbf{X}_*)}{(\lambda_u - 1)\mathbf{W}_u \partial_r \mathbf{X}_*}. \tag{5.54}$$

Using $r = r_* + \delta r_k$ to compute $\mathbf{X}_{k+1}$ places the iterate close to the rail that leads to the fixed point, owing to the contraction induced by the tangent map $\partial_{\mathbf{X}}\mathcal{F}(\mathbf{X}_*, r_*)$ along its stable eigen-direction. The algorithm is then applied again and again to ensure control. If noise kicks the system out of the neighbourhood of the fixed point, application of the control is adjourned and the trajectory is computed using the nominal value of the control parameter r_* until the trajectory comes back close to the fixed point, which is bound to happen since it explores the attractor in an ergodic fashion. Since the stable manifold guides the trajectory towards the fixed point, it is possible to determine the relevant neighbourhood essentially in terms of the distance to its linear approximation as given by value of the scalar product $\mathbf{W}_u(\mathbf{X}_k - \mathbf{X}_*)$.

Let us now consider maps in dimension d larger than 2. On general grounds, the tangent space at the fixed point, can be decomposed into an unstable subspace and a stable subspace. If the unstable subspace is one-dimensional, nothing is changed since the condition of orthogonality to $\mathbf{W}_u$ defines a subspace with dimension $d - 1$. Corrections computed from (5.54) involving the scalar product $\mathbf{W}_u(\mathbf{X}_k - \mathbf{X}_*)$ again drive the system toward the nominal fixed point. In fact the proposed method can be seen as a special case of a general method of use in linear control theory, where (5.54) is replaced by a more general rule:

$$\delta r_k = -\mathbf{K}(\mathbf{X}_k - \mathbf{X}_*).$$

The linear dynamics around the fixed point can indeed be written in full generality as

$$\mathbf{X}_{k+1} - \mathbf{X}_* = \mathcal{A}(\mathbf{X}_k - \mathbf{X}_*) + \mathbf{B}\delta r_k,$$

where $\mathcal{A} = \partial_{\mathbf{X}}\mathcal{F}_*$ and $\mathbf{B} = \partial_r\mathcal{F}_*$. Linear control theory in state space introduced in §A.4, p. 391 tells us that the system can be stabilised, i.e. that one is able to find a time-independent feedback that damps out any small perturbation, when the matrix $[\mathbf{B}, \mathcal{A}\mathbf{B}, ..., \mathcal{A}^{d-1}\mathbf{B}]$ has maximal rank (d). One then must choose $\mathbf{K}$ so that

$$\mathbf{X}_{k+1} - \mathbf{X}_* = (\mathcal{A} - \mathbf{B}\mathbf{K})(\mathbf{X}_k - \mathbf{X}_*)$$

defines a convergent iteration (eigenvalues of the tangent map < 1 in absolute value).

5.5 Conclusions

As a whole, one can say that the problem of the transition to turbulence in strongly confined systems is, at least at a conceptual level but also most often at a practical quantitative level, well understood in terms of temporal chaos and a small set of universal scenarios.

This framework is, by construction, that of discrete systems introduced in Chapter 2. It can also legitimately be applied to continuous systems in case of strong confinement effects, by virtue of the distinction between driving and enslaved modes and the adiabatic elimination of the latter, thus offering a vast field where a detailed comparison of theory, tools and concepts, and laboratory or numerical experiments is meaningful.

The next chapter is devoted to the study of the case when confinement effects are no longer sufficiently strong to restrict the effective dynamics to the interaction of such a small number of discrete modes.

5.6 Exercises

5.6.1 *Fractals*

Apply formula (5.5) for a topologically connected set, usual line, surface, volume, and observe that it yields its topological dimension as expected. Then compute the fractal dimension of the following objects:

1) von Koch snow flake (Fig. 5.25): starting with a triangle with sides of unit length, replace each side by a broken line formed with 4 segments of

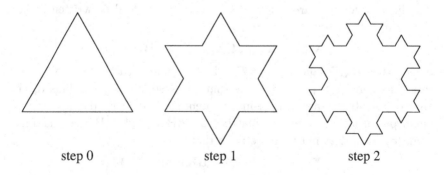

Fig. 5.25 von Koch snow flake.

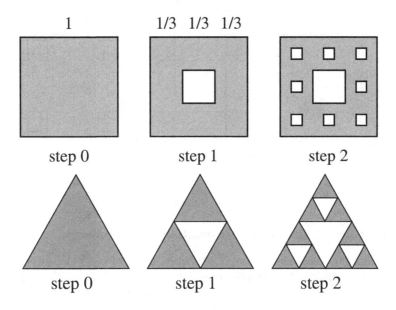

Fig. 5.26   Two-dimensional Sierpinski sets.

length 1/3 as shown in Figure 5.25 and repeat indefinitely the construction on each so obtained segment.

2) Sierpinski sets (Fig. 5.26): On the plane, start with the unit square and suppress the central square with side 1/3; repeat on each of the eight remaining squares; and so on. Same rule but starting with a triangle and dropping the central part. In three-dimensional space, start with a cube cut it into 27 cubes of side 1/3, suppress seven cubes, the six in the middle of the sides and the central one, repeat indefinitely to get a fractal sponge.

3) Cantor dust (Fig. 5.27): Take the square and keep the elements shown, repeat indefinitely. Observe that the fractal dimension can be an integer and that here the result could have been obtained by noticing that, at every step, the whole set is in one-to-one correspondence with a continuous interval by projection along the indicated direction.

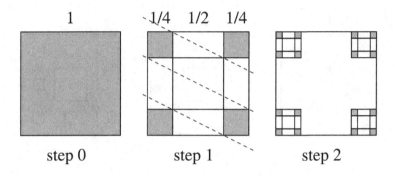

Fig. 5.27 Cantor dust.

5.6.2 *Curry–Yorke model (transition QP → chaos)*

Consider a two-dimensional map[40] expressed in terms of (X, Y) interpreted as the components of a complex number $Z = X + iY$, and defined in two steps:
(1) rotation of $Z_k = X_k + iY_k = |Z|_k \exp(i\varphi_k)$ by an angle φ:
$$\varphi_{k+1/2} = \varphi_k + \bar\varphi \quad (\text{mod } 2\pi)$$
and stretching of the modulus $|Z|_{k+1/2}$:
$$|Z|_{k+1/2} = (1+\eta)\log(1+|Z|_k)\,,$$
where parameter η controls the stability of the fixed point at the origin (unstable when $\eta > 0$).
(2) nonlinear transformation expressed in Cartesian coordinates:
$$X_{k+1} = X_{k+1/2}\,, \qquad Y_{k+1} = Y_{k+1/2} + X_{k+1/2}^2\,.$$
The model can be understood as the Poincaré map of a time-continuous system with a limit cycle bifurcating towards a two-periodic regime at $\eta = 0$ and may serve to illustrate the breakdown of a torus into a chaotic attractor. Draw the attractors obtained numerically for $\bar\varphi = 2$ (rotation $\simeq 114.6°$ close to strong resonance at $2\pi/3 = 120°$) and various values of $\eta > 0$.

[Answer: Fig. 5.28. For $\eta = 0.27$ the attractor is a smooth loop (section of a smooth torus corresponding to a two-periodic regime for the time continuous system). For $\eta = 0.48$, the attractor is a singular curve with fractal structure generated by stretching and folding in a regime of developed chaos. These corrugations appear around $\eta = 0.40$ which is thus close to the border of chaos.]

---

[40] J. Curry, J.A. Yorke, "A transition from Hopf bifurcation to chaos: computer experiment with maps on $\mathbb{R}^2$," Springer Notes in Mathematics **668** (1977) 48ff.

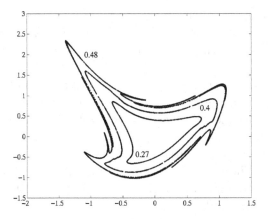

Fig. 5.28 The Curry–Yorke map $\bar{\varphi} = 2$ and several values of η. Component X (Y) in the horizontal (vertical) direction.

5.6.3 Permanent vs. transient chaos

Consider the one-dimensional so-called *tent map* $\mathcal{F}(X; \alpha)$ defined by

$$0 \leq X_k \leq 1/2 \qquad X_{k+1} = 2\alpha X_k ,$$
$$1/2 \leq X_k \leq 1 \qquad X_{k+1} = 2\alpha(1 - X_k) .$$

1) Suppose first $\alpha = 1$, draw the graph of $\mathcal{F}$ and check that interval $\mathbb{I}_0 \equiv [0, 1]$ is invariant in the sense that it contains its image by $\mathcal{F}$. Find the fixed points of the map and show that they are all unstable. Draw the graph of the iterated map $\mathcal{F} \circ \mathcal{F}$, i.e. $X \mapsto \mathcal{F}(\mathcal{F}(X))$. Find its fixed points and obtain from them a periodic orbit of $\mathcal{F}$ with period 2. Show that this orbit is also unstable. The orbit of an initial condition chosen at random is usually (with probability 1) chaotic. Compute its Lyapunov exponent using (5.2).

2) Suppose now that $0.5 < \alpha < 1$ and draw the graph of $\mathcal{F}$. Find the image $\mathbb{I}_1$ of $\mathbb{I}_0$ and the image $\mathbb{I}_2$ of $\mathbb{I}_1$, so that the definition interval of the map can be restricted to $\mathbb{I}_1$. Show that all trajectories are unstable by computing the Lyapunov exponent.

3) Consider the case $1 < \alpha$. The interval $\mathbb{I}_0$ is no longer invariant. Draw the corresponding graph of $\mathcal{F}$.

3a) Find the set $\mathbb{J}_1$ of initial conditions that escape at the first iteration. Determine the width of this interval as a function of α.

3b) Construct the pre-image $\mathbb{J}_2$ of this set, i.e. the set of points such that $\mathcal{F}(\mathbb{J}_2) = \mathbb{J}_1$. Notice that it is made of two disconnected parts $\mathbb{J}_2^1$ and $\mathbb{J}_2^2$. Find the total width of the set of initial conditions that escape in two iterations.

3c) Generalise the argument to trajectories escaping in k iterations. Assuming that the probability to have an initial condition in some interval $\mathbb{K}$ is proportional to its length, show that the probability of having a transient of length k is proportional to α^{-k}, i.e. transients are Poisson distributed.

3d) Observe that the initial conditions that do not escape belong to a Cantor set. When $\alpha = 3$, the construction rule is exactly that of the triadic Cantor set. From the formula, compute the Lyapunov exponent of a trajectory starting right on this set (and remains in it). Such an unstable invariant set is called a *chaotic repeller*.

5.6.4 Chaotic advection in periodic 2-d flows using maps

The flows considered here have abstract definitions that can be viewed as naive approximations of experimental situations. They have been extensively studied but remain of interest as training exercises.

The first case is the so-called *tendril-whorl* flow.[41] It is made of a discontinuous periodic succession of extensional flow and rotational flow, with period T that we re-scale to unity, $T_e = \alpha$ and $T_r = (1 - \alpha)$, being the times during which each type of flow is applied. The extensional part is written in Cartesian coordinates as:

$$\tfrac{\mathrm{d}}{\mathrm{d}t}X = v_x = -\epsilon X, \qquad \tfrac{\mathrm{d}}{\mathrm{d}t}Y = v_y = \epsilon Y, \qquad t \in [0, \alpha] \pmod 1 \quad (5.55)$$

and the rotational part in polar coordinates as:

$$\tfrac{\mathrm{d}}{\mathrm{d}t}R = v_r = 0, \qquad R\tfrac{\mathrm{d}}{\mathrm{d}t}\Theta = v_\theta = -\omega(R), \qquad t \in [\alpha, 1] \pmod 1 \quad (5.56)$$

Integrating (5.55–5.56) over a period is a trivial trivial matter once the initial conditions $X(t = 0)$ and $Y(t = 0)$ are given since variables are separated. Show that the Poincaré map is the compound of the extensional part $\mathbf{f}_e(X, Y) = (X/\beta, \beta Y)$ with $\beta = \exp(\epsilon\alpha)$ and the rotational part $\mathbf{f}_r(R, \Theta) = (R, \Theta + \Delta\Theta)$ with $\Delta\Theta = -(1 - \alpha)\omega(R)/R$. Authors in Note 41 have chosen $\omega(R)$ such that $\Delta\Theta = bR\exp(-R)$.

---

[41] D.V. Khakhar *et al.*: "Analysis of chaotic mixing in two model systems," J. Fluid Mech. **172** (1986) 419–451.

The second system is the *blinking vortex* flow introduced by Aref.[42] Around a vortex, the velocity field is given by

$$v_r = 0, \qquad v_\theta = \Gamma/2\pi r \qquad (5.57)$$

where the strength of the vortex Γ is just the circulation of the velocity around it. The flow generated by two such vortices separated by a fixed distance $2a$ is considered. It is further assumed that the the vortices are co-rotating (same Γ) and that they are 'on' and 'off' with a constant total period $2T$, blinking alternatively, i.e. during one half-period T, one is 'on' while the other is 'off' and conversely. The motion of a test particle in the flow amounts to a rotation of angle $\Delta\Theta$ around one or the other of the vortex cores alternatively, with $\Delta\Theta = \Gamma T/2\pi R_i^2$, where R_i is the distance of the particle to the active vortex. Trajectories thus depend on the initial position of the particle and a single parameter $\mu = \Gamma T/2\pi a^2$ (show it using a dimensional argument).

As a homework, study these systems using the tools of dynamical systems theory (determination of fixed and periodic points, linear stability analysis around those points, etc.) and trace a few trajectories. Consult references cited for more information.

---

[42] H. Aref, "Stirring by chaotic advection," J. Fluid Mech. **143** (1984) 1–21. See also the second part of the article mentioned in Note 41.

Chapter 6

Nonlinear Dynamics of Patterns

Chaos theory represents an important conceptual advance and offers an appropriate framework to account for experiments in confined geometry. However, as already pointed out in §1.4.2, p. 27, it becomes rapidly inapplicable when the effective dimension of the dynamics increases, which is the case for extended systems (large aspect ratio). As already indicated, eigenmodes are then quasi-degenerate with wavevectors typically such that

$$k_{n+1} - k_n = 2\pi/\ell \ll k_c,$$

ℓ being the lateral extension (p. 127). Many modes may be unstable close to the threshold. A large part of the interaction between them can be understood in terms of *linear* interference accounting for spatial modulations brought to a regular uniform reference state at the limit of a laterally unbounded system, e.g. parallel straight rolls for convection. At the *nonlinear* stage, focusing on the modulations directly leads one to the *envelope* formalism that gives a satisfactory account of confinement effects and defects at lowest order, a first step toward understanding *spatiotemporal chaos*.

For simplicity, we mostly limit this presentation to the case of cellular stationary instabilities. The case of dissipative waves, just briefly introduced in §6.4.4, is left to more advanced studies. We consider first, in §6.1, the determination of uniform quasi-one-dimensional structures, periodic in a single space direction. Then we generalise the theory to the case of quasi-two-dimensional patterns, with square or triangular/hexagonal plan-forms, in §6.2 and §6.3. Their most universal instability modes in terms of long wavelength perturbations are next investigated in §6.4. We conclude the chapter by §6.5 where we present some other modelling approaches that may help one analysing space-time chaotic regimes in extended systems.

6.1 Quasi-one-dimensional Cellular Structures

6.1.1 Steady states

Consider an instability mechanism which, like plain Rayleigh–Bénard convection, generates stationary structures locally periodic in a single space direction, say x. At the linear stage we have:

$$\mathcal{L}(\partial_x, \partial_t; R)\mathbf{V} = 0, \qquad (6.1)$$

the solutions of which are searched for in the form:

$$\mathbf{V} = \bar{\mathbf{V}} \exp(ikx + st), \qquad (6.2)$$

which leads to the dispersion relation:

$$\mathcal{L}(ik, s; R) = 0. \qquad (6.3)$$

We are particularly interested in the neighbourhood of the critical conditions $R \approx R_c$, corresponding to a marginal mode $k = k_c$ for which $\mathcal{R}e(s) = \sigma = 0$. ($\mathcal{I}m(s) \equiv 0$ along a stationary instability branch.) At threshold, we have $\mathcal{L}(ik_c, 0; R_c) = 0$, so that the linear problem (6.1) has a non-trivial solution:

$$\mathbf{V}_c = \bar{\mathbf{V}}_c \exp(ik_c x) + \text{c.c.}, \qquad (6.4)$$

where $\bar{\mathbf{V}}_c$ accounts for the structure of the critical normal mode and c.c. means 'complex conjugate'.

The nonlinear problem extending (6.1) reads:

$$\mathcal{L}(\partial_x, \partial_t; R)\mathbf{V} = \mathcal{N}(\mathbf{V}, \mathbf{V}), \qquad (6.5)$$

where $\mathcal{N}(\mathbf{V}, \mathbf{V})$ represents the higher order terms that were neglected in the linearisation procedure. The notation suggests a formally quadratic nonlinearity, as in hydrodynamics. The solution is searched for as an expansion in powers of a small parameter ϵ:

$$\mathbf{V} = \epsilon \mathbf{V}_1 + \epsilon^2 \mathbf{V}_2 + \epsilon^3 \mathbf{V}_3 \ldots, \qquad (6.6)$$

and, like in the case of nonlinear oscillators (§2.3.2.3), the control parameter is also expanded:

$$R = R_c + \epsilon R_1 + \epsilon^2 R_2 + \ldots \qquad (6.7)$$

Isolating the distance to threshold in the expression of the linear part on the l.h.s. of (6.5) we can write:

$$\mathcal{L} = \mathcal{L}_c - (R - R_c)\mathcal{M}, \qquad (6.8)$$

where $\mathcal{M}$ is the opposite of the formal derivative of $\mathcal{L}$ with respect to R evaluated at threshold. Inserting expansions (6.6, 6.7) into (6.5) and taking (6.8) into account, we get a series of linear problems:

$$\mathcal{L}\mathbf{V}_1 = 0 \tag{6.9}$$
$$\mathcal{L}\mathbf{V}_2 = R_1\mathcal{M}\mathbf{V}_1 + \mathcal{N}(\mathbf{V}_1, \mathbf{V}_1), \tag{6.10}$$
$$\mathcal{L}\mathbf{V}_3 = R_2\mathcal{M}\mathbf{V}_1 + R_1\mathcal{M}\mathbf{V}_2 + \mathcal{N}(\mathbf{V}_2, \mathbf{V}_1) + \mathcal{N}(\mathbf{V}_1, \mathbf{V}_2), \tag{6.11}$$

of the general form:

$$\mathcal{L}\mathbf{V}_k = \mathbf{F}_k. \tag{6.12}$$

The first problem is homogeneous. Since the critical conditions are fulfilled, it has a non-trivial solution $\mathbf{V}_1 \propto \mathbf{V}_c$. The higher order problems ($k > 1$) are all inhomogeneous and depend on the solutions computed at previous orders ($k' < k$). Unknown free quantities R_k introduced through (6.7) are fixed by the condition that the r.h.s. of (6.12) do not contain resonant terms, exactly like in the Poincaré–Lindstedt calculation, p. 67ff.

At a formal level, let the relevant scalar product be denoted as $\langle\ldots|\ldots\rangle$ and the adjoint $\mathcal{L}^\dagger$ to $\mathcal{L}$ defined by (§A.3):

$$\langle \mathbf{W}|\mathcal{L}\mathbf{V}\rangle = \langle \mathbf{V}|\mathcal{L}^\dagger\mathbf{W}\rangle^*,$$

the conditions (Fredholm alternative) fixing the unknown parameters in (6.7) read:

$$\langle \tilde{\mathbf{V}}|\mathbf{F}_k\rangle = 0.$$

where $\tilde{\mathbf{V}}$ is the solution generating the kernel of $\mathcal{L}^\dagger$, such that $\mathcal{L}^\dagger\tilde{\mathbf{V}} = 0$. See §A.3.2 for a reminder.

From (6.9) one gets $\mathbf{V}_1 \propto \mathbf{V}_c$ and when applied to (6.10), the condition for $k = 2$ yields R_1. Once this condition is fulfilled, a particular solution can be found, to which one adds an arbitrary solution to the homogeneous problem to obtain the most general solution. To fix this solution uniquely, one may ask it to be orthogonal to $\hat{\mathbf{V}}_c$:

$$\langle \mathbf{V}_c|\mathbf{V}_2\rangle = 0,$$

so that it appears as a true correction to the first order solution in the sense of the scalar product. Once $\mathbf{V}_2$ is determined, it is reported in (6.11), where the sole unknown on the r.h.s. is R_2, and so on.

When $R_1 \neq 0$, one can truncate the expansion at lowest order, which gives:

$$R = R_c + \epsilon R_1, \qquad \mathbf{V} = \epsilon \mathbf{V}_c,$$

and, after elimination of ϵ between the two equations

$$\mathbf{V} = \frac{(R - R_c)}{R_1}\mathbf{V}_c,$$

so that the bifurcation is in fact *two-sided*, the solution exists for both $R < R_c$ and $R > R_c$, and its amplitude varies linearly. This situation was encountered in Exercise 4.4.4, Eq. (4.48) describing what was called a *transcritical* bifurcation where two solutions exchanged their stability. This was shown to occur in the absence of '$A \mapsto -A$' symmetry.

In fact, it often happens that $R_1 = 0$ for symmetry reasons. This is the case of Rayleigh–Bénard convection with symmetric top/bottom boundary conditions within the Boussinesq approximation. The '$A \mapsto -A$' symmetry then results from the translation invariance by $\lambda_c/2$ in the direction perpendicular to the roll axis. When $R_1 = 0$, the lowest non-trivial truncation of the expansion is one order higher:

$$R = R_c + \epsilon^2 R_2, \qquad \mathbf{V} = \epsilon \mathbf{V}_c + \epsilon^2 \mathbf{V}_2,$$

and thus, neglecting $\epsilon^2 \mathbf{V}_2$ when compared to $\epsilon \mathbf{V}_c$ for ϵ sufficiently small, we obtain:

$$\mathbf{V} \simeq \pm\sqrt{(R - R_c)/R_2}\,\mathbf{V}_c$$

The bifurcation is now *one-sided*. Bifurcated states are to be found either for $R > R_c$ when $R_2 > 0$ (*supercritical*) or for $R < R_c$ when $R_2 < 0$ (*subcritical*).

Explicit calculation shows that Rayleigh–Bénard convection between good-conducting plates is supercritical. It is of course possible to continue the expansion and determine a more accurate solution by going to next order $k+1$ since everything is known at order k (R_{k-1}) or can be determined ($\mathbf{V}_{k-1}$) and that the compatibility condition contains only R_k as unknown.

6.1.2 Amplitude equations

We now turn to a variant of the same calculation that brings back a problem similar to the one studied in §4.1, reintroducing time in a way similar to the method of multiple scales, introduced in §2.3.2.4, p. 70.

We consider the emergent stationary dissipative structure and make it explicit that the bifurcation is supercritical by defining a new small parameter ε through:

$$R = R_c + \varepsilon^2. \qquad (6.13)$$

We again search the solution as an expansion:
$$\mathbf{V} = \varepsilon \mathbf{V}_1 + \varepsilon^2 \mathbf{V}_2 + \ldots, \tag{6.14}$$
but we no longer assume that it is time-independent. In order to account for this new feature we introduce a *slow* time scale t_1 in addition to the natural time scale that we now denote t_0 by setting:
$$\partial_t = \partial_{t_0} + \varepsilon^2 \partial_{t_1}. \tag{6.15}$$
The order of t_1 in ε results from the choice (6.13) and anticipates the fact that, close to the threshold, the growth rate of perturbations varies as $R - R_c$. Here, since the instability is stationary, the action of ∂_{t_0} is trivial.[1] Let us come back to (6.5) and expand also the ∂_t present in $\mathcal{L}$. With respect to system (6.9,...), in addition to the assumptions $R_1 = 0$ and $R_2 = 1$ inherent in (6.13) the first important modification enters Eq. (6.11) where a term ∂_{t_1} appears. We can thus rewrite it as:
$$\mathcal{L}\mathbf{V}_3 = \mathcal{M}\mathbf{V}_1 + \mathcal{N}(\mathbf{V}_2, \mathbf{V}_1) + \mathcal{N}(\mathbf{V}_1, \mathbf{V}_2) - \mathcal{Q}\partial_{t_1}\mathbf{V}_1, \tag{6.16}$$
where $\mathcal{Q}$ is the opposite of the operator obtained by differentiating $\mathcal{L}$ with respect to ∂_t formally.

The solution at order ε reads
$$\mathbf{V}_1 = A_1 \bar{\mathbf{V}}_c \exp(ik_c x) + \text{c.c.}$$
where A_1 is now a function of the slow variable t_1.

At order ε^2, (6.10) is left unchanged with the introduction of time. The compatibility condition that determined R_1 in the previous approach, is now trivially fulfilled by assumption. Now, non-resonant terms of the form $\exp(ink_c x)$ with $n = 0$ and $n = \pm 2$ are present in (6.10), stemming from the evaluation of $\mathcal{N}(A_1 \exp(ik_c x) + \text{c.c.}, A_1 \exp(ik_c x) + \text{c.c.})$. A particular solution can thus be found in the form:
$$\mathbf{V}_{2,\text{part}} = |A_1|^2 \mathbf{V}_{20} + (A_1^2 \exp(2ik_c x) \mathbf{V}_{22} + \text{c.c.}),$$
where the first subscript indicates the order in ε and the second one the harmonic generated by the nonlinear couplings. A solution of the homogeneous problem, $A_2 \bar{\mathbf{V}}_c \exp(ik_c x) + \text{c.c.}$, must be added to this particular solution in order to obtain the full solution at order ε^2.

At order ε^3, it is easily seen that (6.16) contains a certain number of resonant terms coming from the evaluation of $\mathcal{N}(\mathbf{V}_2, \mathbf{V}_1) + \mathcal{N}(\mathbf{V}_1, \mathbf{V}_2)$.

---

[1] Things would be different for an oscillatory instability, in which case we would have $\partial_{t_0} = -i\omega_c$, but the approach can easily be extended thanks to what we learned in Chapter 2.

Because $\mathbf{V}_{2,\text{part}}$ contains harmonics 0 and ± 2 and $\mathbf{V}_1$ harmonics ± 1, these terms contain harmonics 0 ± 1, i.e. ± 1, and $\pm 2 \pm 1$, producing ± 3 and ± 1. Other resonant terms come from $R_2 \mathbf{V}_1$ (with $R_2 = 1$ by definition), and $\mathcal{Q}\,\partial_{t_1}\mathbf{V}_1$. Instead of giving R_2 as in the previous approach, the compatibility condition now reads:

$$0 = A_1 \langle \tilde{\mathbf{V}} | \mathcal{M} \bar{\mathbf{V}}_c \rangle + \langle \tilde{\mathbf{V}} | \mathcal{N}(\mathbf{V}_2, \bar{\mathbf{V}}_c) + \mathcal{N}(\bar{\mathbf{V}}_c, \mathbf{V}_2) \rangle - \partial_{t_1} A_1 \langle \tilde{\mathbf{V}} | \mathcal{Q} \bar{\mathbf{V}}_c \rangle . \quad (6.17)$$

One gets easily convinced that the second term is of the form $|A_1|^2 A_1$ by counting the powers of $\exp(ik_c x)$, so that (6.17) effectively reads:

$$\tau_0 \partial_{t_1} A_1 = a A_1 - g |A_1|^2 A_1, \quad (6.18)$$

where τ_0, a, and g are constants that can be evaluated by computing the scalar products in (6.17). Returning to notations introduced in Chapter 3, especially in (3.23), p. 95, coefficient a can be identified with R_c^{-1} since the linear growth rate was defined there as $\sigma = \tau_0^{-1}(R - R_c)/R_c$.

For a stationary instability, it is easily shown that if the system is symmetrical under the change $x \mapsto -x$, then (6.18) has real coefficients since this symmetry implies symmetry under complex conjugation: $A \exp(ik_c x) \mapsto A^* \exp(-ik_c x)$. This is no longer the case for oscillatory instabilities and dissipative waves which are much more complicated in this respect, see §6.4.4 below.

The expansion can be continued. It is then observed that the freedom introduced by R_k in (6.6) for the computation of the time-independent solutions (6.7) is now replaced by the introduction of successive amplitudes A_k at each order. The Fredholm alternative now governs these A_k as functions of the slow time t_1.

At lowest order we have just $A = \varepsilon A_1$. Coming back to the natural time t using (6.15), observing that all terms of (6.18) are of order ε^3, we get:

$$\tau_0 \partial_t A = a(R - R_c) A - g|A|^2 A \quad (6.19)$$

called the *amplitude equation*. As announced, this result is in line with the argument developed at the beginning of Chapter 4 leading to (4.7), p. 131.

6.2 Dissipative Crystals

We now generalise the previous calculation for the case of systems that are isotropic in the (x, y) plane perpendicular to direction z singled out by the instability mechanism. The linearised problem now reads:

$$\mathcal{L}(\boldsymbol{\nabla}_\perp, \partial_t; R)\mathbf{V} = 0, \qquad \boldsymbol{\nabla}_\perp \equiv (\partial_x, \partial_y) \quad (6.20)$$

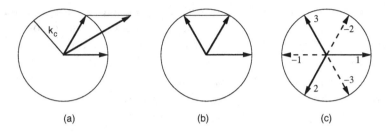

Fig. 6.1 Spatial resonance at second order: (a) Non-resonant combination $\mathbf{k} = \mathbf{k}_1 + \mathbf{k}_2$ with $|\mathbf{k}| \neq k_c$. (b) Resonant combination $\mathbf{k} = \mathbf{k}_1 + \mathbf{k}_2$ with $|\mathbf{k}| = k_c$. (c) Regular superposition of three wavevectors making a three-branch star at $2\pi/3$ forming a resonant set with their opposites (dashed arrows).

and, owing to the orientation degeneracy, its solution can be taken as a superposition of plane waves with wavevectors in different directions

$$\mathbf{V}_1 = \sum_{\mathbf{k}_j} \bar{\mathbf{V}}_{\mathbf{k}_j} \exp(i\mathbf{k}_j \cdot \mathbf{x}_h) + \text{c.c.} , \qquad (6.21)$$

with $|\mathbf{k}_j| = k_c$ for all j.

The expansion in powers of ϵ is performed like in §6.1.1 and again leads to (6.9, ...). Let us work sequentially as before. Though the equation at order ϵ^2 remains formally the same as in the one-dimensional case, the two-dimensional character of the problem implies novelty in the determination of resonant terms: nonlinearity $\mathcal{N}(\mathbf{V}_1, \mathbf{V}_1)$ that was then only able to generate the non-resonant harmonics 0 and 2, can now produce resonant combinations as shown in Figure 6.1(b).

We can now rewrite (6.10) in the form

$$\mathcal{L}\mathbf{V}_2 = R_1 \sum_{\mathbf{k}_j} \mathcal{M}\bar{\mathbf{V}}_{\mathbf{k}_j} \exp(i\mathbf{k}_j \cdot \mathbf{x}_h)$$
$$+ \sum_{\mathbf{k}_{j'}, \mathbf{k}_{j''}} \mathcal{N}(\bar{\mathbf{V}}_{\mathbf{k}_{j'}}, \bar{\mathbf{V}}_{\mathbf{k}_{j''}}) \exp(i(\mathbf{k}_{j'} + \mathbf{k}_{j''}) \cdot \mathbf{x}_h) , \quad (6.22)$$

so that when $\mathbf{k}_{j'} + \mathbf{k}_{j''}$ falls right on the critical circle, we must compensate this term with a term in $R_1 \neq 0$ for some well chosen $\mathbf{k}_j$. In the case of a formally quadratic nonlinearity, it is thus generically expected that solutions at order ϵ^2 exist in the form of a regular superposition of three wavevectors at angles $2\pi/3$ and their opposites, Figure 6.1(c). Such solutions bifurcate transcritically as already shown and only special circumstances can suppress them by cancelling the scalar products involving the last terms on the r.h.s. of (6.22) for symmetry reasons, e.g. the top-bottom symmetry in convection.

Let us suppose that the nonlinearity does not generate resonant terms at second order. The solution is still *a priori* made of a superposition of linear modes (6.21). So, let us consider two pairs of wavevectors $\pm\mathbf{k}_1$ and $\pm\mathbf{k}_2$ and start with:
$$\mathbf{V}_1 = \bar{\mathbf{V}}_{\mathbf{k}_1} \exp(i\mathbf{k}_1\cdot\mathbf{x}_h) + \bar{\mathbf{V}}_{\mathbf{k}_2} \exp(i\mathbf{k}_2\cdot\mathbf{x}_h) + \text{c.c.}\,.$$
The special solution at second order then formally reads:
$$\mathbf{V}_2 = \sum_{\pm} \mathbf{V}^{(\pm)}_{\mathbf{k}_1,\mathbf{k}_2} \exp(i(\mathbf{k}_1 \pm \mathbf{k}_2)\cdot\mathbf{x}_h) + \text{c.c.}\,.$$
Inserting these expressions in (6.11), we obtain terms with space dependence in the form:
$$\exp(i(\mathbf{k}_{j'} + \mathbf{k}_{j''} + \mathbf{k}_{j'''})\cdot\mathbf{x}_h) \quad \text{with} \quad \mathbf{k}_{j'}, \mathbf{k}_{j''}, \mathbf{k}_{j'''} = \pm\mathbf{k}_1 \text{ or } \pm\mathbf{k}_2\,,$$
so that we cannot avoid the generation of resonant terms through the relation:
$$\mathbf{k}_{j'} + \mathbf{k}_{j''} + \mathbf{k}_{j'''} = \mathbf{k}\,,$$
with $\mathbf{k}$ lying right on the critical circle, as shown in Figure 6.2.

The difficulty in the computation only comes from the fact that this relation can be fulfilled in many ways. Among all possible combinations, those involving a single pair of wavevectors are immediately identified, which were already present in the one-dimensional case. Other resonant combinations are readily discovered, Fig. 6.2(a), with nontrivial contribution to R_2 depending quantitatively on the angles made by the wavevectors pairs.

In general one is first interested in the coefficient $R_2^{(N)}$ associated with a superposition of N pairs of wavevectors forming angles of π/N between neighbours, Fig. 6.2(b). Apart from stability considerations to be examined in §6.3 below, a superposition with an arbitrary number of pairs can be considered. However, obtaining *simple* periodic patterns implies either $N = 1$ for *rolls*, $N = 2$ for *squares*, or $N = 3$ for *hexagons* as demonstrated for standard two-dimensional crystals.[2]

In general, a superposition with more than three regularly disposed pairs of wavevectors forms a *multi-periodic* pattern called a *quasi-crystal*. The latter may degenerate into a periodic *super-lattice* with large periods when certain commensurability relations are fulfilled. See Chapters 5 and 6 of [Rabinovich *et al.* (2000)]. This is in complete parallel with the locking phenomenon for quasi-periodic regimes studied in §4.2.3, p. 146. A natural extension of the temporal setting would suggest the existence of *chaotic crystals*.[3]

---

[2]See, Chapter 13 of L.D. Landau and E.M. Lifshitz, *Statistical Physics* (Butterworth–Heinemann, 1980).

[3]A.C. Newell and Y. Pomeau, "Turbulent crystals in macroscopic systems," J. Phys. A: Math. Gen. **26** (1993) L429–L434.

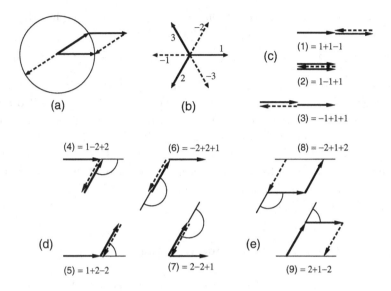

Fig. 6.2 Spatial resonance at third order: (a) An example of resonant superposition; (b) N pairs of wavevectors making an angle π/N between nearest neighbours, here with $N=3$. (c) Combinations with a single pair of wavevectors $\pm\mathbf{k}$ (rolls). (d, e) Combinations with two pairs of wavevector forming a parallelogram, degenerate (d) or not (e).

6.3 Short Term Selection of Patterns

What has just been said relates only to the existence of stationary nonlinear dissipative structures but the approach leading to the amplitude equation (6.19) for a roll pattern, characterised by a single complex amplitude A, can be reproduced for more complex patterns with a superposition of plane waves $\exp(i\mathbf{k}_j\cdot\mathbf{x}_\mathrm{h})$, with amplitudes A_j and their complex conjugates. Symmetry and resonance considerations (Exercise 6.6.1) lead to phenomenological amplitude equations generalising (6.19). Assuming formally cubic nonlinearity (or formally quadratic but such that relevant resonances at second order are killed for symmetry reasons), one obtains:

$$\tau_0 \partial_t A_j = a(R-R_\mathrm{c})A_j - g_0|A_j|^2 A_j - \sum_{j'} g_{jj'}|A_{j'}|^2 A_j \qquad (6.23)$$

where coefficients g_0 and $g_{jj'}$ respectively account for interactions of type (c) and (d, e) displayed in Figure 6.2.

When formally quadratic interactions do not kill the second order resonance, sets of three pairs of wavevectors, $\mathbf{k}_j$, $j=1,2,3$, such as in Figure 6.1(b) have to be considered with the corresponding amplitudes A_j,

$j = 1, 2, 3$. The resonance condition $\mathbf{k}_1 = -\mathbf{k}_2 - \mathbf{k}_3$ then implies the presence of a term $A_2^* A_3^*$ in the equation for A_1, which leads to:

$$\tau_0 \partial_t A_1 = a(R - R_{\rm c})A_1 - \tilde{g} A_2^* A_3^* - g[|A_1|^2 + \mu(|A_2|^2 + |A_3|^2)]A_1 \quad (6.24)$$

and two other equations obtained by circular permutation of the subscripts. Coefficients g, μ and $\tilde{g}$ in (6.24) are to be determined from a detailed nonlinear calculation or just introduced using a phenomenological argument.

Regular configurations correspond to specific fixed points of the amplitude equations (6.23) or (6.24) with $|A_j| = A$ for all j. Patterns selected by the nonlinearity can be discussed from the stability of these fixed points in the strictly temporal setting of Chapters 2 and 4. The case of squares is examined in Exercise 6.6.1.

6.4 Modulations and Envelope Equations

The previous analyses all referred to uniform cellular structures with wavelengths equal to the critical wavelength $\lambda_{\rm c}$. By assumption, confinement effects are weak in extended geometry and thus may prove unable to control the development of such *ideal* structures. *Natural* patterns that develop are therefore usually disordered, with local vectors $\mathbf{k}$ such that $|\mathbf{k}| \approx k_{\rm c}$ but with slowly variable lengths and/or orientations not everywhere perfectly aligned with the directions defining a unique underlying reference pattern. Defects can also perturb the regular ordering of the individual cells. An example from convection in a large Prandtl number fluid was displayed in Figure 3.10, p. 108. Imperfect patterns are usually called *textures*.

The problem is approached within the framework of *envelope equations* adding a spatial significance to the strictly temporal setting which, up to now, rules the amplitude of the *order parameter* accounting for the bifurcation. This allows the description of slow modulations to regular patterns and universal instabilities attached to them.

The derivation again starts with a solution to the nonlinear problem as a superposition of modes but with amplitudes that can be *slowly* varying in time and space, now called *envelopes*. The derivation rubs out the specificity of the primary instability, so that the result is expected to bear a universal content: all patterns with the same symmetries behave in the same way. Here we restrict ourselves to a heuristic approach[4] mostly based

---
[4] For technical details consult the seminal paper by A.C. Newell, "Envelope equations," Lectures in Appl. Math. **15** (1974) 157–163; or else: S. Fauve, "Pattern forming instabilities" in [Godrèche and Manneville (1998)]; A.C. Newell, Th. Passot, and J. Lega,

on the general properties of the linear *dispersion relation* in the neighbourhood of the threshold discussed earlier, §3.1.6, p. 95, completed by *symmetry* considerations. We consider first the case of a one-dimensional cellular instability before proceeding to several extensions.

6.4.1 *Quasi-one-dimensional cellular patterns*

At distance $r = (R - R_c)/R_c$ from threshold R_c, the growth rate of a normal mode with wavevector $k = k_c + \delta k$ is generically given by (3.26), i.e.:

$$\tau_0 \sigma = r - \xi_0^2 \delta k^2 \,. \tag{6.25}$$

Slightly above threshold ($0 < r \ll 1$) the wavepackets serving to build the modulated pattern are made of unstable wavevectors ($\sigma > 0$) in the band $k \in [k_c - \Delta k, k_c - \Delta k]$ with $\Delta k = \xi_0^{-1}\sqrt{r}$. The space modulations are thus slow when compared to variations at the scale of the wavelength λ_c ($\Delta k / k_c \ll 1$). Two time variables, t_0 and t_1, were introduced in order to obtain the amplitude equations. In the same way, two space variables are defined, a fast one and a slow one, x_0 and x_1, respectively. Like the slow time variable t_1, the slow space variable is related to the distance to threshold r by (6.25) from which we guess:

$$\partial_{x_1 x_1} \sim \delta k^2 \sim r \sigma \sim \partial_{t_1} \,.$$

A systematic expansion in powers of a small parameter ε should therefore rest on the assumptions:

$$r = \varepsilon^2, \quad \partial_t \mapsto \partial_{t_0} + \varepsilon^2 \partial_{t_1}, \quad \partial_x \mapsto \partial_{x_0} + \varepsilon^2 \partial_{x_1},$$

and

$$V = \varepsilon A_1(x_1, t_1) \exp(i k_c x_0) + \text{c.c.} + \ldots$$

where the distinction between the *carrier wave* at k_c and the *modulation* is made explicit.

The translation in physical space (slow x_1) is easily obtained by formally performing an inverse Fourier transform $i \delta k \mapsto \partial_{x_1}$ which leads to

$$\tau_0 \partial_{t_1} A_1 = A_1 + \xi_0^2 \partial_{x_1 x_1} A_1 \,. \tag{6.26}$$

After having unfolded the space dependence, we have to add the contribution of the nonlinearity previously obtained in (6.18). Back to the natural variables with $A = \varepsilon A_1$, (6.26) reads:[5]

$$\tau_0 \partial_t A = r A + \xi_0^2 \partial_{xx} A - g |A|^2 A \tag{6.27}$$

"Order parameter equations for patterns," Annu. Rev. Fluid Mech. **25** (1993) 399–453; [Cross and Hohenberg (1993)] or else [Rabinovich *et al.* (2000)].

[5] As already noticed, coefficient a in (6.18) is absorbed in the definition of r.

where it can be seen that each term is of order $r^{3/2}$, which is in fact the lowest significant order. This equation is nothing but (1.29) introduced p. 29 under the generic name of *Ginzburg–Landau equation*. Here we get the version with real coefficients; below, in §6.4.4, we shall find the complex version (1.30).

In the case of convection, Segel[6] has shown that one could account for lateral boundary effects inhibiting the instability mechanism by imposing:

$$A(x_{\rm b}, t) = 0 \qquad (6.28)$$

at the position of the lateral wall $x_{\rm b}$. In a semi-infinite medium, setting the origin at the wall, one easily determines the profile of the modulation by integrating the second order differential equation

$$\xi_0^2 \frac{{\rm d}^2}{{\rm d}x^2} A + rA - gA^3 = 0\,.$$

(Since the equation is invariant under the change $A \mapsto A \exp(i\varphi)$ one can choose the phase A that makes it real.) By identification one finds

$$A = A_0 \tanh(x/\xi\sqrt{2})$$

with $A_0 = \sqrt{r/g}$ and where $\xi = \xi_0/\sqrt{r}$, often called the *coherence length*, diverges in the vicinity of the threshold.

6.4.2 Two-dimensional modulations of quasi-1D patterns

Let us now proceed to several extensions of (6.27) and first consider the case of an instability that still favours rolls but in an effective two-dimensional medium which is rotationally invariant in its plane. An argument due to Newell and Whitehead[7] shows that, x being the direction of the local wavevector and y the perpendicular direction, (i) modulations along y are slow but faster than along x ($\partial_y \sim \mathcal{O}(r^{1/4}) \gg \partial_x \sim \mathcal{O}(r^{1/2})$ for $r \ll 1$) and (ii) rotational invariance implies the replacement of ∂_x by $\partial_x - (i/2k_{\rm c})\partial_{yy}$ in (6.27). This substitution leads to what is known as the Newell–Whitehead–Segel (NWS) equation:

$$\tau_0 \partial_t A = rA + \xi_0^2 \left[\partial_x + \frac{1}{2ik_{\rm c}}\partial_{yy}\right]^2 A - g|A|^2 A\,, \qquad (6.29)$$

in the original variables x, y, and t, and where each term is again easily seen to be of order $r^{3/2}$.

---

[6] L.A. Segel, "Distant side-walls cause slow amplitude modulation of cellular convection," J. Fluid Mech. **38** (1969) 203–224.

[7] A.C. Newell and J.A. Whitehead, "Finite bandwidth, finite amplitude convection," J. Fluid Mech. **38** (1969) 279–303.

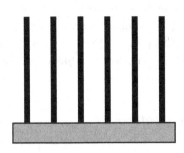
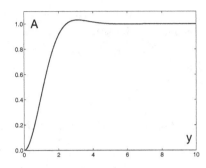

Fig. 6.3 Left: Pattern with rolls arriving perpendicular to a lateral wall. Right: Corresponding solution of (6.31).

The origin of the substitution can be understood by considering the operator $\left(\nabla_\perp^2 + k_c^2\right)^2$ from which periodic structures with space periodicity close to $\lambda_c = 2\pi/k_c$ emerge in media with rotational invariance in the (x, y) plane. Under Fourier transform, for a wavevector $\mathbf{k} = (k_c + \delta k_x)\hat{\mathbf{x}} + \delta k_y \hat{\mathbf{y}}$, at lowest significant order in $(\delta k_x, \delta k_y)$ one obtains:

$$\left[-(k_c + \delta k_x)^2 - \delta k_y^2 + k_c^2\right]^2 \simeq \left[-2k_c \delta k_x - \delta k_y^2\right]^2$$
$$= \left[(2ik_c)(i\delta k_x) + (i\delta k_y)^2\right]^2,$$

from which the indicated replacement derives by just performing the inverse Fourier transform $(i\delta k_x \mapsto \partial_x, i\delta k_y \mapsto \partial_y)$.

Brown and Stewartson[8] have shown that the boundary condition at a wall inhibiting the instability and perpendicular to the y direction (Figure 6.3, left) reads:

$$A(x, y_b, t) = 0 \quad \text{and} \quad \partial_y A(x, y_b, t) = 0 \tag{6.30}$$

The y-dependence of the envelope in the vicinity of the lateral wall, depicted in the right panel of the figure, has been obtained by numerical integration of the fourth-order differential equation

$$(\xi_0^2/4k_c^2)\frac{d^4}{dy^4}A = rA - gA^3 \tag{6.31}$$

with boundary conditions (6.30) at $y = 0$ and $A \to A_0 = \sqrt{r/g}$ for $y \to \infty$. The value of the second derivative at $y = 0$ is obtained by multiplying (6.31) with $\frac{d}{dy}A$ which can be integrated. The so-obtained first integral is then evaluated at $y \to \infty$ where all derivatives of A are zero, which fixes

---

[8] S.N. Brown and K.S. Stewartson: "On finite amplitude Bénard convection in a cylindrical container," Proc. R. Soc. Lond. A **360** (1978) 455–469.

$\frac{d^2}{dy^2}A$ at $y=0$. Writing (6.31) as a four-dimensional first-order system, one finally obtains the solution by integrating it as an initial value problem, by means of a shooting method in which the value of the third derivative of A at $y=0$ is the sole unknown initial condition to be adjusted so that $A(y) \to A_0$ as $y \to \infty$.

6.4.3 Quasi-two-dimensional cellular patterns

It is not difficult to extend the formalism to treat patterns with several pairs of wavevectors at a phenomenological level. We shall consider here only the case of a square pattern simply obtained by noticing that the x direction for one of the wavevectors is the transverse direction to the other and reciprocally. Combining results already obtained we get

$$\tau_0 \partial_t A_1 = rA_1 + \xi_0^2 \left[\partial_x + \frac{1}{2ik_c}\partial_{yy}\right]^2 A_1 - g(|A_1|^2 + \mu|A_2|^2)A_1, \quad (6.32)$$

$$\tau_0 \partial_t A_2 = rA_2 + \xi_0^2 \left[\partial_y + \frac{1}{2ik_c}\partial_{xx}\right]^2 A_2 - g(|A_2|^2 + \mu|A_1|^2)A_2. \quad (6.33)$$

When $\mu > 1$, the calculation developed in Exercise 6.6.1 shows that rolls are preferred locally.[9] Exercise 6.6.3 then shows that a system of rolls with a wavevector too far from k_c also becomes unstable against the formation of rectangles owing to the growth of a system of rolls with wavevector k_c but at right angles with it. This is the *cross-roll* instability, and the way it is shown to exist implies its *universal character*: all roll patterns in rotationally invariant media may experience it. On the other hand, owing to their local stability properties, roll systems at right angles may coexist in different but contiguous regions of space if their wavevectors are sufficiently close to k_c, forming a stationary *grain boundary*, Figure 6.4 (left). A grain boundary parallel to the y axis is obtained by solving (6.32, 6.33) in the special case $\partial_y \equiv 0$ which leaves one with a differential system in x. For example if both underlying wavevectors are equal to k_c, one gets:

$$0 = rA_1 + \xi_0^2 A_1'' - g(|A_1|^2 + \mu|A_2|^2)A_1,$$

$$0 = rA_2 - \frac{\xi_0^2}{4k_c^4} A_2'''' - g(|A_2|^2 + \mu|A_1|^2)A_2,$$

where the primes indicate differentiation with respect to x. The solution illustrated in Figure 6.4 (right) has here been rapidly and accurately obtained

---

[9] In essence, when $\mu < 1$, everything occurs as if $\mu = 0$, the two systems of rolls ignore each other and grow, so that squares are obtained everywhere in space; when $\mu > 1$, they cannot but feel each other and one kills the other, at least locally.

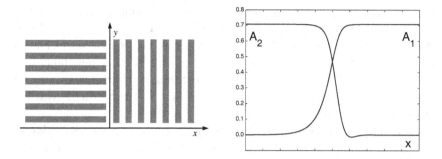

Fig. 6.4 Left: Roll system with a grain boundary. Right: Solution for the corresponding amplitudes.

by searching it as the asymptotic solution of a partial differential problem in x and t using a simple numerical scheme of the kind described in §B.2.1. The mathematical solution can however be obtained by analytical means (matched asymptotic expansions).[10]

The envelope formalism just introduced can account for the essentials of the dynamics of *scalar* textures such as those observed in convection at large Prandtl numbers. In particular, it explains the orientation of rolls at lateral boundaries, the presence of curvature and well ordered *grains* and other *defects* such as those present in Figure 3.10, p. 108. All this follows from the fact that the NWS equation (6.29) derives from a potential in the sense of §2.1.3, p. 47, that is to say, can be written as

$$\tau_0 \partial_t A = -\frac{\delta \mathcal{G}}{\delta A^*} \qquad (6.34)$$

where the right hand side is a notation representing the functional differentiation of

$$\mathcal{G}(A, A^*) = -r|A|^2 + \xi_0^2 \left| \left[\partial_x + \frac{1}{2ik_c} \partial_{yy} \right] A \right|^2 + \frac{1}{2} g|A|^4, \qquad (6.35)$$

with respect to A^*, with $g \in \mathbb{R}$, hence $\mathcal{G} \in \mathbb{R}$.

Functional differentiation in (6.34) is understood in the sense of *variation calculus*. In the change $A \mapsto A + \delta A$:
- variations of $|A|^2$ and $|A|^4$ expand as $A\,\delta A^* + A^*\delta A$ and $2|A|^2(A\,\delta A^* + A^*\delta A)$, which immediately gives the corresponding terms in (6.29) upon isolation of terms in δA^*;

---

[10] P. Manneville and Y. Pomeau: "A grain boundary in cellular structures near the onset of convection," Phil. Mag. A **48** (1983) 607–621.

– terms involving derivatives need a slightly more complicated treatment; for example, the variation of $\partial_x A\, \partial_x A^*$ gives $\partial_x A\, \partial_x \delta A^* + \partial_x \delta A\, \partial_x A^*$ but $\partial_x \delta A^*$ is not independent of δA^* so that an integration by parts has to be performed to isolate the latter, which yields $-\delta A^* \partial_{xx} A$ plus boundary terms; the other terms can be treated in the same way (integration by parts for the terms arising from $|\partial_{yy} A|^2$) so that all the partial derivatives of (6.29) with respect to space are also recovered;
– boundary conditions (6.28, 6.30) for rolls parallel or perpendicular to the lateral boundaries may be used to cancel boundary terms arising from integration by parts in case of a rectangular domain; otherwise contributions from the boundary term can simply be neglected when compared to bulk contributions in the general case.

The system then evolves so as to minimise the potential

$$G(t) = \int \mathcal{G}(A(x,y,t), A^*(x,y,t))\,\mathrm{d}x\,\mathrm{d}y$$

over the domain considered since $\frac{\mathrm{d}}{\mathrm{d}t} G \leq 0$. Using the same argument as in §2.1.3, one obtains that the solution that achieves this local minimum is time independent. On this basis, selected patterns are those corresponding to stationary solutions of (6.29) at local minima of $\mathcal{G}$ (local maxima or saddles are unstable solutions). Favoured textures can be found by comparing the contributions of the different causes of inhomogeneity (lateral walls, defects) to the potential increase with respect to the uniform solution.

In this context, the slow residual time dependence often observed in experiments can be attributed to higher order terms omitted in (6.29) which only holds at lowest significant order, at least in the scalar case. However, the interpretation of space-time chaos (weak turbulence) is at any rate more complex as soon as one leaves this simple framework, in particular for convection at low Prandtl numbers for which the most relevant field is no longer the temperature but the velocity.

6.4.4 Oscillatory patterns and dissipative waves

Up to now, we have considered only stationary instabilities. The approach in terms of envelope equations extends also in the more difficult case of oscillatory instabilities, $\omega_c \neq 0$, in particular for waves when $k_c \neq 0$. Difficulties arise from the fast time dependence, drastically destroying the relaxational property by introducing complex coefficients in the evolution equations, like in the strictly temporal setting of standard Hopf bifurcation, §4.2.1.2, p. 138, and not mildly by higher order corrections as above.

Here, we just give a few results practically without derivation, inviting the reader to consult references mentioned in Note 4. The primary role is again played by the linear dispersion relation introduced in §3.1.6, p. 95. Considerations about resonance developed in §4.1.3, are then used to deal with nonlinearity directly derived from the relevant *normal forms*.

For an oscillatory instability with $k_c = 0$ or for a dissipative wave with $k_c \neq 0$ but in a reference frame moving at the group velocity (p. 96), we have:

$$\tau_0 \partial_t A = (r - i\tilde{\alpha})A + \xi_0^2(1 + i\alpha)\partial_{xx}A - (g' + ig'')|A|^2 A.$$

This equation accounts for the space unfolding of the Hopf bifurcation (4.17) as derived from the dispersion relation of the corresponding instability (3.26, 3.27), pp. 95–96. Coefficient $\tilde{\alpha}$ is the critical pulsation in units of τ_0^{-1}, i.e. $\omega_c = \tilde{\alpha}/\tau_0$; it can be eliminated by changing to a 'rotating frame' $A(x,t) \mapsto A(x,t)\exp(-i\omega_0 t)$.

The next coefficient, α, is a measure of linear dispersion in units of ξ_0^2. Finally, β is the nonlinear dispersion coefficient re-scaled by g, which accounts for saturation effects from nonlinearity.

Upon further re-scaling of time, length, and amplitude, one obtains the universal form of the *complex Ginzburg–Landau equation* (CGL):[11]

$$\partial_t A = A + (1 + i\alpha)\partial_{xx}A - (1 - i\beta)|A|^2 A, \qquad (6.36)$$

already introduced as (1.30) in Chapter 1. This formulation is not restricted to the one-dimensional case introduced here but can be extended to higher dimensions, ∂_{xx} simply being replaced by the Laplacian when the physical system is isotropic in space. See Figure 1.13, p. 30, for illustrations of different regimes obtained in the two-dimensional case.

The structure of the equations governing one-dimensional waves propagating in opposite directions (complex envelopes $A_{1,2}$) can be obtained in the same way by symmetry and resonance considerations which lead to:

$$\partial_t A_1 + v_g \partial_x A_1 = A_1 + (1 + i\alpha)\partial_{xx}A_1$$
$$- [(1 - i\beta)|A_1|^2 + (\mu' + i\mu'')|A_2|^2]A_1,$$
$$\partial_t A_2 - v_g \partial_x A_2 = A_2 + (1 + i\alpha)\partial_{xx}A_2$$
$$- [(1 - i\beta)|A_2|^2 + (\mu' + i\mu'')|A_1|^2]A_2,$$

after appropriate re-scaling. Notice however that, since the first order partial derivative ∂_x is expected to be of order $r^{1/2}$ before re-scaling, the group

---
[11] W. van Saarloos, "The complex Ginzburg–Landau equation for beginners," in [Cladis and Palffy-Muhoray (1995)].

velocity must also be of order $r^{1/2}$ (i.e. sufficiently small), to ensure the consistency of the set of equations at order $r^{3/2}$.

Solutions with $A_2 = 0$ and $A_1 \neq 0$ (or the reverse) account for *propagating waves* moving to the right (or left), i.e. $\equiv A_1(x - vt)$ (or $\equiv A_2(x + vt)$). In contrast, condition $|A_1| = |A_2|$ describes a superposition of right and left waves forming a *standing wave*. A calculation analogous to that in Exercise 6.6.1 shows that the system prefers propagating waves when $|\mu'| > 1$ (one wave "feels" the presence of the other and "kills" it) and standing waves when $|\mu'| < 1$ (each wave does not "feel" the presence of the other much and therefore "accept" cohabitation).

When propagating waves are preferred, defects similar to grain boundaries may separate homogeneous domains with opposite kinds of wave in each. These defects are called *sources* or *sinks* according to whether the waves travel away from the defect or toward it.

6.4.5 Universal long-wavelength instabilities

One of the interests of the envelope formalism is to offer a framework for the study of universal secondary instabilities of dissipative structures linked to the symmetry properties of patterns at the limit of a laterally unbounded system. For stationary cellular structures, two such secondary modes are the *Eckhaus instability* against local compression/expansion and the *zigzag instability* against local torsion of the rolls, as sketched in Figure 6.5.

These two universal modes are called *phase instabilities* because they relate to modulations of the position of the cells. Within the envelope approach, the solution is searched for in the form $A = |A|\exp(i\phi)$, and while the modulus $|A|$ describes the intensity of the response to the primary instability mechanism, the phase serves to specify the absolute position of the pattern in the laboratory frame. The universal instability linked to the amplitude is the cross-roll instability mentioned earlier, p. 240, and further studied in Exercise 6.6.3. Contrasting with the amplitude that has a finite relaxation time, the phase is dynamically neutral as long as the solution is uniform.

A slightly irregular pattern can be described by a phase modulation which decays when the pattern is stable or gets amplified when it is unstable. Typically, at least for scalar dissipative structures, the phase of the envelope is governed by a diffusion equation:

$$\partial_t \phi = D_\| \partial_{xx} \phi + D_\perp \partial_{yy} \phi, \tag{6.37}$$

Fig. 6.5 Initial aspect of perturbation associated with the Eckhaus instability (left) and the zigzag instability (right).

where $D_\parallel$ and $D_\perp$ are diffusion coefficients along the wavevector of the structure, or perpendicular to it. As long as these coefficients are both positive, the phase perturbation relaxes and the roll pattern is stable against (infinitesimal) phase perturbations. The instability is observed when one of these coefficients changes its sign. The Eckhaus instability corresponds to $D_\parallel$ becoming negative, and the zigzag instability when it is $D_\perp$.

The phase diffusion coefficients depend on the wavelength of the underlying pattern. Exercise 6.6.2 is a first approach to the Eckhaus instability resting on (6.27), the zigzag instability would be studied in the same way using (6.29). The result is generally presented as a stability diagram in the parameter plane $(\delta k, r)$ where δk measures how the wavevector of the underlying pattern departs from the critical wavevector and r is the control parameter. Figure 6.6 displays the results at lowest significant order.

The Eckhaus instability is a *side-band instability* that develops "far from" k_c and close to the marginal stability curve, in a region where the instability mechanism is not very efficient and the amplitude of the solution, $A_0 = \sqrt{(r - \xi_0^2 \delta k^2)/g}$, is small. The system, which "prefers" a solution with a more optimal wavevector, evolves so as to amplify the phase modulation by compressing certain regions and expanding others (Fig. 6.5, left). Assuming a phase fluctuation in a pattern with an initial wavelength which is too short ($k_i \gg k_c$), expanded regions will have a more favourable local wavevector while the local wavelength will get even shorter in compressed regions. The primary mode will completely disappear at the location of the most compressed cells since its amplitude drops to zero on the marginal stability curve. The pattern will then lose a pair of rolls and after relaxation of the phase fluctuations, the average wavelength will be longer and thus more optimal $k_f < k_i$. If the obtained wavevector is still in the unstable

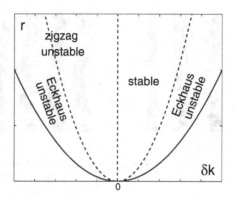

Fig. 6.6 Stability diagram of roll patterns against universal phase modes of compression/expansion (Eckhaus) and torsion (zigzag). As obtained from equation (6.29) valid at lowest order, the marginal stability curve is given by $r = \xi_0^2 \delta k^2$. Rolls are unstable against the Eckhaus mode in the region $\xi_0^2 \delta k^2 < r < 3\xi_0^3 \delta k^2$. The zigzag instability domain is for $\delta k < 0$. Stable rolls have wavevectors in the domain $0 < \delta k$, $r > 3\xi_0^3 \delta k^2$.

domain ($k_f \gg k_c$, the process repeats. If not, it will stop and the system will evolve toward a uniform pattern with a wavelength inside the stable domain.

The zigzag instability sets in when $\delta k < 0$, i.e. for wavelengths larger than critical. This can also be easily interpreted by noticing (Fig. 6.5, right) that, when measured perpendicularly to the local axis of the rolls, the wavelength is shorter than that measured along the x axis by a factor equal to the cosine of the angle between the local periodicity direction and the x axis. The instability again develops so as to make the (too long) initial wavelength shorter and thus closer to the critical value. In contrast with the case of the Eckhaus instability which changes the number of rolls, the zigzag instability saturates as it amplifies. The process ends in wide regularly spaced domains of 'zig' and 'zag' straight rolls connected by barrow bands of strong bending.

Up to now, we have only considered the initial development of these universal phase instabilities and just sketched the ultimate fate of the unstable state. Analytically, one can go a little further and complete the phase diffusion equation (6.37) by appropriate nonlinear terms. In this context, let us mention the *Kuramoto–Sivashinsky equation*[12] (KS) to be used in the numerical experiments described at the end of Appendix B. This equation

---

[12] Y. Kuramoto, "Phase dynamics of weakly unstable periodic structures," Prog. Theor. Phys. **71** (1984) 1182–1196.

describes the dynamics of the phase of nearly uniform solutions to the complex Ginzburg–Landau equation (6.36) in a narrow neighbourhood of the threshold of the *Benjamin–Feir* instability.

On general grounds this instability, which closely corresponds to the Eckhaus instability of one-dimensional stationary patterns, develops when the phase diffusion coefficient $D = 1 + \alpha\beta$ becomes negative (*Newell's criterion*, see Note 4. By symmetry considerations one can check that the equation governing the phase, at most quadratic in ϕ and up to order four in ∂_x, reads:

$$\partial_t \phi = D\partial_{xx}\phi - K\partial_{xxxx}\phi + g_0(\partial_x \phi)^2 + g_1(\partial_x \phi)(\partial_{xxx}\phi)$$
$$+ g_2(\partial_{xx}\phi)^2 + g_3(\partial_x \phi)^2(\partial_{xx}\phi), \qquad (6.38)$$

where each coefficient can be derived from those in (6.36). In addition to $D = 1 + \alpha\beta$, one gets[13] $K = \frac{1}{2}\alpha^2(1 + \beta^2) > 0$, $g_0 = \beta - \alpha$, $g_1 = 2g_2 = \alpha g_3 = -2\alpha(1 + \beta^2)$.

In practice, for $D \approx 0$ but negative, the order of magnitude of the space gradient is fixed by the competition between the two linear terms and all nonlinear terms except the first one can be neglected. After appropriate re-scaling, Equation (6.38) can be reduced to the KS equation which reads in universal form:[14]

$$\partial_t \phi + \partial_{xx}\phi + \partial_{xxxx}\phi + \tfrac{1}{2}(\partial_x \phi)^2 = 0 \qquad (6.39)$$

or, after differentiating it with respect to x and setting $\psi = \partial_x \phi$:

$$\partial_t \psi + \psi \partial_x \psi + \partial_{xx}\psi + \partial_{xxxx}\psi = 0. \qquad (6.40)$$

The nonlinearity $\psi \partial_x \psi$ present in this last expression is reminiscent of the advection term of hydrodynamics. It already appeared in the *Burgers equation* (Exercise 1.5.1, p. 35) producing shocks (Exercise 6.6.4, part 1) and the *Korteweg–de Vries* equation producing solitons (Exercise 2.5.6). The KS equation in one or the other form, (6.39) or (6.40), is a particularly simple model of *phase turbulence*.[15]

As long as D remains small, the modulus of the envelope is enslaved to the phase gradient, stays close to its nominal value A_0, and therefore

---

[13] J. Lega, *Défauts topologiques associés à la brisure d'invariance de translation dans le temps*, PhD Dissertation, Nice University, 1989 (in French).

[14] The equation also appears in problems of front propagation as shown by G.I. Sivashinsky, "On self-turbulization of a laminar flame," Acta Astronautica **6** (1979) 569–591; hence the joint names for the equation.

[15] For a brief review with references, see: H. Chaté and P. Manneville, "Phase turbulence," in [Tabeling and Cardoso (1994)].

remains bounded away from zero. This is no longer the case when the phase instability is more intense. The field $|A|$ then revolts and recovers a dynamics of its own. It then explores a larger range of values that extends down to zero. At places where $|A| = 0$, the phase is no longer defined and phase defects nucleate, with 2π-jumps of ϕ. The CGL equation then enters new regimes whose precise nature depends on the value of α and β in (6.36) and the space dimension. Strong space-time chaos called *defect mediated turbulence* then sets in. In two dimensions, it is characterised by the permanent birth of defects in pairs, that further dissociate and move around before merging.[16]

The theory of the transition to chaos is rooted in the idea of dimensional reduction leading to effective dynamical systems in terms of ordinary differential equations. To summarise this section, one can say that the envelope formalism is the required adaptation of this idea when confinement effects are too weak to legitimate the approach in terms of isolated modes and corresponding discrete amplitudes. Pattern selection strictly relies on this reduction for uniform solutions.

Defects and universal secondary instabilities involve modulations to some ideal reference situation. As long as the system remains sufficiently close to a stable regularly ordered pattern, the relevant instability modes relates to the phase of envelope and a supplementary reduction is possible by adiabatic elimination of the envelope modulus. Localised defects also play a role, either because they are present in the initial conditions (when the pattern emerges) or as the result of secondary instabilities that do not saturate. Sometimes tools borrowed from the theory of dynamical systems can still be used, e.g. to determine special solutions to the envelope equations.

It should also be noticed that, while the envelope formalism can be made rigourous within the framework of multiple scale methods, an interesting alternative is the derivation of generic models by phenomenological arguments resting on resonance and symmetry considerations. Further, the numerical simulation of such models has proved crucial to the understanding of space-time chaos. Such simulations usually do not require considerable investment and the reader is encouraged to practice them using the simple numerical methods presented in Appendix B, in order to get a personal

---

[16] P. Coullet, L. Gil, J. Lega, "Defect-mediated turbulence," Phys. Rev. Lett. **62** (1989) 1619–1622. A complete phase diagram is given in the one-dimensional case by H. Chaté, "Disordered regimes of the one-dimensional complex Ginzburg–Landau equation," in [Cladis and Palffy-Muhoray (1995)].

intuition of the problem while taking advantage of the published material listed in, e.g. [Rabinovich et al. (2000)] or to be found on the Internet.

6.5 What Lies Beyond?

Before closing this chapter, in line with the introductory section §1.4.2, p. 27, let us point out the interest of approaches to the modelling of extended systems by analogy, specially useful when the minimum space-time coherence necessary to apply the envelope formalism is absent from the system. As a matter of fact, when the space-time coherence is limited, there is some advantage to consider the continuous system as an aggregate of subsystems coupled to each other. The local dynamics is accounted for at the scale of the subsystem while the space extension arises through the coupling between the subsystems usually arranged at the node of a lattice.

A system can thus be discretised at several levels. At first, one can just discretise the physical space and get lattices of differential systems, e.g. lattices of Hopf oscillators or Lorenz systems. These systems are then coupled by some rule. For example, nearest-neighbour diffusive coupling of identical Hopf oscillators in one dimension would yield

$$\tfrac{\mathrm{d}}{\mathrm{d}t} Z_n = (r - i\omega) Z_n - g|Z_n|^2 Z_n + D(Z_{n+1} - 2Z_n + Z_{n-1}),$$

where subscript n indicates the space position of an individual oscillator and the discrete version of the diffusion operator, here $D\partial_{xx}$ is easily recognised.

Next step, time can also be discretized, which gives *coupled map lattices* already introduced as (1.31), p. 30. Time and space being indicated by superscript k and subscript n, respectively, in one dimension and again for diffusive coupling, one will start with

$$\mathbf{X}_n^{k+1} = (1 - 2D)\boldsymbol{\mathcal{F}}\left(\mathbf{X}_n^k\right) + D\left[\boldsymbol{\mathcal{F}}\left(\mathbf{X}_{n+1}^k\right) + \boldsymbol{\mathcal{F}}\left(\mathbf{X}_{n-1}^k\right)\right],$$

which is preferable to the seemingly more straightforward formulation:

$$\mathbf{X}_n^{k+1} = \boldsymbol{\mathcal{F}}\left(\mathbf{X}_n^k\right) + D\left[\mathbf{X}_{n+1}^k - 2\mathbf{X}_n^k + \mathbf{X}_{n-1}^k\right],$$

since it is immediately checked that the latter may not preserve the invariant domain of each subsystem while the former does (if $\mathbf{X} \in \mathcal{D}$ implies $\boldsymbol{\mathcal{F}}(\mathbf{X}) \in \mathcal{D}$, then $\mathbf{X}_n^k \in \mathcal{D}$ implies $\mathbf{X}_n^{k+1} \in \mathcal{D}$ for all n). The appropriateness of the model relies entirely on the skill of the modeller while choosing the local map and the type of coupling.[17]

---

[17] See for example [Kaneko (1993)] and for a specific application H. Chaté and P. Manneville, "Spatiotemporal intermittency" in [Tabeling and Cardoso (1994)].

Up to now, the local phase space was still a continuous set. One can make a last step by considering *cellular automata* where each subsystem can be in one of a finite set of states. Boolean automata have states labelled by bits 0 and 1, with evolution rules functions of the configuration of the neighbourhood of each cell. Such systems can already have complicated dynamics in spite of very simple definitions [Wolfram (1986)].

Lattice gas automata are automata specially adapted to hydrodynamics. They describe fluids at the level of individual molecules but governed by simplified evolution rules. "Living" at the nodes of a regular lattice, the molecules can be in one of a discrete set of motion states (speed and orientation), jump from node to node and change their state of motion when they meet at some given node according to rules given in collision tables. At first, when massively parallel computers began to appear, they were presented as an alternative to the direct numerical simulation of hydrodynamics equations. Nowadays, they are rather considered as useful models when the local dynamics is rich or complicated (chemical reactions), or when the boundaries have a complex topology or special physical properties (surface catalysis). An interesting review of lattice gas automata as applied to complex hydrodynamics is [Rothman and Zaleski (1997)].

The present chapter was mostly devoted to stationary patterns in weakly confined systems, with only a few words about waves. The latter will be central to the study of open flow instabilities in the next chapter.

6.6 Exercises

6.6.1 *Pattern selection*

Consider patterns described by a superposition of plane waves:

$$V = \sum_j \tfrac{1}{2}[A_j \exp(i\mathbf{k}_j \cdot \mathbf{x}_h) + \text{c.c.}],$$

namely, a roll pattern with a single pair of wavevectors and a square with two pairs at right angles. The wavelength of the participating modes is $\lambda_c = 2\pi/k_c$.

1) Show that translational invariance along wavevector $\mathbf{k}_j = k_c \hat{\mathbf{x}}_j$ (unit vector $\hat{\mathbf{x}}_j$), $x_j \mapsto x_j + x_j^{(0)}$, $j = 1, 2$, implies a (gauge) invariance for the corresponding complex amplitude $A_j \mapsto A_j \exp(i\phi_j^{(0)})$.

Taking these symmetries into account, show that the supercritical bifurcation towards a uniform pattern is governed at lowest order by

$$\tfrac{d}{dt} A_1 = r A_1 - g_{11}|A_1|^2 A_1 - g_{12}|A_2|^2 A_1 \tag{6.41}$$

$$\tfrac{d}{dt} A_2 = r A_2 - g_{22}|A_2|^2 A_2 - g_{21}|A_1|^2 A_2 \tag{6.42}$$

with $g_{ij} \in \mathbb{R}$. Use the rotational invariance to reduce the number of interaction coefficients to g (self-interaction) and $g' = \mu g$ (cross-interaction). Show that g can be eliminated by appropriate re-scaling of A_1 and A_2.

2) Determine the potential $G(A_1, A_1^*, A_2, A_2^*)$ from which (6.41, 6.42) derive in the sense of (6.34), here simply

$$\tfrac{d}{dt} A_j = -\partial G / \partial A_j^*$$

and conclude that permanent regimes are time-independent.

3) Find all fixed points (A_{1*}, A_{2*}) of system (6.41, 6.42) and, apart from the trivial solution, give their explicit expression and physical significance (take advantage of the gauge invariance to specify the phase of the envelopes so as to have real amplitudes).

4) Compute the value of the potential and study its curvature at the different fixed points to guess their stability properties. Then perform the explicit stability analysis by inserting $A_j = A_{j*} + a_j$ in the system and linearising. Even though the A_{j*} are real quantities, the most general perturbations are not, thus take $a_j = u_j + i v_j$. Show that the fourth order linear system obtained splits into two subsystems, producing two neutral modes (explain their origin) and two non-trivial modes that can be stable or unstable, according to the value of μ and the considered fixed point. Use the result to interpret phase portraits displayed in Figure 6.7. What happens when $\mu = 1$?

6.6.2 Eckhaus instability

Consider a roll pattern described by an envelope evolving according to:

$$\partial_t A = r A + \partial_{xx} A - |A|^2 A. \tag{6.43}$$

The Eckhaus instability relates to the phase ϕ of the complex field A.

1) Determine the amplitude $\tilde{A}^{(0)}$ of a *phase-winding* solution corresponding to uniform rolls with wavevector $k = k_c + \delta k$, $A(x,t) = \tilde{A}^{(0)} \exp(i \delta k\, x)$.

The stability of this solution against long wavelength perturbations (phase modes) is studied by assuming that the pre-factor of $\exp(i \delta k\, x)$ can be a function of space and time. From (6.43), derive the equation governing $\tilde{A}(x,t)$ defined through $A(x,t) = \tilde{A}(x,t) \exp(i \delta k\, x)$.

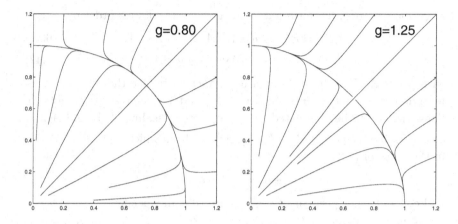

Fig. 6.7 Phase portrait of system (6.41, 6.42) for $\mu = 0.80$ (left) and $\mu = 1.25$ (right).

2) Insert $\tilde{A}(x,t) = \tilde{A}^{(0)} + a(x,t)$ in this new equation, and show that the linearised equations governing the real and imaginary parts of $a(x,t) = u(x,t) + iv(x,t)$ read:
$$\partial_t u = -2(r - \delta k^2)u + \partial_{xx} u - 2\delta k \partial_x v,$$
$$\partial_t v = 2\delta k \partial_x u + \partial_{xx} v.$$

Derive the dispersion relation for the Fourier normal modes with growth rate s and wavevector q taken in the form $u = \bar{u} \exp(st) \cos(qx)$ and $v = \bar{v} \exp(st) \sin(qx)$.

3) Show that the roots are real and that the phase-winding solution is unstable when
$$2(r - 3\delta k^2) + q^2 \leq 0,$$
derive from this that the instability occurs first for $q \to 0$ (long wavelength) and recover the result displayed in Figure 6.6.

6.6.3 Cross-roll instability

In the conditions of Exercise 6.6.1, assume $\mu > 1$ (rolls preferred to squares), consider a phase winding solution with $k_x = k_c + \delta k$, and study its stability against rolls at right angles with $k_y = k_c$.

From (6.32, 6.33), show that:
$$\tfrac{d}{dt} A_1 = r A_1 + \partial_{xx} A_1 - g(|A_1|^2 + \mu |A_2|^2) A_1,$$
$$\tfrac{d}{dt} A_2 = r A_2 - g(|A_2|^2 + \mu |A_1|^2) A_2,$$

is an appropriate starting point.

Show that the condition for instability has the same form as the Eckhaus condition but with a different pre-factor to be determined. Compute the value of μ that makes the cross-roll instability more dangerous than the Eckhaus instability and interpret the limit $\mu \to 1$.

6.6.4 Dynamical systems and nonlinear waves

6.6.4.1 The Burgers equation

Consider the Burgers equation:
$$\partial_t v + v \partial_x v = \nu \partial_{xx} v \,.$$

1) Show that it is invariant through a Galilean change of frame, i.e. that the equation for $\tilde{v}(\tilde{x}, t)$ with $\tilde{x} = x - Vt$ and $\tilde{v} = v - V$ is the same as that governing $v(x,t)$. (The corresponding symmetry of the NS equation is discussed in Note 10, p. 281.)

2) Derive the differential equation governing a solution that propagates at speed V without deformation ($\tilde{x} = x - Vt$ is the sole independent variable) and integrate it once with respect to x. Consider a solution such that $v_\pm$ for $\tilde{x} \to \pm\infty$, derive its speed V from the first integral just obtained, and derive the analytical shape of the corresponding solution.
[Solution: a hyperbolic tangent.]

6.6.4.2 The Kuramoto–Sivashinsky equation

Consider variant (6.40) of the KS equation:
$$\partial_t v + v \partial_x v + \partial_{xx} v + \partial_{xxxx} v = 0 \,.$$

1) Derive the linear dispersion relation governing infinitesimal perturbations $\delta v = \bar{\delta v} \exp(st + iqx)$ to the trivial solution $v \equiv 0$. Show the instability of those belonging to a range of q to be determined.

2) Write down the ordinary differential equation in $\tilde{x} = x - Vt$ governing a solitary wave solution propagating without deformation in a frame moving at speed V and derive a first integral. By identification, find a solution in the form
$$\tilde{v} = \alpha \tanh(\kappa \tilde{x}) + \beta \tanh^3(\kappa \tilde{x})$$
and compare the value of κ with the wavevector $q_{\max}$ of the perturbation with maximum growth rate $s_{\max}$.

6.6.4.3 *Flow down an inclined plane*

The Benney equation is a partial differential equation governing the long wavelength perturbations to a fluid film of uniform thickness flowing down an inclined plane. After appropriate re-scaling of length, time, and film thickness, one obtains[18]

$$\partial_t h + \tfrac{2}{3}\partial_x \left[h^3 + \left(\alpha h^6 - h^3\right)\partial_x h + h^3 \partial_{xxx} h\right] = 0$$

where α is the control parameter.

1) Derive the linearised equation governing infinitesimal perturbations to a time-independent uniform solution $h = h_0$. Show that solution $h \equiv h_0 = 1$ is linearly unstable for $\alpha > \alpha_c = 1$.

2) A hydraulic jump is a solution such that $h \to 1$ for $x \to -\infty$ and $h \to h_\infty$ for $x \to +\infty$. A solitary wave precisely corresponds to $h_\infty = 1$. Write down the ordinary differential equation in $\tilde{x} = x - Vt$ governing a solution that propagates without deformation in a frame moving at speed V, integrate this equation to find a first integral and determine the value K of this first integral as a function of c when $x \to -\infty$. Then derive the relation between V and $h_\infty < 1$.

---

[18] See: A. Pumir, P. Manneville, Y. Pomeau, "On solitary waves running down an inclined plane," J. Fluid Mech. **135** (1983) 27–50, and cited references.

Chapter 7

Open Flows: Instability and Transition

Convection studied in Chapter 3 is the prototype of systems becoming turbulent by steps when progressively driven far from equilibrium upon increasing the temperature gradient. All along this progression, the physical problem is clearly ruled by a mechanical cause, buoyancy-induced advection, counteracted by thermodynamic dissipation processes, viscous friction and thermal diffusion. Turning to open flows, one could thus try to stay within the same framework and start with the no-flow equilibrium situation, then consider weakly out-of-equilibrium *laminar* regimes mostly controlled by viscous dissipation, and further increase the shear[1] to observe the transition to *turbulence* after some cascade of instabilities. This approach can indeed be followed in some cases (e.g. wakes) but, in most situations, open flows turn out to be representative of strongly out-of-equilibrium systems in which mechanical evolution largely pre-empts the relaxation trends of thermodynamic origin.

This has been recognised early since the historical tradition gives the first place to the study of inviscid flows, for which velocity gradients are not rubbed out by viscous friction. A simple illustration of this change of perspective can be found in considering the damped linear oscillator $m\frac{d^2}{dt^2}X + \eta\frac{d}{dt}X + kX = 0$: the high-viscosity limit $\eta^2 \gg 4km$, yielding over-damped behaviour (stable node at $X = \frac{d}{dt}X = 0$) typical of thermodynamic relaxation, is not a good starting point to understand weakly damped oscillatory behaviour of inertial origin $\eta^2 \ll 4km$ (spiral point), whereas the ideal oscillator $\eta = 0$ (elliptic point) is more relevant.

Here, we restrict our attention to one-directional and time-independent base flows (§7.1). Inertial effects due to rotation or time dependence (e.g.

---

[1]The Reynolds number introduced in Chapter 1, p. 9, compares mechanical advection to supposedly stabilising viscous effects.

alternating with $\partial_t \equiv i\omega$) will not be considered. Linear stability theory (§7.2) was first developed for inviscid flows in the second half of the Nineteenth Century, pointing out the possibility of purely mechanical instabilities (§7.2.2), namely the *Kelvin–Helmholtz instability* of inflexional flow profiles, and *a contrario* the existence of mechanically stable flow profiles of more delicate study.

The inviscid case then served as a natural starting point for the study of the effects of viscous dissipation considered as a perturbation ($R < \infty$ but large). While viscosity tends to stabilise purely mechanical instabilities as expected, it can destabilise mechanically stable flows according to mechanisms that are not much intuitive, producing the so-called *Tollmien–Schlichting waves*. During the second third of the Twentieth Century, these mechanisms were elucidated by analytical means resting on asymptotic analyses in the limit $R \to \infty$, which gave the behaviour of the asymptotes to marginal stability curves and, to a limited extent, estimates of instability thresholds. Precise numerical results were obtained later using computers. We shall not consider this theoretical and numerical issue in detail but just present some results (§7.2.3) and redirect the interested reader toward specialised literature, the chapter by Huerre and Rossi in [Godrèche and Manneville (1998)] to begin with, and [Drazin (2002); Schmid and Henningson (2001); Drazin and Reid (1981); Betchov and Criminale (1967)], among others, for more detailed presentations.

Another source of difficulty is linked to the presence of the upstream-downstream flow of matter which, by the way, also transports the perturbations. The latter can thus be carried away and the instability said to be *convective*, or instead develop *in situ* in spite of global transport, in which case it is *absolute*.[2] The technical approach to this problem is delicate and relies on a mastery of complex variable analysis that goes much beyond the prerequisites for this course. Accordingly, we shall only introduce the subject at a qualitative level (§7.2.4).

After the description of primary instabilities we shall turn to the situation beyond threshold and examine the main physical scenarios of transition to turbulence, the understanding of which is still incomplete owing to specific difficulties. After an abstract presentation of the characteristics of secondary instability modes (§7.3.1), we shall focus on the most typical cases representing the types of flow pointed out at the linear stages:

---

[2]The terms *convective* and *absolute* are the accepted ones. We avoided to using the word "convective" all along Chapter 3 dealing with "convection" which is "absolute" in the absence of (sufficiently strong) through-flow.

the *wake* of a blunt body for mechanically unstable flows, and the *boundary layer* along a flat plate for mechanically stable flows. The case of the wake, which bifurcates toward the classical Bénard–von Kármán *vortex street* (see Figure 1.10, p. 21) through a supercritical Hopf bifurcation, is examined first (§7.3.2) since it can be approached with the same tools as those described earlier. Wall-bounded flows such as the boundary layer considered next (§7.3.3) are more difficult to understand, leaving us with still partially open problems. One known source of difficulties is that, at the Reynolds numbers relevant to the transition, the base flows may remain stable against delocalised infinitesimal disturbances, but become unstable against localised finite-amplitude perturbations. Turbulent spots arising from the breakdown of such perturbations can therefore coexist (in physical space) with laminar flow, see Figure 1.11, p. 22. The global stability of wall-bounded flows, the role of transient structures and the spatiotemporal competition between laminar and turbulent domains are thus examined in a separate section (§7.3.4) with special emphasis on the case of plane Couette flow known to be linearly stable for all Reynolds numbers.

7.1 Base Flow Profiles

7.1.1 *Strictly one-dimensional flows*

The *Navier–Stokes* (NS) equation reads:
$$\rho(\partial_t \mathbf{v} + \mathbf{v} \cdot \boldsymbol{\nabla}\mathbf{v}) = -\nabla p + \mu \nabla^2 \mathbf{v}, \tag{7.1}$$
where ρ is the density, p the pressure and μ the dynamic viscosity. It is completed by the *continuity* equation for an incompressible flow:
$$\boldsymbol{\nabla} \cdot \mathbf{v} = 0. \tag{7.2}$$
We consider only the case of constant density fluids so that we can divide (7.1) by ρ and introduce the kinematic viscosity $\nu = \mu/\rho$, which is the diffusivity of velocity fluctuations (homogeneous to $[L]^2[T]^{-1}$). The quantity p/ρ does not play a dynamical role and can be eliminated by appropriate analytical manipulations, i.e. applying the curl operator to (7.1) while taking (7.2) into account.

Let us examine the case of a time-independent one-directional flow whose direction defines the x axis. Here we denote u, v, and w the velocity components along x, y, and z, respectively. Introducing $\mathbf{v} \equiv u(x,y,z)\,\hat{\mathbf{x}}$ in the continuity equation, we simply get:
$$\partial_x u = 0, \tag{7.3}$$

i.e. u is independent of x but still a function of the transverse coordinates (y, z). The advection term $\mathbf{v} \cdot \nabla \mathbf{v}$ then cancels identically, while the equations for v and w simply read $\partial_z p = \partial_y p \equiv 0$, i.e. the pressure is a function of x only. Differentiating the x component of the NS equation with respect to x yields:
$$-\partial_{xx} p + \mu(\partial_{xx} + \partial_{yy} + \partial_{zz}) \partial_x u = 0$$
and, using (7.3),
$$\partial_{xx} p = 0,$$
so that, for a straight tube with length L, we obtain:
$$p(x) = p_0 - Gx \quad \text{with} \quad G = \Delta p / L,$$
where p_0 is the pressure at $x = 0$ and $\Delta p = p_{\text{up}} - p_{\text{down}}$ is the pressure head. The velocity component $u(y, z)$ is then a solution to the Poisson equation
$$\left(\partial_{y^2} + \partial_{z^2} \right) u = -G/\mu, \qquad (7.4)$$
with the usual no-slip boundary condition at the wall. Of course $u(y, z)$ depends on the shape of the tube's section. The case of a tube with circular cross-section is standard (Hagen–Poiseuille flow). The base flow is the axisymmetric solution to (7.4) in cylindrical coordinates:
$$\mu \left(\partial_{\varrho\varrho} + \varrho^{-1} \partial_\varrho \right) u + G = 0,$$
where ϱ denotes the radial coordinate. Integrating this equation, requiring the regularity of the solution on the tube axis $\varrho = 0$, and using the no-slip condition at $\varrho = \bar{\varrho}$, the interior radius of the tube, we get
$$u = U_{\max} \left(1 - (\varrho/\bar{\varrho})^2 \right) \qquad \text{with} \qquad U_{\max} = \frac{G \bar{\varrho}^2}{4\mu}. \qquad (7.5)$$
The classical expression of the flow rate is obtained by integrating over ϱ:
$$Q = \int_0^{\bar{\varrho}} u(\varrho) 2\pi \varrho \, d\varrho = \frac{\pi G \bar{\varrho}^4}{8\mu}.$$

For simplicity we further restrict ourselves to the consideration of plane base flows, i.e. flows that develop in channels such that the velocity is uniform in along one of the transverse coordinates, called *spanwise* and traditionally labelled z. The velocity profile u of course remains a function of the *cross-stream* coordinate y, i.e. $u = U(y)$. This configuration is depicted in Figure 7.1 (left). The profile of the flow driven by a pressure gradient G between two parallel plates moving along the x axis at speeds U_1 and U_2 is then given by the classical plane Couette–Poiseuille expression:
$$U(y) = -\tfrac{1}{2} G y^2 + b y + c,$$
where constants b and c are determined from the no-slip conditions:
$$U(y_i) = U_i, \qquad i = 1, 2.$$
This profile is displayed in Figure 7.1 (right).

7. Open Flows: Instability and Transition   259

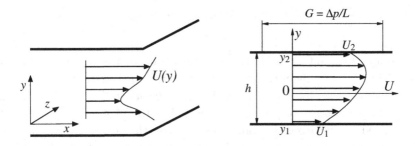

Fig. 7.1 One-directional flow between parallel planes. Left: Geometry and notation conventions for the axes. Right: Plane Couette–Poiseuille profile.

7.1.2 More general velocity profiles

The class of base profiles to be studied can be enlarged by relaxing the condition of strict parallelism. Let us, for example, consider a flow that is slowly evolving downstream. The continuity equation (7.2), i.e. $\partial_x u = -\partial_y v$, implies a slight divergence of the stream-lines: $v \neq 0$ but small as long as ∂_x remains small in a region where $\partial_y = \mathcal{O}(1)$. The boundary layer close to a wall, the plane jet, the wake downstream a body in a stream illustrated in Figure 7.2 are examples of nearly-parallel flows.

Taking into account the orders of magnitude of the fields u, v, and their x and y derivatives, the time-independent ($\partial_t \equiv 0$) two-dimensional ($\partial_z \equiv 0$) NS equation and continuity equations at lowest order read:

$$u\partial_x u + v\partial_y u = -\rho^{-1} dp_0/dx + \nu \partial_{xx} u, \qquad (7.6)$$

$$\partial_x u + \partial_y v = 0, \qquad (7.7)$$

where dp_0/dx is a possible applied pressure gradient (so-called *boundary layer approximation*). The solution is then searched by standard *similarity*

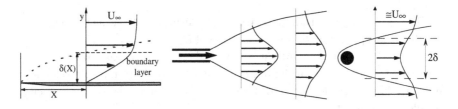

Fig. 7.2 Example of flows developing downstream. From left to right: The boundary layer, the jet, the wake.

variable methods (see [Tritton (1990)], Chapter 11) in the form:
$$u = u_{\max}(X)\bar{u}(y/\delta(X)) \tag{7.8}$$
where $u_{\max}(X)$ and $\delta(X)$ are the relevant scales for the streamwise speed and the transverse distances at the point with x-coordinate X.

Let us consider first the case of the *boundary layer* that develops along a wall in the absence of pressure gradient, i.e. the region where a stream with speed U_∞ far from the wall accommodates the no-slip condition. The thickness δ of the layer varies with the distance X from the leading edge. Considering the equation for u, it is immediately observed that the advection term, which is no longer strictly zero, must be compensated by the dominant dissipative term, hence:
$$u\partial_x u \sim \nu \partial_{yy} u. \tag{7.9}$$
The distance X from the leading edge is the only streamwise length scale available, while δ is obviously the relevant cross-stream scale at that position. Dimensionally, this yields:
$$\partial_x u \propto U_\infty/X, \qquad \partial_{yy} u \propto U_\infty/\delta^2,$$
and, upon insertion of these relations in (7.9):
$$\delta(X) = \sqrt{\nu X/U_\infty}. \tag{7.10}$$

The local velocity profile $u = U_\infty f'(\zeta)$, where $\zeta = y/\delta(X)$ is the similarity variable and the prime denote differentiation with respect to ζ, is obtained as the solution of a third-order nonlinear differential equation in ζ for the stream-function f:
$$ff'' + 2f''' = 0, \tag{7.11}$$
with boundary conditions $f = f' = 0$ at $\zeta = 0$, $f' \to 1$ for $\zeta \to \infty$. The corresponding *Blasius velocity profile* is displayed in Figure 7.3 (left). It starts with very weak curvature for $\zeta < 1$, and quickly tends to its asymptote for $\zeta \to \infty$.

The thickness Δ of the boundary layer is usually defined as the distance at which the speed equals a given fraction of U_∞, for example $\Delta_{0.99}$ defined by $u(\Delta_{0.99}) = 0.99\, U_\infty$ is about $5\,\delta(X)$, where $\delta(X)$ is defined by (7.10). Since 0.99 is somewhat arbitrary, other quantities with more physical significance have been defined in order to characterise boundary-layer velocity profiles experimentally:
$$\delta_1 = \frac{1}{U_\infty}\int_0^\infty (U_\infty - u(y))\,dy \quad \text{and} \quad \delta_2 = \frac{1}{U_\infty^2}\int_0^\infty (U_\infty - u(y))u(y)\,dy,$$

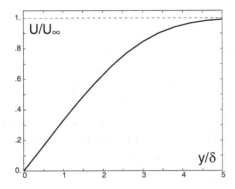 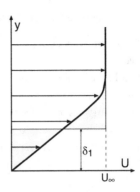

Fig. 7.3 Left: Blasius profile. Right: Sketch illustrating the definition of the displacement thickness δ_1 (shaded domains have the same surface).

called the *displacement thickness* and the *momentum thickness*, respectively. Figure 7.3 (right) illustrates the definition of the displacement thickness. For the Blasius profile, one finds $\delta_1 = 1.721\,\delta(X)$ and $\delta_2 = 0.664\,\delta(X)$.

In the case of a plane jet (Figure 7.2, middle), the maximum streamwise speed decreases as X increases due to spreading and flow-rate conservation, so that $u_{\max}(X)$ introduced in (7.8) now depends on X. The similarity argument leads to:

$$u_{\max} \propto M^{2/3}/(\nu X)^{1/3} \quad \text{and} \quad \delta \propto (\nu X)^{2/3}/M^{1/3},$$

where $M = \int_{-\infty}^{+\infty} u^2 dy$. Equation (7.11) is replaced by another slightly different equation which can be explicitly integrated yielding a stream-function $f \propto \tanh(\zeta)$. The base flow speed $u = f'$ accordingly reads:

$$u_{\text{jet}}(X) = \left(\frac{3M^2}{32\nu X}\right)^{1/3}\left\{1 - \tanh^2\left[y\left(\frac{M}{48\nu^2 X^2}\right)^{1/3}\right]\right\}. \tag{7.12}$$

The far wake of a blunt obstacle (Figure 7.2, right) can be modelled using a similar profile by considering the slowing-down behind it as a jet in the reversed direction corresponding to a *velocity defect*), i.e. a profile in the form:

$$u_{\text{wake}}(X) = U_{\min} + (U_\infty - U_{\min})\tanh^2(y/\delta),$$

where $U_{\min}$ is the minimal speed in the plane of the wake passing through the obstacle.

A similar approach can be developed for flows that are slowly time-dependent, e.g. a boundary layer along an impulsively started plate.

7.1.3 Extension to arbitrary profiles

Velocity profiles considered up to now were all controlled by viscous effects. However, on physical grounds, viscosity is expected to be dominant at low Reynolds numbers R only. Let us remind that, given a typical shear $\Delta U/\Delta \ell$ measured by a Reynolds number $R = \Delta U \Delta \ell/\nu$, the viscosity can impose its law only as long as the viscous time $\tau_\nu = \Delta \ell^2/\nu$ stays much shorter than the advection time $\Delta \ell/\Delta U$. This is precisely what makes the inviscid limit $\nu \to 0$ relevant, as already pointed out. As a first approximation, any velocity profile $U(y)$ can be thus accepted provided the Reynolds number is sufficiently large.

The stability theory that we are now ready to develop focuses on strictly parallel and time-independent flows. As applied to the extended class of flows considered above, its results are thus expected to give hints on what happens only when the predicted critical wavelengths and angular frequencies of the unstable modes are sufficiently large when compared to the imposed evolution scales in space and/or time.

7.2 Linear Stability

7.2.1 General framework

7.2.1.1 Setting

Let us consider a base flow $\mathbf{V}_0 \equiv \{(U(y), 0, 0); p_0(x)\}$ to which we add a perturbation $\mathbf{V} \equiv \{(u, v, w); p\}$, function of (x, y, z, t). The NS equations governing the perturbation are easily obtained by inserting $\mathbf{V}_0 + \mathbf{V}$ into (7.1, 7.2). These equations are written here directly in dimensionless form using the scales characterising the unperturbed profile $U(y)$, typically $\Delta U = U_{\max} - U_{\min}$ where $U_{\max} = \max_y U(y)$ and $U_{\min} = \min_y U(y)$ and $\Delta \ell = y_2 - y_1$ for a channel flow, or some appropriately defined reference thickness δ in other cases. The corresponding time and pressure scales are $\Delta t = \Delta \ell/\Delta U$ and $\Delta p = \rho(\Delta U)^2$. The Reynolds number is defined as $R = \Delta U \Delta \ell/\nu$ accordingly. The NS equations for the perturbation then read:

$$\partial_t u + [(U+u)\partial_x + v\partial_y + w\partial_z](U+u) = -\partial_x p + R^{-1}\nabla^2 u, \quad (7.13)$$

$$\partial_t v + [(U+u)\partial_x + v\partial_y + w\partial_z]v = -\partial_y p + R^{-1}\nabla^2 v, \quad (7.14)$$

$$\partial_t w + [(U+u)\partial_x + v\partial_y + w\partial_z]w = -\partial_z p + R^{-1}\nabla^2 w, \quad (7.15)$$

$$\partial_x u + \partial_y v + \partial_z w = 0. \quad (7.16)$$

Neglecting all terms quadratic in u, v, and w, we get

$$\partial_t u + U\partial_x u + v\partial_y U = -\partial_x p + R^{-1}\nabla^2 u, \quad (7.17)$$

$$\partial_t v + U\partial_x v = -\partial_y p + R^{-1}\nabla^2 v, \quad (7.18)$$

$$\partial_t w + U\partial_x w = -\partial_z p + R^{-1}\nabla^2 w, \quad (7.19)$$

$$\partial_x u + \partial_y v + \partial_z w = 0, \quad (7.20)$$

which can be formally written as

$$\mathcal{L}(\partial_t, \nabla; r)\mathbf{V} = \mathbf{0}. \quad (7.21)$$

At this stage, it can be useful to eliminate the pressure field and reduce (7.17–7.20) to a system of two equations for two unknowns. This reduction is the subject of Exercise 7.4.1 in which the perturbation is expressed in terms of the cross-stream velocity component v and the component of the vorticity normal to the boundary, $\Omega_y = (\nabla \times \mathbf{v})_y = \partial_x w - \partial_z u$.

7.2.1.2 Reduction to two dimensions

The solution to the linearised problem (7.21) is, as usual, sought for as a superposition of elementary modes in the form

$$\mathbf{V} = \tilde{\mathbf{V}}(y)\exp[i(\mathbf{k}_h \cdot \mathbf{x}_h - \omega t)], \quad (7.22)$$

where subscript "h" means "horizontal", i.e. in the plane of the flow. Physical perturbations would be obtained by taking the real part of these complex modes.

The relevance of the Fourier modes $\exp(i\mathbf{k}_h \cdot \mathbf{x}_h)$ is related to the fact that the problem is autonomous with respect to variables x and z. When studying convection, we took normal modes in the form $\exp(i\mathbf{k}_h \cdot \mathbf{x}_h + st)$, with real wavevectors $\mathbf{k}_h$ and a complex growth rate s, the real part of which accounted for the amplification/decay properties and the imaginary part for a possible oscillatory behaviour. Here the introduction of the time dependence as $\exp(-i\omega t)$ is a matter of convenience and the two formulations are equivalent when taking $\omega \in \mathbb{C}$. We see in particular that the real part of $\omega = \omega_r + i\omega_i$ now accounts for the oscillation while the imaginary part ω_i gives the growth rate ($s = \sigma - i\omega = -i(\omega_r + i\omega_i) = \omega_i - i\omega_r$). The pertinence of choice (7.22) comes from the fact that, in contrast with plain Rayleigh–Bénard convection producing stationary roll patterns, here we shall mainly have to deal with waves.

As before, the existence of a nontrivial solution to the stability problem (7.21) is subjected to a compatibility condition expressed as a relation between $\mathbf{k}_h$ and ω function of the control parameter R, the *dispersion relation* that can be formally written as

$$\mathcal{L}(-i\omega, i\mathbf{k}_h; R) = 0, \qquad (7.23)$$

where $\mathcal{L}$ is a scalar functional relation whereas $\mathcal{L}$ in (7.21) was a linear operator. Typically, the relation $\mathcal{L} = 0$ expresses the existence condition for nontrivial solutions to an algebraic homogeneous linear system, see later §7.2.2.3, p. 272 and Exercise 7.4.4.

In fact, relation (7.23) can be read in different ways without any assumption on the real or complex character of the wavevector and the angular frequency. The approach taken when assuming $\mathbf{k}_h$ real and ω complex is usually called *temporal* and was particularly relevant to the case of convection. The *spatial* approach assumes ω real and k complex, which is more appropriate to open flows since it allows one to describe the streamwise growth of perturbations in a flow forced by a time-periodic, spatially localised perturbation with externally controllable angular frequency ω_f, as currently done in experiments.

We shall come back to this issue in Section 7.2.4 but, for the moment, let us focus on the temporal approach. Noticing that we shall have to deal with waves, we introduce their (complex) phase velocity c and take normal modes (7.22) in the form

$$\mathbf{V}(x,y,z,t) = \{\tilde{\mathbf{v}}(y), \tilde{p}(y)\} \exp\left[i\big(k_x(x-ct) + k_z z\big)\right],$$

that we insert in (7.17–7.20). According to this definition, the modes evolve in time as $\exp(st)$ with

$$s = -i\omega = -ik_x c = -ik_x(c_r + ic_i) = k_x c_i - ik_x c_r,$$

so that the flow is unstable when $k_x c_i > 0$, hence $c_i > 0$ with $k_x > 0$ by convention.

At this stage, it is useful to take advantage of the Squire theorem, subject of Exercise 7.4.2. It stipulates that the most dangerous perturbations with wavevectors $\mathbf{k}_h = (k_x, k_z)$ and $k_x \neq 0$ are those with $k_z = 0$ and $w \equiv 0$. More precisely it shows that these *two-dimensional* modes are the most amplified ones in the inviscid case and have the lowest threshold when viscosity is taken into account.

For such reduced two-dimensional perturbations $(\tilde{u}, \tilde{v}, \tilde{p})$ with $k_x \equiv k$, one readily obtains the following simpler system

$$ik(U-c)\tilde{u} + \tilde{v}U' = -ik\tilde{p} + R^{-1}\left(\tfrac{d^2}{dy^2} - k^2\right)\tilde{u}, \qquad (7.24)$$

$$ik(U-c)\tilde{v} = -\tilde{p}' + R^{-1}\left(\tfrac{d^2}{dy^2} - k^2\right)\tilde{v}, \qquad (7.25)$$

$$ik\tilde{u} + \tilde{v}' = 0, \qquad (7.26)$$

where the prime denotes differentiation with respect to the cross-stream variable y, for example $U' \equiv dU/dy$.

7.2.1.3 Orr–Sommerfeld and Rayleigh equations

Two-dimensional incompressible flow are usually solved by introducing a stream-function ψ such that:

$$u = \partial_y \psi \quad \text{and} \quad v = -\partial_x \psi.$$

The continuity condition is then automatically fulfilled.

Dropping the tildes indicating perturbations, we introduce the normal modes as:

$$\psi(x, y, t) \mapsto \psi(y) \exp(ik(x - ct)) \tag{7.27}$$

in (7.24, 7.25), which leads to

$$ik(U - c)\psi' + (-ik\psi)U' = -ikp + R^{-1}\left(\tfrac{d^2}{dy^2} - k^2\right)\psi', \tag{7.28}$$

$$ik(U - c)(-ik\psi) = -p' + R^{-1}\left(\tfrac{d^2}{dy^2} - k^2\right)(-ik\psi). \tag{7.29}$$

The pressure is eliminated by differentiating (7.28) with respect to y, multiplying (7.29) by ik and subtracting the second equation from the first, which leads to the *Orr–Sommerfeld* equation:

$$(U - c)\left(\tfrac{d^2}{dy^2} - k^2\right)\psi - U''\psi = \frac{1}{ikR}\left(\tfrac{d^2}{dy^2} - k^2\right)^2 \psi. \tag{7.30}$$

This is a fourth-order differential equation in y, to which four boundary conditions must be applied, two at each end of the cross-stream domain, $y = y_1$ and $y = y_2$. The boundary conditions on ψ derive from those on the velocity perturbation at a solid boundary, i.e. $u = v = 0$, which gives:

$$0 = k\psi(y_1) = k\psi(y_2) = \psi'(y_1) = \psi'(y_2). \tag{7.31}$$

In infinite domains, these conditions are replaced by the requirement that ψ remains bounded as $y_1 \to -\infty$ and/or $y_2 \to +\infty$.

It is immediately remarked that the inviscid limit '$\nu \to 0$' is singular since when $R \to \infty$, the small parameter R^{-1} is a factor of the terms containing the fourth-order derivative. As a matter of fact, in this limit the Orr–Sommerfeld equation simply reads:

$$(U - c)\left(\tfrac{d^2}{dy^2} - k^2\right)\psi - U''\psi = 0, \tag{7.32}$$

and is called the *Rayleigh equation*. For this second order differential equation, the no-slip condition $u = 0$ is no longer applicable, but condition $v = 0$ remains, hence

$$k\psi(y_1) = 0 = k\psi(y_2). \qquad (7.33)$$

On general grounds, the solutions to the Orr–Sommerfeld equation are built on four elementary functions and those of the Rayleigh on only two. When R is sufficiently large, out of these four functions, two will derive from those found for the corresponding Rayleigh equation. The other two, of viscous origin, will remain localised in regions of strong velocity gradient.

7.2.2 Inviscid flows

Let us start with the Rayleigh equation (7.32) and assume that we have found a complex solution $(\psi; c)$. Since the equation has real coefficients, it follows that $(\psi^*; c^*)$ is also a solution. So, by complex conjugation, an unstable solution ($c_i > 0$) corresponds to every stable solution ($c_i < 0$), and reciprocally. This property comes from the time-reversal symmetry characteristic of the absence of dissipation. A sufficient stability condition is therefore the absence of solution with $c_i \neq 0$, so that we face three possibilities: either (7.32) has no solution other than the trivial one, or it admits only neutral modes ($c_i = 0$), or it has also unstable modes ($c_i \neq 0$). The problem is therefore to find the conditions under which the Rayleigh equation has unstable modes.

7.2.2.1 The Rayleigh theorem and related properties

A necessary condition for the existence of unstable solutions to Equation (7.32) is that the base flow profile displays inflection points (U'' passes through zero and change its sign).[3] The derivation of this result due to Rayleigh (1880) is a good example of a *global* method in the sense of §1.3, p. 9, and Exercise 2.5.2.

Let us assume the existence of an unstable mode ($c_i \neq 0$), then Equation (7.32) can be divided by $(U - c)$ that never cancels. This leads to:

$$\left(\tfrac{\mathrm{d}^2}{\mathrm{d}y^2} - k^2\right)\psi = \frac{U''}{U - c}\psi.$$

---

[3] Here we assume implicitly that U'' is not identically zero, which would be the case of plane Couette flow or, more generally, of linear-by-part velocity profiles, for which (7.32) has to be solved directly, see later p. 271.

Multiplying by ψ^* and integrating the result by parts over $[y_1, y_2]$ using boundary conditions $\psi(y_1) = \psi(y_2) = 0$, we get:

$$-\int_{y_1}^{y_2} \left(|\psi'|^2 + k^2|\psi|^2\right) dy = \int_{y_1}^{y_2} \frac{U''}{U-c} |\psi|^2 dy. \tag{7.34}$$

Expanding the right hand side yields:

$$\int_{y_1}^{y_2} \frac{U''}{U-c} |\psi|^2 dy = \int_{y_1}^{y_2} \frac{U''(U - c_r + ic_i)}{|U-c|^2} |\psi|^2 dy.$$

Next, separating real and imaginary parts, and focusing on the latter, we find:

$$c_i \int_{y_1}^{y_2} \frac{U''}{|U-c|^2} |\psi|^2 dy = 0.$$

It remains to be noticed that the assumption $c_i \neq 0$ implies $\int_{y_1}^{y_2}(\ldots) dy = 0$, which is possible only if the integrand goes through zero and changes its sign, and implies the same property for U'', hence the theorem. This is a necessary but *not* sufficient condition for instability.

Stated another way, the Rayleigh theorem stipulates that unstable flow profiles ($U'' = 0$ somewhere in the cross-stream domain) display an extremum of the vorticity $\Omega_z(y) = -U'$. A more refined condition is obtained by working with the real part of the integrated Rayleigh equation (7.34). Following Fjørtoft,[4] one can show that a monotonic velocity profile with an infection point may have unstable modes only if $|\Omega_z(y)|$ displays a maximum at the inflection point. This criterion, is derived in Exercise 7.4.3 and illustrated in Figure 7.4.

Other results can be shown about neutral and unstable modes, if they exist. For example the *Howard semicircle theorem* stipulates that the (complex) phase speeds of solutions to the Rayleigh equation, neutral or unstable, lie in the upper half-disk of the c-complex plane, centred on the real axis at $U_{av} = \frac{1}{2}(U_{min} + U_{max})$ with diameter $\Delta U = (U_{max} - U_{min})$, where $U_{min} = \min_y U(y)$ and $U_{max} = \max_y U(y)$, Figure 7.5. For a derivation by a global method similar to that for the Rayleigh theorem, consult [Schmid and Henningson (2001)], p. 23ff.

Given a neutral mode with phase velocity c_r it may be remarked that, since $U_{min} \leq c_r \leq U_{max}$ there is a point y_c in the flow where $U(y_c) = c_r$. At such a point, called the *critical level*, the coefficient of the second order derivative in the Rayleigh equation is zero, which makes it singular. This

---
[4] R. Fjørtoft, "Application of integral theorems in deriving criteria of stability for laminar flows and for the baroclinic circular vortex," Geofys. Publ. Oslo **17** (1950) 1–52.

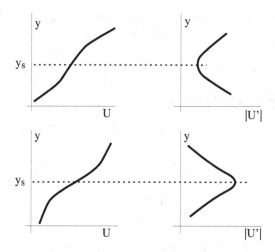

Fig. 7.4 Velocity profiles $U(y)$ (left) and vorticity profiles $|U'(y)|$ (right) corresponding to stable (top) and unstable (bottom) flows according to Fjørtoft criterion.

singularity expresses a particular resonance between the base flow and the perturbation, since the latter travels at exactly the same speed as the fluid particles at that level. We shall not enter a detailed discussion of this phenomenon, leaving it to more theoretically oriented works such as [Drazin and Reid (1981)] or [Schmid and Henningson (2001)], but just notice that it raises delicate analytical difficulties when y_c is different from the level y_s of the inflection point determined by solving $U''(y_s) = 0$. Instead, we now turn to a typical flow with an inflection point in its velocity profile, the *mixing layer* already mentioned in Chapter 1 (Figure 1.9, left).

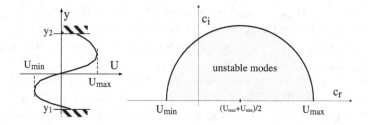

Fig. 7.5 Domain of unstable modes for an inviscid flow according to the Howard semi-circle theorem.

7.2.2.2 Kelvin–Helmholtz instability

When two flows merge at the end of a splitting plate, they form a *mixing layer* (Figure 7.6). Let U_1 (U_2) be the speed of the fluid when $y < 0$ ($y > 0$) and large and forget about the boundary layers that form on each side of the plate. At some distance of the trailing edge, viscosity smoothes out the velocity profile $U(y)$ that can be locally characterised by its vortical content. The vorticity thickness of the mixing layer is then defined as:

$$\delta_{\text{vort}} = \frac{U_2 - U_1}{U'_{\text{max}}}$$

where $|U'_{\text{max}}|$ is the absolute value of the maximum of the spanwise vorticity component $\Omega_z = -U'(y)$. Close to the end of the plate, δ_{vort} thickens as the square root of the downstream distance X (viscous diffusion).

At a given point X, the mixing layer can thus be seen as a quasi-parallel flow characterised by speeds U_1 and U_2, thickness δ_{vort}, and kinematic viscosity ν. Two dimensionless control parameters can be built with these quantities: the Reynolds number:

$$R = \frac{|U_2 - U_1| \delta_{\text{vort}}}{\nu}$$

and the velocity ratio:

$$\varrho = \frac{U_2 - U_1}{U_2 + U_1}.$$

Obviously, this flow is prone to a mechanical instability in the Rayleigh–Fjørtoft context. Locally, the profile is close to a hyperbolic tangent that can serve to model it analytically:

$$U(y) = U_{\text{av}} + \tfrac{1}{2}\Delta U \tanh(2y/\delta_{\text{vort}}).$$

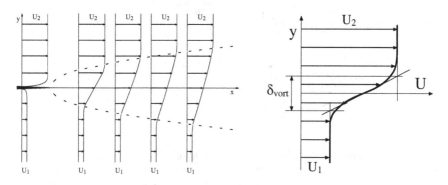

Fig. 7.6 Mixing layer and definition of the vorticity thickness.

The spanwise vorticity component is then

$$|\Omega_z| = \frac{\Delta U}{\delta_{\text{vort}} \cosh^2(2y/\delta_{\text{vort}})}.$$

In the limit $\nu \to 0$, which implies $\delta_{\text{vort}} \to 0$, we get $|\Omega_z| = \Delta U \, \delta_D(y)$, where $\delta_D(y)$ is the Dirac distribution. Stated another way, seen as an extreme case of inflexional profile, a velocity discontinuity can be interpreted as a vortex sheet.

The Kelvin–Helmholtz (KH) instability that we now study is typical of inviscid vortex sheets. Let us consider a base flow profile displaying a velocity discontinuity at $y = 0$:

$$\begin{aligned}
\text{region (1)} \quad y < 0, \quad U(y) = U_1 = U_{\text{av}} - \tfrac{1}{2}\Delta U, \\
\text{region (2)} \quad y > 0, \quad U(y) = U_2 = U_{\text{av}} + \tfrac{1}{2}\Delta U,
\end{aligned} \qquad (7.35)$$

and discuss the instability mechanism following Batchelor's approach [Batchelor (1967), p. 511], sketched in Figure 7.7:

The unperturbed vorticity sheet is represented by a regular distribution of equal intensity vortices in (a). In (b) a periodic infinitesimal modulation of the intensity is assumed, where points equivalent in a translation by one wavelength are labelled with the same letters and a higher local vorticity

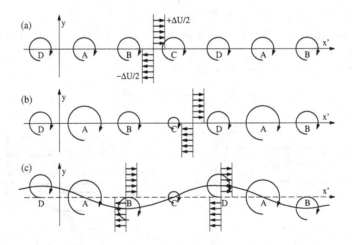

Fig. 7.7 Mechanism of the KH instability. The frame is moving at the average speed U_{av} (coordinate $x' = x - U_{\text{av}}t$) so that the flow profile displays a jump from $-\tfrac{1}{2}\Delta U$ for $y < 0$ to $+\tfrac{1}{2}\Delta U$ for $y > 0$. (a) Unperturbed vorticity sheet. (b) Modulation of the vorticity intensity. (c) Sinusoidal deformation of the interface and amplification of the modulation due to the advection of the vorticity by the base flow.

is represented by an oriented circle with a larger diameter. Intuitively, the differential effect of vortices at A and C is to move the interface at B and D in the directions indicated in (c). At lowest order the effect of the base flow is to carry vorticity along the flow, to the left when the interface has been moved downwards and to the right in the opposite case. Typically, at B it is carried from B toward A and at D from D toward A, thus increasing the intensity of the vortex at A and depleting the vortex at C. The initial modulation is thus increased, hence the instability.

All along this discussion, the velocity ratio ϱ does not show up since the mechanism is analysed in a frame moving at the average speed $U_{\text{av}} = \frac{1}{2}(U_2 + U_1)$. The physical conditions for the formation of the mixing layer have been, in some sense, evacuated. In practice, the trailing edge of the splitting plate has a fixed position in the laboratory frame and, accordingly, a *spatial stability analysis* is relevant (§7.2.4 below). When $\varrho \ll 1$, downstream transport is dominant and perturbations are amplified by the mechanism but are blown away so that the instability is expected to be *convective*: the flow plays the role of a *noise amplifier*. In the opposite case $\varrho \gg 1$, i.e. $U_2 \simeq -U_1$, the instability has time to develop on the spot before being evacuated, it is then *absolute*: the flow behaves as a genuine *self-sustained oscillator*. The precise value of ϱ at which the behaviour change takes place depends on the base profile. For the simplest continuous linear-by-part approximation, this happens at $\varrho = 1$ (Exercise 7.4.4) and for the smooth hyperbolic tangent profile at $\varrho \simeq 1.3$, consult Huerre and Rossi in [Godrèche and Manneville (1998), p. 169ff] for details and references to original work.

7.2.2.3 The Rayleigh equation with linear-by-part velocity profiles

Let us now take an analytical viewpoint on the KH instability, with just the assumption that it is legitimate to replace the actual base profile by a linear-by-part approximation, splitting the cross-stream interval $[y_1, y_2]$ into subintervals such that

$$U^{(j)}(y) = \alpha^{(j)} + \beta^{(j)} y \qquad \text{for} \qquad y \in [y_j, y_{j+1}].$$

The second derivative of such a profile is identically zero, except at discontinuity points where it is not defined. The Rayleigh equation then simply reads:

$$\left(\tfrac{\mathrm{d}^2}{\mathrm{d}y^2} - k^2\right)\psi = 0. \tag{7.36}$$

Over each subinterval, the solution to (7.36) is given by:
$$\psi^{(j)} = \sum_{(\pm)} A_{\pm}^{(j)} \exp(\pm kx). \tag{7.37}$$

In order to find the full solution, it remains to express boundary conditions (7.33) at y_1, y_2 and to match the different partial solutions (7.37) at the discontinuity points of U and/or U'. The matching conditions are obtained from the continuity of pressure in (7.24)

$$\left. \psi U' - (U-c)\tfrac{\mathrm{d}}{\mathrm{d}y}\psi \right|_j = \left. \psi U' - (U-c)\tfrac{\mathrm{d}}{\mathrm{d}y}\psi \right|_{j+1}, \tag{7.38}$$

and the continuity of the cross-stream velocity component v at the surface of discontinuity. This second condition derives from the definition of v as the cross-stream velocity at a point of the interface between domains (j) and $(j+1)$, with coordinate Y, hence $v = \tfrac{\mathrm{d}}{\mathrm{d}t}Y$ (Exercise 7.4.4, first item). It reads:

$$\left. \frac{\psi}{U-c} \right|_j = \left. \frac{\psi}{U-c} \right|_{j+1}. \tag{7.39}$$

Assuming that the interval $[y_1, y_2]$ has been cut into n pieces, $2n$ constants $(A_\pm^{(j)}, j = 1, \ldots, n)$ have to be determined. One is left with two boundary conditions and $2(n-1)$ matching conditions, which indeed makes a system of $2n$ homogeneous linear equations. The dispersion relation is obtained from the condition that this system has nontrivial solutions.

Let us analyse the KH instability along these lines using the approximate, discontinuous base flow profile (7.35). Solutions to the Rayleigh equation (7.36) are taken in the form (7.37). Requiring that ψ remain bounded as $y \to \pm\infty$ implies

$$A_{(-)}^{(1)} = 0 = A_{(+)}^{(2)},$$

while the matching conditions (7.38, 7.39) at $y=0$ read

$$(+k)\left(U^{(1)} - c\right) A_{(+)}^{(1)} = (-k)\left(U^{(2)} - c\right) A_{(-)}^{(2)},$$

$$\frac{1}{U^{(1)} - c} A_{(+)}^{(1)} = \frac{1}{U^{(2)} - c} A_{(-)}^{(2)},$$

so that we are left with a homogeneous linear system of two equations for two unknowns. The corresponding compatibility condition:

$$(U_1 - c)^2 + (U_2 - c)^2 = 0,$$

is solved to yield:

$$c = U_{\mathrm{av}} \pm i\tfrac{1}{2}\Delta U,$$

The instability of the flow directly derives from $c_i = \pm\frac{1}{2}\Delta U \neq 0$. Since there is no specific scale in this problem (unbounded medium, infinitely thin vorticity sheet) the instability condition is independent of the wavevector k and all modes are unstable. Their growth rates $\sigma = kc_i$ are larger and larger as k increases. Things are different for a confined layer $-\infty < y_1 < 0 < y_2 < +\infty$, as can be seen by solving Exercise 7.4.4 where several other linear-by-part profiles are also considered.

7.2.3 *Viscous flows*

7.2.3.1 *Instability and viscous dissipation*

For the KH instability, the destabilising process is of mechanical origin and viscosity plays a normal stabilising role. But the Rayleigh theorem shows that some flows do not have unstable modes at the inviscid limit. In order to learn what happens to them, we have to return to the Orr–Sommerfeld equation (7.30).

The problem is difficult from an analytical point of view owing to the singular character of the limit $R \to \infty$. As a matter of fact, whereas the absence of inflection point y_s such that $U''(y_s) = 0$ forbids the existence of unstable inviscid modes, neutral modes may exist, for which the Rayleigh problem is singular: at the critical level y_c, the phase speed c_r is equal to the speed $U(y_c)$ of the flow while $U''(y_c) \neq 0$. These neutral modes are the best candidates for becoming unstable when perturbed by viscous effects. Here we restrict ourselves to a qualitative presentation of the results, often obtained from a numerical solution of the Orr–Sommerfeld equation, and refer to e.g. [Drazin and Reid (1981)] for the relevant theory.

Let us first indicate why viscous dissipation could play a role in the destabilisation of a mechanically stable flow. This hint is obtained by adopting a global point of view and considering the evolution of the kinetic energy contained in a specific two-dimensional mode with wavevector $k = 2\pi/\lambda$ and phase speed c_r. Let us compute (7.17)×u + (7.18)×v:

$$u\partial_t u + v\partial_t v + uvU' + U(v\partial_x v + u\partial_x u) =$$
$$-u\partial_x p - v\partial_y p + R^{-1}[u(\partial_{xx} + \partial_{yy})u + v(\partial_{xx} + \partial_{yy})v], \quad (7.40)$$

and determine the average kinetic energy (per unit length in the spanwise direction) contained in a domain of length λ and height $(y_2 - y_1)$, in the frame moving at speed c_r labelled by coordinate $\bar{x} = x - c_r t$. The quantity of interest is:

$$K_{\text{pert}} = \left\langle \tfrac{1}{2}\left(u^2 + v^2\right)\right\rangle,$$

where $\langle \ldots \rangle$ denotes $[1/\lambda(y_2 - y_1)] \int_{y_1}^{y_2} dy \int_{\bar{x}_0}^{\bar{x}_0+\lambda} (\ldots) d\bar{x}$.

Averaging (7.40), we obtain:

$$\tfrac{d}{dt}K_{\text{pert}} + \tfrac{1}{2}\langle U\partial_x \left(u^2 + v^2\right)\rangle + \langle uvU'\rangle =$$
$$-\langle(u\partial_x p + v\partial_y p)\rangle + R^{-1}\langle u(\partial_{xx} + \partial_{yy})u + v(\partial_{xx} + \partial_{yy})v\rangle.$$

The second term on the l.h.s. cancels upon streamwise integration by virtue of the periodicity in $\bar{x}$: from the fact that U is independent of $\bar{x}$, for every y we get $\int_{\bar{x}_0}^{\bar{x}_0+\lambda} U\partial_x \left(u^2 + v^2\right) dx \equiv \int_{\bar{x}_0}^{\bar{x}_0+\lambda} \partial_x \left[U \left(u^2 + v^2\right)\right] dx \equiv U \left(u^2 + v^2\right)\big|_{\bar{x}_0}^{\bar{x}_0+\lambda} = 0$. On the right hand side we have $-u\partial_x p - v\partial_y p = -\partial_x(up) - \partial_y(vp) + p(\partial_x u + \partial_y v)$. The two first contributions cancel upon integration, either owing to streamwise periodicity or to cross-stream boundary conditions at y_1 and y_2, and the last one from the continuity condition $\partial_x u + \partial_y v = 0$. Finally we get:

$$\tfrac{d}{dt}K_{\text{pert}} = -\langle U'uv\rangle - R^{-1}\langle((\partial_x u)^2 + (\partial_y u)^2 + (\partial_x v)^2 + (\partial_y v)^2)\rangle, \quad (7.41)$$

where the viscous terms in R^{-1} have been rearranged (integration by parts + boundary conditions). This is the specific form taken by the *Reynolds–Orr equation* governing the evolution of the kinetic energy contained in the perturbation in the linear regime.[5]

Equation (7.41) symbolically reads:

$$\tfrac{d}{dt}K_{\text{pert}} = P - D, \quad (7.42)$$

where P is a production term and D, that contains all the terms in R^{-1}, accounts for viscous dissipation. Term D is manifestly always positive and works so as to decrease the energy contained in the perturbation. The flow can thus be unstable only if the production term P is positive and sufficiently large.

Let us consider perturbations in the form

$$\{u,v\} = \tfrac{1}{2}\big(\{\tilde{u}(y), \tilde{v}(y)\} \exp[ik(x - ct)] + \text{c.c.}\big)$$
$$= \tfrac{1}{2}\big(\{\tilde{u}(y), \tilde{v}(y)\} \exp(ik\bar{x}) + \text{c.c.}\big) \exp(kc_i t),$$

with growth rate kc_i. It can then be checked that every term in (7.42) varies as $\exp(2kc_i t)$ that can be factored out, leaving us with a discussion

---

[5]This equation is valid without modification in the nonlinear regime because advection conserves energy. The nonlinear terms dropped upon linearisation on the l.h.s. read $u^2\partial_x u + uv\partial_y u + vu\partial_x v + v^2\partial_y u$, further rewritten as $\tfrac{1}{2}\partial_x(u(u^2+v^2)) + \tfrac{1}{2}\partial_y(v(u^2+v^2))$ using the continuity condition. All these terms disappear in the averaging process. Even more, the result is not restricted to the two-dimensional case but also holds for three-dimensional perturbations, either localised in space or delocalised but periodic in the (x, z) plane.

of the sign of

$$P \propto - \int U'(y)[\tilde{u}(y)\tilde{v}^*(y) + \text{c.c.}]\,\mathrm{d}y,$$

which is not determined in advance. On the contrary, this sign strongly depends on the shape of u and v, solutions to the full Orr–Sommerfeld equation (7.30). The asymptotic analysis in the limit '$R \to \infty$' points out the role of the variation of the stream-function $\tilde{\psi}$ close to the critical level y_c, and shows the existence of specific unstable modes called *Tollmien–Schlichting (TS) waves*. Rather then starting a long and delicate discussion that would allow one to compute the behaviour of the marginal stability curve in this limit (for an introduction, consult Huerre and Rossi in [Godrèche and Manneville (1998)]) we just present the main results obtained in part analytically and in part numerically.

7.2.3.2 *Results*

Marginal stability curves are conventionally presented in the plane (R, k) with R along the horizontal axis and k vertically, in contrast with convection-like instabilities (see Figure 3.3, p. 92). Whatever the origin of the instability, mechanical or viscous, they take the shape of loops with different aspects in the limit '$R \to \infty$', as shown in Figure 7.8.

• When the flow is *mechanically unstable* (velocity profile with an inflection point and a vorticity maximum at the inflection point), the threshold is *low* and a whole band of wavevectors remains unstable as $R \to \infty$, see Figure 7.8 (left). Viscosity then plays its usual stabilising role.

• When the base profile is *mechanically stable* (no inflection point), the threshold is *high* and the marginal stability curve has the shape of a hairpin that pinches as $R \to \infty$, see Figure 7.8 (right). Unstable modes are the TS waves resulting from the viscous mechanism alluded to above.

Plane Poiseuille flow in a channel is typical of flows without inflection point. With the maximum speed $U_{\max}$ on the centreline of the flow as velocity unit and the half-gap $h/2 = (y_2 - y_1)/2$ as length unit, the Reynolds number is defined as $R = U_{\max} h / 2\nu$ and the dimensionless base velocity profile over the interval $[-1, +1]$ reads $U = 1 - y^2$. The marginal stability curve looks like that in Figure 7.8 (right), with $R_c \simeq 5772 \gg 1$ and $k_c \simeq 1.02$ (critical wavelength $\lambda_c = (h/2)(2\pi/1.02) \simeq 3h$).[6] The phase speed

---

[6]S.A. Orszag, "Accurate solution of the Orr–Sommerfeld stability equation," J. Fluid Mech. **50** (1971) 689–703.

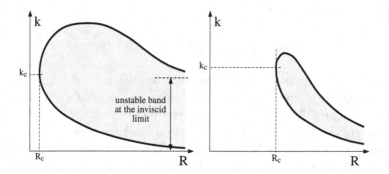

Fig. 7.8 Marginal stability curves typical of plane shear flows. In each case, the unstable domain corresponds to the interior of the loop. Left: Mechanically unstable flows; low threshold and finite-width band of unstable wavevectors when $R \to \infty$. Right: Mechanically stable flows (no inflection point); high thresholds and narrowing band of unstable wavevector when $R \to \infty$.

of waves at threshold is $c_{\rm r,c} \simeq 0.264$ (i.e. about $1/4$ of $U_{\max}$). To these numerical results one can add the asymptotic behaviour of the marginal stability branches obtained analytically (see e.g. [Drazin and Reid (1981)] for a detailed presentation and results summarised in their Figure 4.11, p. 190):

$$\text{Upper branch} \to R^{1/3} \simeq 8.44\,(k^2)^{-11/3}, \qquad c_{\rm r} \simeq \tfrac{4}{15}k^2$$
$$\text{Lower branch} \to R^{1/3} \simeq 5.96\,(k^2)^{-7/3}, \qquad c_{\rm r} \simeq 0.611 k^2\,.$$

7.2.4 Instability and downstream transport

7.2.4.1 Theory vs. experiments

Let us come back to the dispersion relation (7.23). The Squire theorem tells us that the wavevector $\mathbf{k}_{\rm h}$ has a single relevant (streamwise) component noted k, so that we can write

$$\mathcal{L}(-i\omega, ik; R) = 0\,. \tag{7.43}$$

The presentation of results in the plane (R, k) corresponds to a *temporal* reading of this relation, further solved by assuming $k \in \mathbb{R}$ and ω complex, $\omega = \omega_{\rm r} + i\omega_{\rm i}$, so that $\omega_{\rm i}$ is the temporal growth rate (from the normal mode assumption: $\exp[i(kx - \omega t)] = \exp(\omega_{\rm i} t)\exp[i(kx - \omega_{\rm r} t)]$).

This approach was satisfactory in the case of convection that develops in an enclosure and emerges from the background noise, in the absence of any forcing (*absolute* instability). It is no longer appropriate for typical

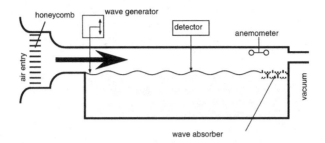

Fig. 7.9 Experimental configuration used in the film *Flow Instabilities* [Mollo-Christensen (1972)]: A wind is blown over a water channel, produced by a vacuum cleaner downstream, and settled by the honeycomb upstream; its speed is measured by an anemometer. A wave generator hits the surface close to the entrance, producing a localised perturbation with given frequency. The height of the waves is measured by an appropriate gauge at a fixed distance from the wave generator.

open flow situations like the one sketched in Figure 7.9, where downstream transport manifestly plays a dominant role (*convective* instability). In such a set-up, a localised forcing, periodic with angular frequency ω_f, is applied at a given point in the flow and the evolution of the perturbation is recorded at downstream stations as a function of the forcing characteristics and the Reynolds number. From the records, a downstream spatial amplification rate can be defined, negative when the perturbation is damped, positive otherwise.

The different behaviours observed, sketched in Figure 7.10, can be analysed in terms of plane waves with angular frequency ω_f, $\exp[i(kx - \omega_\text{f} t)]$, modulated by a slowly decaying or growing envelope, $\exp(\mu x)$, with μ small, hence $\exp[i(kx - \omega_\text{f} t)] \exp(\mu x)$. The two exponential factors can be recombined as $\exp[i(k - i\mu)x - \omega_\text{f} t]$ to form a *complex* wavevector with real part k and imaginary part $-\mu$.

We are thus lead to a *spatial* reading of the dispersion relation (7.43) solved for $\omega = \omega_\text{f}$ real and k complex. This simplistic presentation is not free from criticism. We have indeed avoided evoking what happens upstream of the forcing point and represent it in Figure 7.10. While $-k_\text{i} > 0$ corresponds to an amplification for $x > 0$ (which conventionally denotes the downstream direction), a mode with $-k_\text{i} < 0$ is amplified upstream and the picture is reversed. The non-equivalence of upstream and downstream directions due to advection has delicate analytical implications that we still mostly skip,

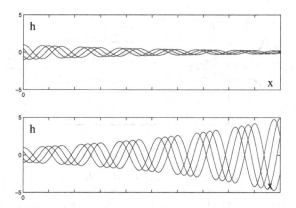

Fig. 7.10 The spatial amplification rate of the waves is extracted from the response of the system to a localised periodic excitation. The marginal value $R_{\mathrm{m}}(\omega)$ of the control parameter separates the domain of damped waves $R < R_{\mathrm{m}}(\omega)$ from that of amplified waves $R > R_{\mathrm{m}}(\omega)$.

asking the reader to turn to specialised references.[7]

Without entering mathematical intricacies, following Gaster[8] we can however relate the main characteristics of the spatial (S) problem to those of the temporal (T) problem. Let us consider the dispersion relation (7.43) and assume that it has been analytically solved for ω as a function of k, both complex, at given R:

$$\omega = \varpi(k; R). \qquad (7.44)$$

The expansion of the dispersion relation around some arbitrary point $\left(k^{(0)}, \omega^{(0)}\right)$ solution to (7.43) reads

$$\omega_{\mathrm{r}} = \omega_{\mathrm{r}}^{(0)} + \frac{\partial \varpi_{\mathrm{r}}}{\partial k_{\mathrm{r}}} \delta k_{\mathrm{r}} + \frac{\partial \varpi_{\mathrm{r}}}{\partial k_{\mathrm{i}}} \delta k_{\mathrm{i}} + \ldots \qquad (7.45)$$

$$\omega_{\mathrm{i}} = \omega_{\mathrm{i}}^{(0)} + \frac{\partial \varpi_{\mathrm{i}}}{\partial k_{\mathrm{r}}} \delta k_{\mathrm{r}} + \frac{\partial \varpi_{\mathrm{i}}}{\partial k_{\mathrm{i}}} \delta k_{\mathrm{i}} + \ldots \qquad (7.46)$$

A nearly neutral T-mode is characterised by:

$$k_{\mathrm{r}}^{(\mathrm{T})} = k_{\mathrm{r}}^{(0)}, \qquad k_{\mathrm{i}}^{(\mathrm{T})} = 0 \quad \text{(by definition)}$$

and the corresponding angular frequency

$$\omega_{\mathrm{r}}^{(\mathrm{T})} = \omega_{\mathrm{r}}^{(0)}, \qquad \omega_{\mathrm{i}}^{(\mathrm{T})} = \omega_{\mathrm{i}}^{(0)} \quad \text{with} \quad |\omega_{\mathrm{i}}^{(\mathrm{T})}| \ll 1.$$

---

[7]e.g. the review article by P. Huerre, "Open shear flow instabilities," in [Batchelor et al. (2000)] and references quoted therein.

[8]M. Gaster, "A note on the relation between temporally increasing and spatially-increasing disturbances in hydrodynamic stability," J. Fluid Mech. **14** (1962) 222–224.

We are looking for the characteristics of an S-mode $\left(k_r^{(S)}, k_i^{(S)}\right)$ with $k_r^{(S)} = k_r^{(T)}$ and angular frequency $\omega_r^{(S)}$ with $\omega_i^{(S)} = 0$ by definition. The T and S modes fulfil the dispersion relation (7.44) and are both nearly neutral, which means in particular $k_i^{(S)} = k_i \neq 0$ with $|k_i^{(S)}| \ll 1$. Inserting these assumptions in (7.45, 7.46) gives at lowest order

$$\omega_r^{(S)} = \omega_r^{(T)} + \frac{\partial \varpi_r}{\partial k_i} k_i^{(S)}, \qquad (7.47)$$

$$0 = \omega_i^{(T)} + \frac{\partial \varpi_i}{\partial k_i} k_i^{(S)}. \qquad (7.48)$$

Now, assuming that the dispersion relation (7.44) is analytic implies the existence of relations between the partial derivatives of the real and imaginary parts of ϖ, the so-called *Cauchy relations*:[9]

$$\text{(a)} \quad \frac{\partial \varpi_r}{\partial k_r} = \frac{\partial \varpi_i}{\partial k_i}, \qquad \text{(b)} \quad \frac{\partial \varpi_r}{\partial k_i} = -\frac{\partial \varpi_i}{\partial k_r}. \qquad (7.49)$$

Using (7.49a) in (7.48) yields:

$$\frac{\omega_i^{(T)}}{k_i^{(S)}} = -\frac{\partial \varpi_r}{\partial k_r}. \qquad (7.50)$$

Strictly at threshold, this relation is an indeterminate ratio $0/0$. Slightly off threshold, it allows us to convert spatial and temporal growth rates into each other. The quantity that shows up on the r.h.s. of (7.50) is the (real) group velocity of the normal modes, which is somewhat natural in view of the discussion on p. 96.

Furthermore, by definition of the threshold condition in the temporal case, ϖ_i reaches a maximum at $k = k_c$ for $R = R_c$. Quantity $\partial \varpi_i / \partial k_r$ is thus small, of order $(k_r - k_c)$, which makes $\partial \varpi_r / \partial k_i$ also small from (7.49b) so that, from (7.47) at lowest significant order:

$$\omega_r^{(S)} = \omega_r^{(T)},$$

i.e. the most dangerous forcing frequency is that corresponding to the most amplified temporal mode, which is intuitively expected.

The stability diagram corresponding to the spatial reading of the dispersion relation displays the forcing frequency on the vertical axis and the Reynolds number on the horizontal axis. Its experimental determination is concretely illustrated in the film *Flow instabilities* [Mollo-Christensen (1972)]. For mechanically stable flows, the marginal curve has again a hairpin shape, as illustrated in Figure 7.11.

---

[9] In order to derive relations (7.49), one takes advantage of the fact that $\varpi = \varpi_r + i\varpi_i$, as a function of $k_r = (k + k^*)/2$ and $k_i = -i(k - k^*)/2$, is only a function of k and not of its complex conjugate k^*, which would of course not be the case of the most general function of the two variables k_r and k_i.

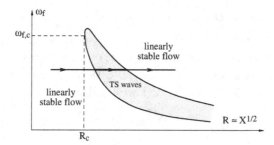

Fig. 7.11 Marginal stability curve of a Blasius boundary layer flow submitted to a localised periodic forcing (vibrating ribbon experiment). As the flow develops downstream, R is a function of the distance X to the leading edge. A path in the (R, ω_f) plane is thus followed by moving the observation point at fixed infinitesimal forcing and perturbations are damped or amplified depending on its position. Far downstream the flow is expected to be stable provided that perturbations have no time to grow enough to invalidate the linear approach.

Flows most often develop downstream, the plane Poiseuille flow being an exception. The Reynolds number is then usually a function of the downstream distance, e.g. the distance to the leading edge for a boundary layer as indicated by the horizontal axis of that figure. A first information about the stability of the flow is then obtained by using the results of the linear theory locally, i.e. by neglecting the downstream dependence of the flow and assuming a plane parallel flow with a velocity profile identical to that at the considered position. Results will be relevant provided that the space-time growth rate of the normal mode remains large when compared to the space-time evolution rate of the flow itself. They can be corrected by developing a multiple-scale approach to account for the effects of non-parallelism or slow time dependence.

Testing the stability of an open flow then comes to explore the plane (R, ω_f) by changing the angular frequency of the forcing. Of course, one should take the predictions of this linear approach with some caution. In particular, the boundary layer could be anticipated to be stable (relaminarisation) from a naive reading of Figure 7.11 but this can happen only if, while crossing the unstable domain, perturbations are kept sufficiently small, so that no secondary instability (or even turbulence) sets in, which somehow leaves the *receptivity* problem open, i.e. how the system extracts the dangerous TS eigenmodes from the random perturbations forming the residual background turbulence (see §7.3.3.2, p. 292).

7.2.4.2 Absolute and convective instabilities

The general stability problem is not only that of the resonant response of the flow to a periodic localised forcing at a fixed position. Previously, we considered temporal modes in the form of delocalised waves $\exp(i(kx - \omega t)) = \exp(\omega_i t)\exp(i(kx - \omega_r t))$. The stability of the flow was then implicitly discussed in a frame moving at the phase speed of the waves: $kx - \omega_r t \equiv k(x - c_r t)$. As long as no reference is made to fixed boundaries, this speed can freely be subtracted using Galilean invariance[10] and the growth/decay properties can be obtained from the sign of ω_i. But neither the spatial approach nor the temporal approach are satisfactory when dealing with the *natural* transition that relates to the evolution of temporally incoherent and spatially localised small perturbations composing the residual turbulence, in the presence of walls or obstacles breaking Galilean invariance. In the most general case, a complete linear-response theory has to be developed. This implies a mastery of complex-variable analysis that goes beyond the prerequisites of this course, so we limit our ambition to giving a sketchy presentation that just keeps its spirit and suggest the reader to consult specialised works, e.g. Note 7, p. 278.

Since phenomena are ultimately detected in the laboratory frame, we have to account for the competition between the downstream transport of perturbations and their amplification by instability mechanisms. Accordingly, as already alluded to before, we shall say that:

• The instability is *convective* when fluctuations are carried downstream. In that case, its effects will be detectable depending on the level of triggering residual turbulence in the flow (or of external perturbations) and the flow behaves as a *noise amplifier*.

• The instability is *absolute* when its mechanism is sufficiently intense that perturbations can go against the stream and invade the whole experimental domain in spite of a general downstream transport by the base flow. The system then behaves as a *self-sustained oscillator*.

In the present context, it will be enough to express the result of the theoretical analysis in terms of the so-called *Briggs–Bers criterion* that,

---

[10] Reference to the laboratory frame affects the *external* characteristics of the flow that break the symmetry of the NS equation with respect to a Galilean transformation: The change to a frame translating at constant speed $\bar{x} = x - Vt$, $\bar{t} = t$ and $v = \bar{v} + V$, i.e. $v(x,t) = \bar{v}(x - Vt, t) + V = \bar{v}(\bar{x}, \bar{t})$, indeed leaves these equations unchanged owing to the special form of the *advection* term: $\partial_t v + v\partial_x v = \partial_t[\bar{v}(x-Vt,t)+V] + [\bar{v}(x-Vt,t)+V]\partial_x[\bar{v}(x-Vt,t)+V] = -V\partial_{\bar{x}}\bar{v} + \partial_{\bar{t}}\bar{v} + [\bar{v}+V]\partial_{\bar{x}}\bar{v} = \partial_{\bar{t}}\bar{v} + \bar{v}\partial_{\bar{x}}\bar{v}$.

in some sense, extends Gaster's approach. Since the 'absolute/convective' discrimination relates to the laboratory frame, let us assume that we can restrict ourselves to the study of wave-packets composed of modes travelling at group velocity $v_g = 0$. Now, v_g is defined as usual through $v_g = \partial \varpi / \partial k$, where ϖ is here the complex function of the complex variable k obtained by solving the dispersion relation (7.43) for ω. Condition $v_g = 0$ is thus an equation for k. Let $k^{(0)}$, be the solution to this equation and $\omega_i^{(0)} = \varpi_i(k^{(0)})$ the corresponding growth rate.[11]

The criterion goes as follows:

- The flow is *absolutely unstable* if $\omega_i^{(0)} > 0$ (amplification on the spot).
- It is *convectively unstable* if $\omega_i^{(0)} < 0$, provided that $\omega_i > 0$ for some k.
- It is linearly *stable* when $\omega_i < 0$ for all k.

These situations are depicted in Figure 7.12 which displays perturbation profiles as functions of space at several successive times. An elementary illustration of the use of the criterion is considered in Exercise 7.4.5.

The preceding discussion is typically *local* in that is applies to a quasi-parallel flow provided that its x dependence can be neglected. In this perspective, when the flow develops downstream, the intensity of the instability mechanism varies in space so that its absolute/convective character may change downstream. In that case only a *global* analysis can help us understand the stability properties of the flow, which is another full story in itself. Finally, in all that precedes, departures from the base flow were tacitly assumed to be wave-like and infinitesimal, making the linear approach legitimate (localised finite-size perturbations were discarded). Besides the identification of important instability mechanisms, a thorough understanding of the transition to turbulence in developing flows has to face difficulties arising from a combination of alternatives: absolute *vs.* convective, local *vs.* global, and linear *vs.* nonlinear.

7.3 Transition to Turbulence

7.3.1 Nonlinear development of instabilities

7.3.1.1 Saturation of the primary mode

In its principle, the approach is parallel to the one followed in Chapter 3 for natural convection. In order to identify the different steps more clearly

---

[11] If there are several solutions $k^{(0)}$, we consider only the one with the largest $\omega_i^{(0)}$.

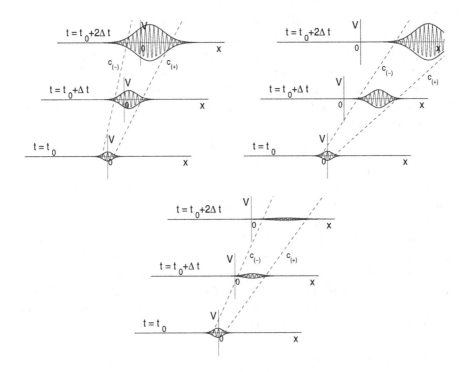

Fig. 7.12 Space-time representation of the growth/decay of a perturbation introduced at $x=0$ and $t=0$. Top-left: The two edges of the wave-packet go in opposite directions and the perturbation grows on the spot, the instability is *absolute*. Top-right: The perturbation grows but is evacuated since the two edges of the wave-packet go in the same direction, the instability is *convective*. Bottom: The perturbation vanishes and the flow is *stable*.

let us introduce subscripts, '0' for the base state, '1' for the primary perturbation, '2' for the secondary modes, etc. The linear stability theory of the parallel base flow $\mathbf{V}_0$ yields a critical primary mode $\mathbf{V}_1$ in the form of a two-dimensional wave with wavevector $\mathbf{k}$ aligned with the flow direction, phase speed c, and no spanwise flow component (Squire theorem). Let A_1 be the amplitude of this mode. The first question relates to the saturation of A_1 beyond threshold, which is dealt with by means of an expansion in powers of the amplitude. This *à la Landau* approach has been introduced in the field of open flows mainly by Stuart.[12] As seen in Chapter 3, convection is a stationary cellular instability essentially governed by an equation for a

---

[12] For a review of early work, consult: J.T. Stuart, "Nonlinear stability theory," Annu. Rev. Fluid Mech. **3** (1971) 1–21.

single variable with real coefficients accounting for a supercritical fork bifurcation. In contrast, here the wavy behaviour of the flow beyond threshold makes it time-oscillatory in the laboratory frame so that a Hopf bifurcation on a two-dimensional centre manifold is expected, (4.17) p.138. With $s_1 = \sigma_1 - i\omega_1$, where σ_1 is the real growth rate of the primary mode (positive when unstable) and ω_1 the angular frequency, changing for a rotating amplitude by setting $A_1 = \tilde{A}_1 \exp(-i\omega_1 t)$, we get:

$$\tfrac{\mathrm{d}}{\mathrm{d}t}\tilde{A}_1 = \sigma_1 \tilde{A}_1 - g|\tilde{A}_1|^2 \tilde{A}_1 ,$$

where the complex coefficient $g = g_\mathrm{r} + ig_\mathrm{i}$ describes the effective interaction of the unstable mode with itself (*via* the coupling with other, adiabatically eliminated, non-critical modes, see §4.1).

Saturation is controlled by the sign of the real part g_r of g. When $g_\mathrm{r} > 0$ the bifurcation is supercritical and the wave saturates at an amplitude that smoothly varies with the distance to threshold. In the opposite case, the wave does not saturate at lowest order, one must continue the expansion (see the case of the plane Poiseuille flow discussed later) but there is a risk that no stable branch exists at finite distance from the base solution, as recognised early by Landau (Note 4, p. 16).

7.3.1.2 *Secondary instabilities*

Let us now sketch what should be the next steps. We assume that the nonlinear problem for the primary mode has been solved and that saturation at some amplitude A_1 has been reached. The stability analysis is then repeated for the flow $\mathbf{V}_{01} = \mathbf{V}_0 + A_1 \mathbf{V}_1$ in a frame moving at the speed c_nl of the saturated wave (coordinate $\bar{x} = x - c_\mathrm{nl} t$). In subscript '01' the '0' indicates that it is the new base flow and the '1' that it arises from the saturation of the primary perturbation.

Concrete computations are particularly involved. The standard linearisation process leads to a partial differential system with coefficients periodic in $\bar{x}$ with period $\lambda = 2\pi/k$. For such linear operators, *Floquet theory* suggests to take secondary normal modes $\mathbf{V}_2$ in a form that explicitly isolates nontrivial space-time departures from a trivial space dependence in $\bar{x}$ with period λ:

$$\mathbf{V}_2(\bar{x}, y, z, t) = A_2 \exp[\,i(q_x \bar{x} + q_z z - \omega_2 t)\,]\, \widetilde{\mathbf{V}}_2(\bar{x}, y) , \quad (7.51)$$
$$\widetilde{\mathbf{V}}_2(\bar{x} + \lambda, y) = \widetilde{\mathbf{V}}_2(\bar{x}, y) .$$

The exponential part accounts for a supplementary modulation introduced by the putative secondary instability, with real wavevector $\mathbf{q} \equiv (q_x, q_z)$ and complex angular frequency $\omega_2 = \omega_{2,\mathrm{r}} + \omega_{2,\mathrm{i}}$ (temporal instability scheme). The main possible cases of this conceptually well-posed problem can be identified from (7.51) without any explicit calculation.

It is first not difficult to show that the value of q_x/k can be restricted to the interval $[0,1[$ since, setting $q_x = (\varsigma + \kappa)k$ with $\kappa \in \mathbb{N}$, one gets:

$$\exp(iq_x x) = \exp[2\pi i(\varsigma + \kappa)k\tilde{x}] = \exp(2i\pi\varsigma k\tilde{x})\exp(2i\pi\kappa k\tilde{x}),$$

where the last exponential factor is periodic in $\tilde{x}$ with period $2\pi/\kappa k$, thus also periodic with period $2\pi/k$, so that it can be incorporated into $\tilde{\mathbf{V}}_2$ that accounts for the trivial part of the space dependence.

Next, possible secondary modes can be classified according to whether (i) $q_z = 0$ or $q_z \neq 0$, (ii) $q_x = 0$ or $q_x \neq 0$, and (iii) when $q_x \neq 0$, whether q_x is commensurate or not with k, i.e. ς rational or irrational. In particular:

• When $q_z = 0$ and $q_x = k/2$ ($\varsigma = 1/2$), the secondary mode is still two-dimensional but has a wavelength twice that of the primary mode, this period doubling instability is called *pairing*.

• When $q_z \neq 0$, the secondary mode is most often called *fundamental* when $q_x = 0$, and *subharmonic* when $q_x = k/2$ (see Figure 7.17).

• When q_x is not commensurate to k (ς irrational), the situation is similar to that of quasi-periodic temporal systems considered in §4.2.3, p. 146, and can be transposed from it.

The physical mechanisms involved in secondary instabilities have been studied both theoretically and experimentally. One generally makes a distinction between *viscous* processes evolving over long time scales (in practice mostly the growth of TS waves) and *inertial* processes developing over short time scales. In this latter context, the *elliptical* instability affecting the core of vortical structures is of great importance. For a detailed discussion with references, see e.g. Kerswell[13] or Huerre and Rossi in [Godrèche and Manneville (1998), p. 269ff].

Rather than considering the natural transition, the comparison of theory with experiments often involves flows in which the spanwise dependence of the secondary mode is forced at a given q_z by some experimental trick, e.g. by grooving the wall in a boundary layer experiment, the splitting plate for a mixing layer, the lips of the slit for a plane jet, etc.

The tertiary instability is in general hardly accessible to theory and small scale turbulence usually appears soon after secondary modes set in.

---

[13] R.R. Kerswell, "Elliptical instability," Ann. Rev. Fluid Mech. **34** (2002) 83–113.

The last step of the transition to turbulence is often interpreted as being due to inflection points in the velocity profiles resulting from the superposition of the primary and secondary modes to the base flow, thus promoting KH instabilities at the origin of the smaller scales. The situation is also made particularly complicated by space and time aspects of the development of transitional flows interfering with downstream transport.

In the rest of this chapter we present salient features of the transition to turbulence at a phenomenological level resting on the difference between unstable and stable flows from the inviscid point of view and the supercritical or subcritical character of the bifurcation. Rather than trying to be exhaustive, we shall focus on the most typical situations in each case.

7.3.2 Inviscidly unstable flows

7.3.2.1 General features

Two-dimensional base flows that are unstable at the inviscid limit display inflection points in their velocity profile (Rayleigh and Fjørtoft theorems). These are mixing layers (Figure 7.6), jets and wakes (Figure 7.2).

The transition to turbulence is characterised by primary instabilities that set in at low Reynolds numbers in the form of two-dimensional spanwise structures as expected from the linear theory. These structures then gently saturate beyond threshold to form developed vortices. Secondary modes are most often streamwise modulations ($q_z = 0$) leading to quasi-periodic behaviour and locking, especially sub-harmonic locking that manifests itself as *vortex pairing* ($q_x = k/2$). In the case of mixing layers, successive pairings may take place before a secondary instability can introduce some three-dimensional dependence. Small scale turbulence enters soon after the introduction of three-dimensional modes, usually arising from the *elliptical instability* previously mentioned (Note 13). The plane jet, understood as two side-by-side mixing layers close to the outlet and merging further downstream, behaves in the same way.

Another general feature of this transition scenario is the persistence of *coherent structures* in the post-transitional regime. They are large vortices with lifetimes long when compared to their turn-over times, resembling KH vortices but now developing over a mean flow profile.

7.3.2.2 Wakes

We focus on the wake of a circular cylinder, a quantitatively well studied system which combines specific features of flows displaying inflection points and features arising from the absolute character of the instability which is responsible for its self-oscillatory nature. Here we borrow mainly from the review given by Williamson.[14]

The distinctive feature of the wake behind a bluff body is the occurrence of a sharp Hopf bifurcation that marks the emergence of very regular periodic *vortex shedding* forming a pattern already depicted in Figure 1.10, p. 21 and known as the *Kármán vortex street*.[15] The origin of this behaviour can be understood by noticing that, at flow rates corresponding to the bifurcation point, a steady recirculation develops just behind the obstacle so that the velocity profile displays a region where downstream transport is compensated. In contrast, away from it, the flow has been smoothed by viscous effects and is closer to the uniform flow blowing from infinity around the body (Figure 7.13). Accordingly, there is a region in the flow at the rear of the obstacle where the instability linked to the presence of inflection points can develop on the spot (hence 'absolute'), producing the oscillations. This situation is reminiscent of what happens in closed systems where confinement effects select specific modes and nonlinearity can be studied in the framework of dynamical systems theory (§3.2.3, §4.2).

Let us consider a cylinder placed perpendicular to a uniform flow with speed U_∞.[16] This experimental configuration is characterised by two dimensionless parameters. The first one is the *Reynolds number*:

$$R = U_\infty d/\nu,$$

where d is the diameter of the cylinder and ν the kinematic viscosity of the fluid. The second parameter is the *aspect-ratio* $\Gamma = L/d$ where L is the length of the cylinder. Here we assume $L \gg d$ (or $\Gamma \to \infty$), so that it seems legitimate to neglect end effects and to start with a two-dimensional base flow. Accordingly, we are left with R as the only parameter.

---

[14] C.H.K. Williamson, "Vortex dynamics in the cylinder wake," Annu. Rev. Fluid Mech. **28** (1996) 477–539. See also M. Provansal, "Wake instabilities behind bluff bodies," in [Mutabazi *et al.* (2006)].

[15] T. von Kármán, "Über den Mechanismus des Widerstandes, den ein bewegter Körper in ein Flüssikeit erzeugt," Nachr. Ges. Wiss. Göttingen, Math. Phys. Klasse (1911) 509–517, (1912) 547–556. Prior experimental observations by Bénard (1908) and a controversy between him and von Kármán about vortex shedding are mentioned by J.E. Wesfreid, "Scientific biography of Henri Bénard (1874–1930)," in [Mutabazi *et al.* (2006)].

[16] The case of a cylinder dragged through a viscous fluid at rest is equivalent; comparable patterns would be obtained by setting the speed of the body to $-U_\infty$.

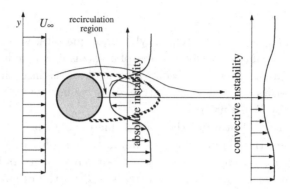

Fig. 7.13 Absolute/convective transition in the wake of a cylinder (and, more generally, of a blunt body).

The flow structure can be understood from visualisations of the velocity field, while the different regimes are most conveniently identified through the variation of global observables. A first one is the oscillation frequency f, or in dimensionless form, the *Strouhal number*

$$S = \frac{fd}{U_\infty}.$$

A second classical measure is the drag, i.e. the force F_D necessary to maintain the cylinder at its position, or rather the dimensionless *drag coefficient* $C_D = F_D / \frac{1}{2}\rho U_\infty^2 d$. Another related quantity is the *base pressure*, the pressure difference between a point at 180 degrees from the upstream stagnation point and a reference pressure, usually the static pressure at infinity. Here we shall just discuss the early stages of the transition using the Strouhal number.

The transition to turbulence behind a cylinder is described at many places besides Williamson's review (Note 14), e.g. in [Tritton (1990)]. Beautiful pictures of the different laminar flow regimes by Taneda can be found in [van Dyke (1982)]. Creeping flow with flow lines smoothly sticking to the cylinder exists up to $R \sim 5$–6, beyond which a steady symmetrical bubble of recirculating fluid sets in. This time independent situation persists up to $R = R_c \simeq 48.5$ beyond which the wake oscillates, periodically emitting vortices parallel to the cylinder axis as illustrated in Figure 7.14. At threshold, the Strouhal number is of order 0.12, so that the streamwise wavelength $\lambda_0 = U_\infty/f$ is of order $8d$.

Beyond threshold, the oscillation frequency increases regularly with the Reynolds number as shown in Figure 7.15. In fact the frequency selection

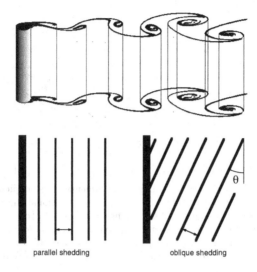

Fig. 7.14 Top: Perspective view of the shedding of parallel vortices, after one of Taneda's pictures in [van Dyke (1982)]. Bottom: Parallel *vs.* oblique shedding.

problem is a difficult one, both experimentally and theoretically:

– On the experimental side, it took some time before it was recognised that the dispersion of the observed frequencies was due to the occurrence of *oblique shedding*, i.e. shedding with the axis of the vortices at an angle θ with the cylinder axis. Assuming that the frequency of parallel shedding is f_0, a simple geometrical construction (Figure 7.14, bottom-right) shows that the frequency of oblique shedding (and with it, the Strouhal number S) is reduced by a factor $\cos\theta$: let λ_θ be the wavelength of the oblique vortex street as measured perpendicularly to the vortex axis and suppose that the instability mechanism makes it identical to that of parallel shedding, then the wavelength measured along the flow direction is longer by a factor $1/\cos\theta$; to get the result one just needs to convert wavelengths into oscillation periods using the speed of the flow U_∞. Re-scaling frequencies accordingly yields the master curve displayed in Figure 7.15 (left).

– The theoretical understanding of the neighbourhood of the threshold is not as obvious as it may seen. At first, the wake can be thought of as a collection of Hopf oscillators, distributed all along the cylinder and coupled to their neighbours. An appropriate local model would then be (4.17), p. 138 which, once unfolded in the spanwise direction z, would yield a complex

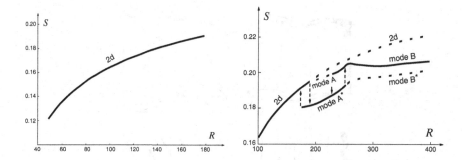

Fig. 7.15 Variation of the Strouhal number S (dimensionless frequency of shedding) as a function of the Reynolds number R. Left: Two-dimensional shedding regime corrected for the effects of oblique shedding. Right: Transition to three-dimensional flow. Modes A* and B* have the same structure than modes A and B but display dislocations; dashed lines correspond to unstable regimes; The transition '2D ↔ mode A*' displays hysteresis; the "natural" route is: 2d → mode A* → mode B. After Williamson, Note 14.

Ginzburg–Landau equation (6.36), p. 243. The dynamics of the wake close to the threshold is in fact reasonably well described by this equation whose coefficients can be fitted against experiments.[17] Chevron patterns observed in case of oblique shedding, finite-size effects, or the presence of dislocations in the vortex system (viewed as phase defects for the complex amplitude), can be understood within this phenomenological approach. But this cannot be the whole story since the approach implicitly assumes that the streamwise (x) structure of the perturbation is frozen, which this is not the case. While the amplitude of the maximum of the wake velocity fluctuations indeed increases as the square-root of the distance to threshold as expected for a Hopf bifurcation, the position of the maximum shifts upstream closer to the cylinder. Progress has been made recently regarding the variation of the streamwise shape of the vortex amplitude and the frequency selection problem by combining convective/absolute, global, and nonlinear issues.[18]

Beyond $R \approx 150$, the wake enters a transition regime where the regular vortex street may be disrupted by several different secondary instability modes, all involving spanwise modulation. The picture is further compli-

---

[17] M. Provansal, C. Mathis, L. Boyer, "Bénard–von Kármán instability: transient and forced regimes," J. Fluid Mech. **182** (1987) 1–22. Th. Leweke, M. Provansal, "The flow behind rings: bluff body wakes without end effects," J. Fluid Mech. **288** (1995) 265–310.
[18] See, e.g. B. Pier and P. Huerre, "Nonlinear self-sustained structures and fronts in spatially developing wake flows," J. Fluid Mech. **435** (2001) 145–174, B. Pier, "On the frequency selection of finite-amplitude vortex shedding in the cylinder wake," J. Fluid Mech. **458** (2002) 407–415, and references cited.

cated by the presence of vortex dislocations. Two main modes have been identified,[19] first experimentally, next in the framework of linear stability analyses.[20] The first one, called 'mode A', arises as a deformation of the primary vortices with wavelength of the order of 0.6–$0.8\lambda_0$, where λ_0 is the primary wavelength at the corresponding value of R. Since the Strouhal number is then about 0.2, the wavelength of mode A is of order 3–$4d$. In contrast the second one, called 'mode B', has a much shorter wavelength of order $0.2\lambda_0$, i.e. of the order of d. The physical origin of these modes cannot be explained in terms of the elementary KH mechanism discussed previously but rather relates to the elliptical instability (Note 13).

The transition regime extends up to $R \approx 260$. Beyond this point the intensity of fine structures increases and the nature of the instabilities involved in the primary vortex shedding mechanism changes gradually. Rather than deriving from the global mode attached to the recirculation bubble, as was the case at lower R, these instabilities have to do first with shear layer that forms downstream the detachment point and, at the highest Reynolds numbers, with the boundary layer that develops along the cylinder itself.

7.3.3 Inviscidly stable flows

7.3.3.1 Plane Poiseuille flow

This prototype of flows without inflection point has been much studied both theoretically and experimentally. The threshold of the instability against TW waves is $R_c \simeq 5772$ (see p. 275). As shown by Herbert[21] the bifurcation is subcritical with unstable nonlinear steady states appearing below threshold, but the bifurcated branch turns back at $R_{nl} \simeq 2900$ (Figure 7.16).

The next step should be the linear stability analysis of two-dimensional saturated TS waves against three-dimensional infinitesimal perturbations, as suggested in §7.2.1. One would thus expect transversally modulated waves later decaying into small scale turbulence. Several difficulties hinder the observation of this scenario. First, the primary instability is convective so that detecting its presence in a finite length channel strongly depends

---

[19] Ch. Williamson, "Three-dimensional wake transition," J. Fluid Mech. **328** (1996) 345–407.
[20] D. Barkley, R.D. Henderson, "Three-dimensional Floquet stability analysis of the wake of a circular cylinder," J. Fluid Mech. **322** (1996) 215–241.
[21] T. Herbert, "Secondary instability of plane channel flows to subharmonic three-dimensional disturbances," Phys. Fluids **26** (1983) 871–874.

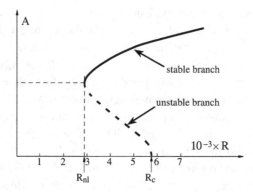

Fig. 7.16 Bifurcation diagram of two-dimensional TW waves in plane Poiseuille flow. After Herbert.[21]

on the level of background noise. Second, the bifurcation is subcritical so that working with a clean base flow inevitably leads to missing the stable bifurcated branch at finite distance from it, whereas triggering appropriate waves, e.g. using vibrating ribbons, is not so easy.

In contrast, a moderately high level of residual turbulence brings about a different scenario (*natural* transition) involving localised finite-size perturbations, called *turbulent spots*, i.e. limited patches of small-scale turbulent flow travelling amid laminar Poiseuille flow, growing in size as they process downstream.[22] These turbulent spots are transient and rapidly decay into laminar flow below $R_g \approx 1000$, but are sustained beyond this value, which is much lower than R_c or even R_{nl}. The value R_g plays the role of a *global stability threshold*, hence the notation, see p. 9. Except for the existence of a linear threshold and a corresponding nonlinear branch, the situation is similar to that taking place in the plane Couette flow to be examined later in §7.3.4.3.

7.3.3.2 *Boundary layer flows*

The main drawback of the Poiseuille flow as a test case for continuous transition scenarios in open flows stems from its strongly subcritical character. Boundary layer flows may seem more favourable from this point of view and, being ubiquitous, have even more practical interest.

The Blasius velocity profile is another example of flow without internal

---

[22]D.R. Carlson, S.E. Widnall, M.F. Peeters, "A flow visualization of transition in plane Poiseuille flow," J. Fluid Mech. **121** (1982) 487–505.

inflection point. As we have seen, the thickness δ of the laminar boundary layer increases as the square root of the distance to the leading edge. Theory predicts the emergence of TS waves at $R = R_c \simeq 519.4$ with $k_c \simeq 0.303$ and $c_{r,c} \simeq 0.397$ (scaling based on δ). The bifurcation seems supercritical or nearly so, but the interpretation is made delicate by the downstream dependence of the flow. Anyway, linear stability has been tested at low residual turbulence levels in vibrating ribbon experiments and the agreement between predictions and observations is satisfactory once corrections due to the slight divergence of the flow are taken into account.

Several tricks have been used in an attempt at controlling the spanwise wavevector q_z of the secondary modes in experiments (regularly spaced grooves or adhesive tapes aligned along the flow direction). In practice the most dangerous secondary modes either do not add streamwise modulation ($q_x = 0$, $\varsigma = 0$, where ς was introduced in §7.3.1.2) or else introduce a spatial subharmonic ($q_x = k/2$, $\varsigma = 1/2$). In the latter case, resonant interactions between a triad formed with the primary mode $(k, 0)$ and two oblique waves ($q_x = k/2, \pm q_z$) may be expected from the quadratic coupling through the advection term $\mathbf{v} \cdot \boldsymbol{\nabla} \mathbf{v}$ (Figure 7.17, top).

The velocity field resulting from the superposition of the base flow and the secondary modes displays streamwise bands with alternatively increased and reduced speed ("peak–valley" alternation) that, upon amplification, take the shape of so-called Λ-vortices, see Figure 7.17, bottom.

A step-by-step transition is thus expected. Laminar flow first becomes unstable against two-dimensional TS waves, the amplitude of which slowly increases over a viscous time-scale. TS-waves then experience growth of secondary modes introducing three-dimensional dependence over a much shorter inertial time-scale, and a final breakdown into small scale turbulence.

In fact the kind of secondary mode that develops the fastest depends on the amplitude A_1 reached by the TS waves (exponential amplification of the primary mode over a limited time, converted into a distance by the downstream flow), which in turn is a function of the level of residual turbulence (*receptivity*). For a detailed presentation consult the reviews by Herbert and/or Kachanov,[23] and also [Schmid and Henningson (2001)].

• At the lowest residual turbulence levels, the amplitude of the TS waves staying below 0.2–0.3% of the speed at infinity, relaminarisation is expected

---

[23]T. Herbert, "Secondary instabilities of boundary layers," Ann. Rev. Fluid Mech. **20** (1988) 487–526; Y.S. Kachanov, "Physical mechanisms of laminar-boundary-layer transition," Ann. Rev. Fluid Mech. **26** (1994) 411–482.

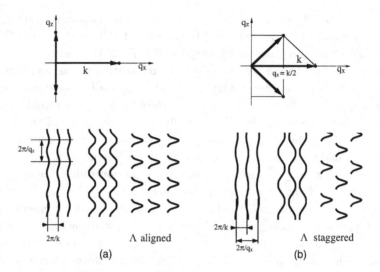

Fig. 7.17 Top: Resonance conditions for secondary modes. Bottom: (a) Aligned Λ-vortices, $q_x = 0$, 'K' mode. (b) Staggered Λ-vortices, $q_x = k/2$, 'H' mode.

as noticed earlier from the consideration of Figure 7.11, p. 280.

- When their amplitude is between 0.3% and 0.6%, the fastest emerging secondary mode is subharmonic, which produces staggered Λ-vortices, this is 'mode H' after Herbert's name, Note 21, p. 291.
- Finally, when their amplitude is larger than 0.6%, the prevalent secondary mode is with $q_x = 0$ and produces aligned Λ-vortices as early observed by Klebanoff and coworkers.[24]

Λ-vortices then breakdown into *spikes* that serve as germs for turbulent spots developing first in the cross-flow direction y and then laterally in the x–z plane while being carried downstream. Merging of the turbulent patches yields a fully developed turbulent boundary layer. Large streamwise rolls called *wall streaks*, with finite lifetime and well-defined wavelength can be identified within the turbulent boundary layer. They experience intermittent breakdown and regeneration. This phenomenon called *bursting* has been studied within the framework of chaotic dynamical systems using some of the tools presented in §5.2.[25]

Figure 7.18 adapted from [Schlichting (1979)] displays a global

---

[24]P.S. Klebanoff, K.D. Tidstrom, L.M. Sargent, "The three-dimensional nature of boundary layer transition," J. Fluid Mech. **12** (1962) 1–34.

[25]N. Aubry, Ph. Holmes, J.L. Lumley, E. Stone, "The dynamics of coherent structures in the wall region of a turbulent boundary layer," J. Fluid Mech. **192** (1988) 115–173.

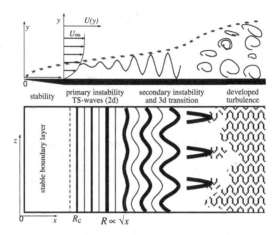

Fig. 7.18 Side view and top view of a 'K mode' transitional boundary layer. After [Schlichting (1979)].

schematic image of these different steps which may be difficult to identify in the absence of control of secondary modes (natural transition) because *turbulent spots* often appear without apparent precursors. Such spots can also be introduced voluntarily by perturbing the laminar boundary layer locally and sufficiently strongly. Of course, when the flow is very clean, one must not conclude that it never become turbulent, but that, when it does, the other scenario involving spots develops. A kind of competition between the standard sequences sketched above and a *bypass* transition not relying on primary (TS) and secondary instability modes can thus be observed in the general case.

7.3.4 Turbulent spots and intermittency

7.3.4.1 Context

In several instances we have mentioned the fact that *turbulent spots*, i.e. bounded regions filled with strongly turbulent flow, could appear, move, and expand within laminar flow. This phenomenon can be understood by noticing that solutions to the Navier–Stokes equation may not all derive continuously from the *thermodynamic branch* (p. 50) by increasing the Reynolds number after supercritical bifurcations, and that there may exist other fully nonlinear solutions belonging to other disconnected branches.

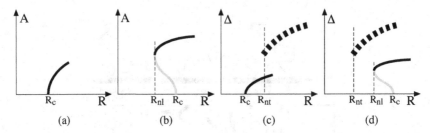

Fig. 7.19 (a, b) Super/sub-critical local bifurcation. The distance to the basic state is measured by the amplitude A of the primary instability mode beyond the corresponding threshold R_c. (c, d) Global bifurcation toward a branch of nontrivial states existing beyond $R > R_{nt}$, disconnected from the basic state, Δ is the measure of the distance to that state.

Figure 7.19 illustrates this situation in a highly simplified way. The bifurcation diagrams (a) and (b) correspond to *local* bifurcations and have the same precise technical meaning as in Chapter 4: variable A on the vertical axis is the amplitude of the bifurcated state. In contrast, the nature of the states on the nontrivial branch in diagrams (c) and (d), resulting from a *global* bifurcation, is left unspecified for the moment. Quantity Δ, which serves as an order parameter, is simply some idealised statistical measure of the distance to the base flow, hiding a multiplicity of (possibly irregular) flow configurations.

From these last two sub-figures, one will just retain (*i*) the possible existence of nontrivial states for $R > R_{nt}$ ('nt' for 'nontrivial') and (*ii*) the fact that depending on cases, the local bifurcation can take place at values of R lower or higher than R_{nt}. Diagram (d) represents a case with $R_c > R_{nl}$ and $R_{nl} > R_{nt}$, i.e. a subcritical, primary, local bifurcation and nontrivial states at even lower values of R, reminiscent of our description of the plane Poiseuille flow transition. Diagram (c) with a supercritical primary bifurcation occurring first $R_c < R_{nt}$, and a nontrivial branch soon becoming relevant, would rather model the case of the Blasius boundary layer, though the exact situation is difficult to appreciate due to the downstream development of the instability.

At any rate, the suggestion is made explicit that the *globally supercritical* cascade toward turbulence can be pre-empted by other mechanisms involved in the so-called bypass transition that can thus be termed *globally subcritical*. In terms of dynamical systems theory, the principal difficulty is then to determine the boundary of the attraction basin of the nontrivial solution, that is to say not only the most dangerous directions in state space, from the base state or related trivial nonlinear solutions, but also the amplitude of the perturbations that actually trigger the transition. Solving Exercise 4.4.3 may contribute to the understanding of that difficulty in an highly simplified case, the connection of that exercise with the present problem is discussed in the article from which it derives, Note 12, p. 164.

The case of plane Couette flow is, from this point of view, the most dramatic since the base flow is linearly stable for all R as shown by Romanov,[26] so that the branch that would correspond to the local bifurcation is pushed at infinity (in R).

This circumstance is shared by the Hagen–Poiseuille flow, the laminar flow in a cylindrical pipe, which is not a plane flow. The parabolic velocity profile (7.5) is also linearly stable for all R,[27] though turbulent flow can of course be observed for R large enough. This flow configuration is mentioned here in view of the historical importance of Reynolds' experiment regarding the definition of the Reynolds number and the problem of the transition to turbulence.[28] The transition occurs without intermediate steps through the formation of *turbulent plugs* equivalent to the turbulent spots observed in plane flows. As suggested in Figure 7.20, a turbulent plug develops close to the tube wall, then invades the section and grows in the streamwise direction to become a *turbulent slug*. These slugs grow in length and thus persist indefinitely, i.e. till the end of the pipe, provided R is large enough. In practice, the head of a slug moves faster than its tails for R greater than about 2300. The slugs become statistically longer and longer beyond this value and, if their birth probability is large enough, the pipe becomes turbulent along most of its length. The transition is thus highly sensitive to the care brought to the experimental conditions (shape of the entrance of the pipe, roughness of the wall, level of residual turbulence). The generation mechanism of the plug depends on the Reynolds number. Laminar flow can be maintained at $R > 10^5$ in experiments designed with

---

[26] V.A. Romanov, "Stability of plane parallel Couette flow" (english transl.) Funktsional'nyi Analiz i Ego Prilozhaniya **7** (1973) 62–73.

[27] H. Salwen *et al.*: "Linear stability of Poiseuille flow in a circular pipe," J. Fluid Mech. **98** (1980) 273–284. Á Meseguer, L.N. Trefethen: "Linearized pipe flow to Reynolds number 10^7," Journal of Computational Physics **186** (2003) 178–187.

[28] O. Reynolds, "An experimental investigation of the circumstances which determine whether the motion of water shall be direct or sinuous, and the law of resistance in parallel channels," Phil. Trans.R. Soc. London **174** (1883) 935-982. Videos of experiments performed with Reynolds' original set-up can be found in [MFM (2007)]. The experiment is repeated in the movie *Turbulence* [Stewart (1972)] where the Reynolds number is modified by mixing water and glycerol in order to vary the viscosity, rather than by changing the flow rate.

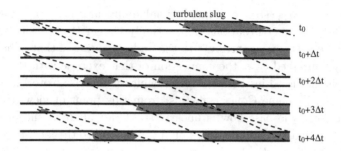

Fig. 7.20 Transition to turbulence by growth of turbulent plugs in a cylindrical pipe. The growth, propagation, and merging of the slugs is suggested by sketches of snapshots taken at regularly spaced successive times.

extreme care but usually the natural transition is observed in pipes of moderate length for $R < 10^4$ due to detachment on sharp angles in the entrance section, or in the range $R \sim 10^4$–10^5 due to the instability of the boundary layer close to the wall of the pipe. At the lowest end of the transitional regime, turbulent plugs are called *slugs*. Below $R \approx 2300$, these slugs turn into coherent structures called *puffs*. The decay of turbulence toward laminar flow happens in the form of *chaotic transients* characteristic of low dimensional dynamical systems, see Exercise 5.6.3, p. 223.[29]

7.3.4.2 Turbulent spots in Blasius and Poiseuille flows

Described first by Emmons (1951), turbulent spots developing in the Blasius boundary layer flow have the shape of an arrow head (Figure 7.21, left).[30] They move rapidly in the flow, about 90% and 50% of the speed at infinity for the head and the tail respectively. They widen while advancing with a spreading angle of order 10°. They also thicken by entraining laminar fluid from the outside of the boundary layer.

In order to partially suppress problems linked to the space dependence of the base flow, Carlson *et al.* (Note 22, p. 292) have studied the development of spots artificially produced in plane Poiseuille flow by local triggering (R is now independent of the streamwise coordinate x). As already mentioned,

---

[29] B. Eckhardt *et al.*, "Turbulence transition in pipe flow," Ann. Rev. Fluid Mech. **39** (2007) 447–468. A.P. Willis *et al.*, "Experimental and theoretical progress in pipe flow transition," Phil. Trans. R. Soc. A 366 (2008) 2671-2684. Articles in: Phil. Trans. R. Soc. A **367** (2009) Feb. 13 (special issue "Turbulence transition in pipe flow: 125th anniversary of the publication of Reynolds' paper").

[30] B. Cantwell, D. Coles, P. Dimotakis, "Structure and entrainment in the plane of symmetry of a turbulent spot," J. Fluid Mech. **87** (1978) 641–672.

Fig. 7.21 Turbulent spots in the Blasius flow at high R (left, after Cantwell *et al.* in [van Dyke (1982)]), and in plane Poiseuille flow (right, after Carlson *et al.*, Note 22).

turbulent patches persist within laminar flow for $R \geq 1000$, i.e. much below the value at which nonlinear TS waves bifurcate ($R_{\rm nl} \simeq 2900$). As seen in Figure 7.21 (right), triggered spots are oval. The head and tail of a spot move at speeds roughly 2/3 and 1/3 of the centreline velocity, respectively. The spreading angle is of the order of 8°. Further downstream, at distances $\sim 150h$ (h = channel's height), a typical turbulent patch has a diameter of order 30–50h and then experiences *spot splitting*: a calm region appears, separating two turbulent spots that run side by side and further grow. Turbulent patches that develop spontaneously within the laminar flow at high levels of residual turbulence evolve similarly but have less regular shapes and, of course, appear randomly in space and time.

As a general rule, turbulent spots display different regions that are well visible in Figure 7.21: (1) the spot as a whole, a turbulent region with well-defined boundaries, that behaves as an obstacle in the flow since it moves somewhat more slowly than the neighbouring laminar flow; (2) on the sides, trains of oblique waves seemingly produced by the motion of this obstacle; (3) an interior filled with small scale turbulence; and (4) a turbulent "wake" made of streamwise elongated streaks. These features are also present in Figure 1.11, p. 22, displaying a mature turbulent spot in plane Couette flow.

7.3.4.3 Plane Couette flow

The globally subcritical character of the transition toward nontrivial flow regimes, that implies coexistence *in state space*, is essential to understand the coexistence of turbulent and laminar flow *in physical space* with well-defined fronts separating them at one and the same value of the control

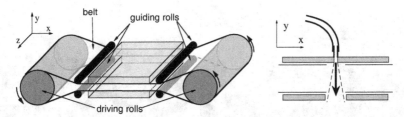

Fig. 7.22 Set-up used to produce high-quality plane Couette flow in view of a quantitative study of turbulent spots. Left: The shear is created in the gap separating two portions of a wide plastic belt kept parallel at a constant distance $2h$ by four adjustable rolls, with counter-plates limiting mechanical vibrations, and entrained at linear speed U by two large rolls. Right: Turbulent spots are generated by a thin jet that goes through the flow when holes in the belt and the counter-plates are aligned. After Daviaud and coll. (CEA Saclay).

parameter.[31] Plane Couette flow, the simplest case of wall-bounded flow, stays linearly stable for all R (Note 26), which guarantees the subcritical character of the transition and avoids any risk of confusion with scenarios based on saturated TS waves. In contrast with Blasius or Poiseuille flows, there is further no problem with the downstream transport of spots when using the apparatus used in Saclay[32] and described in Figure 7.22 since, by construction, the mean advection cancels exactly for symmetry reasons. In spite of the difficulty to produce a high-R flow with a low level of residual turbulence, plane Couette flow thus represents an ideal case from many points of view.[33]

Traditionally, the Reynolds number is defined as $\tilde{R} = Uh/\nu$ where U and $-U$ are the speeds of the walls inducing the shear and $2h$ the width of the gap between them. It turns out physically more significant to define R from the shear rate U/h, the inverse of the advection time, and the viscous time over the gap, $\tau_{\rm v} = (2h)^2/\nu$, since experiments show that the structures involved in the transition occupy the full gap. The Reynolds number will accordingly be defined as $R = 4Uh/\nu = 4\tilde{R}$, which permits easier comparisons with the cases of boundary layer and Poiseuille flows.

---

[31] Y. Pomeau: "Front motion, metastability and subcritical bifurcations in hydrodynamics," Physica D **23** (1986) 3–11.

[32] F. Daviaud et al.: "Subcritical transition to turbulence in plane Couette flow," Phys. Rev. Lett. **69** (1992) 2511–2514; O. Dauchot and F. Daviaud, "Finite amplitude perturbations and spot growth mechanism in plane Couette flow," Phys. Fluids **7** (1995) 335–343.

[33] Interesting information can also be obtained from direct numerical simulations of Navier–Stokes equation, e.g. A. Lundbladh and A.V. Johansson, "Direct simulations of turbulent spots in plane Couette flow," J. Fluid Mech. **229** (1991) 499–516.

Turbulent spots in plane Couette flow display characteristics similar to those in other wall-bounded flows. It is now well established that the plane Couette flow is globally stable for[34] $R < R_g \simeq 1300$. This value compares well with that in the plane Poiseuille flow case (~ 1000 for sustained spots), which is easily understandable when noticing that the latter can be viewed as two juxtaposed Couette flows.[35]

The expression 'globally stable' is to be taken with precisely the meaning defined in Figure 1.2, p. 9. Experiments leading to the result are principally of two kinds:[36] either *quench* experiments ('Q') where a fully turbulent state, prepared at some large initial R, $R_{\text{init}} \gg R_g$, is suddenly decreased to some final value R_{fin}, or *spot triggering* experiments ('S') achieved as sketched in Figure 7.22 (right). In each case, the relative surface occupied by the turbulent flow F_t, called the *turbulent fraction*, is extracted from video recordings of the flow pattern, Figure 1.11, p. 22, after appropriate thresholding of the images. As illustrated in Figure 7.23, these experiments show that for $R \simeq 1120$ all perturbations relax rapidly, whatever their initial structure, shape, and amplitude. For R between $R_u \simeq 1250$ and $R_g \simeq 1300$, turbulent patches exist and live sufficiently long for the average measure of their surface as a function of time to make sense. This lifetime diverges as R gets closer and closer to R_g from below. Finally, turbulent patches persist indefinitely with finite probability for $R > R_g$.

It can be noticed that even when turbulence is sustained (Figure 7.23, bottom-right) some spots may relax immediately (label 'S2') though prepared in the same way as those that do not abort ('S1'). This feature is the trace of both the globally subcritical character of the bifurcation, and the sensitivity to initial condition expected from a basically chaotic process. Numerical simulations confirm this point by showing that, for Reynolds numbers in the corresponding range, the border of the attraction basin of the laminar flow is fractal.[37] Furthermore, in this regime, the average value of the turbulent fraction does not depend on the type of experiment (Q or S), which qualifies it as an adequate order parameter.

---

[34] For a review of experimental results, see P. Manneville and O. Dauchot, "Patterning and transition to turbulence in subcritical systems: the case of plane Couette flow," in *Coherent structures in classical systems*, D. Reguera, J.M. Rubi, L.L. Bonilla, eds. (Springer Verlag, 2001), pp. 58–79.

[35] For Poiseuille flow, as defined on p. 275, $R_P = U_{\max} h_P / 2\nu$, where h_P is the gap. With the correspondence $U_{\max} \mapsto 2U_C$, $h_P/2 \mapsto 2h_C$, one gets $R_P = 2U_C 2h_C / \nu = R_C (\equiv 4\tilde{R})$.

[36] S. Bottin, F. Daviaud, P. Manneville, O. Dauchot, "Discontinuous transition to spatiotemporal intermittency in plane Couette flow," Europhys. Lett. **43** (1998) 171–176.

[37] A. Schmiegel and B. Eckhardt, "Fractal stability border in plane Couette flow," Phys. Rev. Lett. **79** (1997) 5250–5253.

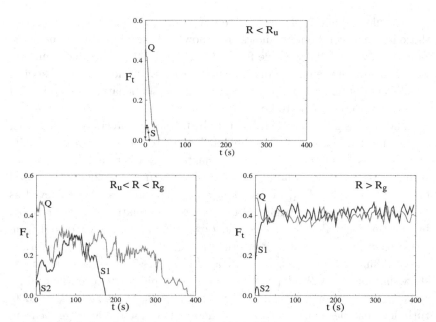

Fig. 7.23 Turbulent fraction function of time for different values of R. 'Q' indicates a *quench* experiment, and 'S' a *spot triggering* experiment. Top: Immediate unconditional relaxation toward laminar flow. Left: Relaxation usually after a long chaotic transient. Right: Sustained turbulence. Adapted from original results, courtesy S. Bottin and O. Dauchot (CEA Saclay, 1998).

The *bifurcation diagram* displayed in Figure 7.24 summarises results on the average turbulent fraction F_t as a function of R. This diagram suggests that F_t increases regularly with R. As a matter of fact, the flow becomes uniformly turbulent beyond $R_t \simeq 1660$. In the range $R_g < R < R_t$ the intensity of turbulence appears to be modulated in the form of alternately laminar and turbulent oblique bands. Just beyond R_g, these bands are fragmented (Figure 7.25, bottom) but, as R increases, for $R \sim 1400$–1450, fragments join together to form continuous bands (Figure 7.25, top). Then the modulation progressively disappears as the regime of uniform turbulence is reached. These results are recapitulated in Figure 7.26.

Turbulent bands are strikingly similar to the celebrated turbulent spirals observed in cylindrical geometry (Taylor–Couette flow, see Exercise 3.3.5, p. 122) when the cylinders rotate in opposite directions and sufficiently fast. Early observations of these spirals date back to Coles' experiments.[38]

---

[38] D. Coles, "Transition in circular Couette flow," J. Fluid Mech. **21** (1965) 385–425.

7. Open Flows: Instability and Transition   303

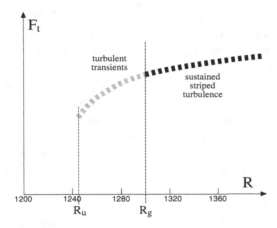

Fig. 7.24 Tentative bifurcation diagram displaying the turbulent fraction F_t (a measure of the distance to the laminar flow) as a function of the Reynolds number: experiments show a change of behaviour from sustained to transient turbulence at $R_g \approx 1300$. This behaviour is typical of extended systems and suggests to reinterpret the ending of the nontrivial branch at R_{nt} in Figure 7.19 in the same way.

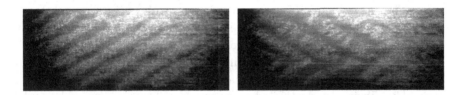

Fig. 7.25 Typical oblique turbulent stripes observed in the range $R_g < R < R_t$ in the plane Couette flow for $R = 1432$ (left) and $R = 1360$ (right). Courtesy A. Prigent (CEA Saclay).

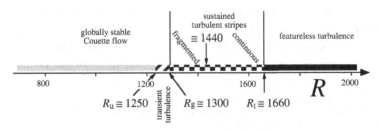

Fig. 7.26 Stability diagram of the plane Couette flow.

A quantitative bifurcation diagram has been established by Andereck et al.[39] for a ratio $\eta = r_o/r_i = 0.883$ where r_o and r_i are the radii of the outer and inner cylinders respectively. Further experiments performed by Dauchot and Prigent[40] with $\eta = 0.983$ show that bands and spirals take place in the same range of Reynolds numbers (weakly dependent on the average rotation rate as far as spirals are concerned), once computed, as in the case of the plane Couette flow, from the shear rate and the viscous time over the gap when curvature effects are small, $\eta \to 1$. A physical explanation for the occurrence of turbulent bands or spirals is still lacking at present. This phenomenon is presently an active numerical and theoretical field of research.[41]

The principal interest of this detailed study is to give a concrete expression to the conceptual presentation given in Figure 7.19, especially in connection with the nature of the states on the postulated disconnected *nontrivial branch* of the bifurcation diagram (compare with Figure 7.24).

7.3.4.4 Mechanisms?

Turbulent spots appear to play an essential role in transitional wall-bounded flows. They have common features that do not involve TS waves and their nonlinear development. First they are present at values of R which are not too large so that the flow as a whole remains relatively coherent and not yet filled with small scale eddies as in fully developed turbulence (Chapter 8). For plane Couette flow this transitional regime extends in the range $1250 < R < 1440$. Trying to identify a limited number of interacting modes and processes thus seems meaningful.

According to the Squire theorem and the conventional TS transition scenario, the most dangerous infinitesimal perturbations are two-dimensional, i.e. periodic in x, functions of y, and independent of z. Disturbances relevant to the non-conventional scenario are thus expected to have finite amplitude and be mostly spanwise, i.e. to depend on z (and y) but not on x. Small perturbations of this kind are damped in the long term but transient linear amplification is not excluded owing to the *non-normal* character of

---

[39] D. Andereck and L.H. Swinney, "Flow regimes in a circular Couette flow system with independently rotating cylinders," J. Fluid Mech. **164** (1986) 155–183.
[40] A. Prigent et al.: "Large-scale finite-wavelength modulation within turbulent shear flows," Phys. Rev. Lett. **89** (2002) 014501.
[41] D. Barkley and L.S. Tuckerman: "Mean flow of turbulent-laminar patterns in plane Couette flow," J. Fluid Mech. **576** (2007) 109–137. Y. Duguet et al.: "Formation of turbulent patterns near the onset of transition in plane Couette flow," J. Fluid Mech. to appear (2010).

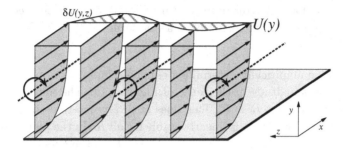

Fig. 7.27 Streamwise vortices and formation of streaks by the *lift-up* mechanism.

the linear stability operator (see §A.3 and Exercise 4.4.3). The required finite amplitude can thus be reached provided that sufficiently large residual turbulence or intentional disturbances are present.

In practice it is not too difficult to imagine a mechanism by which streamwise structures can be sustained in the flow. As a matter of fact, a streamwise vortex $[0, v(y,z), w(y,z)]$, even transient and weak, induces a redistribution of streamwise velocity $\delta U(y,z)$ by displacing fluid with speed $U(y)$ to a different cross-stream coordinate y. This perturbation δU corresponds to a slowing down or a speeding up, depending on the local direction of displacement of the basic flow lines by the vortex (see Figure 7.27). The presence of long, well visible, *streamwise streaks* in the flow is thus not surprising. The superposition of the base flow and the streamwise component of the induced flow now displays inflection points that may make it unstable to KH-like mechanism producing x-periodic waves, which in turn can trigger the transition.

The question is then to determine the initial amplitude of streamwise vortices able to do the job (to find the boundary of the attraction basin of the base state in dynamical systems terms). The answer has been searched mostly by direct numerical simulation of the NS equations and comparison of scenarios starting with different initial perturbations. For a detailed presentation with references, consult [Schmid and Henningson (2001), Chap. 9]. This points to the *receptivity* problem, i.e. the way the flow converts residual turbulence or controlled disturbances into effective perturbations.

In boundary layers, perturbations in the form of oblique waves (an anticipation of secondary instability modes of TS waves) seem to be most efficient. In the case of plane Couette or Poiseuille flows, with Reynolds numbers independent of the streamwise distance, the result is often presented in the form of a diagram relating the initial amplitude required for transition and the Reynolds number in the form

$$\Delta_{\text{init}} \propto R^{-\gamma},$$

where γ is an empirically determined exponent and Δ is a measure of the perturbation amplitude in the sense of Figure 7.19. The above relation is supposed to hold for $R \gg 1$. This does not raise questions in the case of plane Couette flow which is linearly stable for all R. In the plane Poiseuille flow case, such a relation can strictly hold only in the subcritical regime, $R < R_c \simeq 5772$, but turbulent patches are already expected beyond $R \sim 1000$, which leave room to the observation of the scaling behaviour. For boundary layers, the relation makes sense only as long as the level of residual turbulence is sufficiently low that the conventional TS transition is not bypassed. Values $1 \geq \gamma$ are usually found, $\gamma = 1$ being obtained from a scaling argument directly applied to the NE equations.

The study of processes leading to the transition can also give ideas about the nature of states belonging to the nontrivial branch since it is sufficient to admit that the instability of the streamwise streaks produces new streamwise vortices, which closes the loop. Such a cycle has indeed been pointed out in numerical simulations. The model studied considered in Exercise 4.4.3 with its variables X and Y representing the amplitude of the streaks and the streamwise vortices respectively, possesses both trivial and nontrivial states, but is too simple to account for a realistic feedback. A more sophisticated model has been proposed by Waleffe[42] coupling four variables. It reads:

$$\frac{d}{dt}U = -\kappa_u R^{-1} U - \sigma_w W^2 + \sigma_u MV,$$
$$\frac{d}{dt}V = -\kappa_v R^{-1} V + \sigma_v W^2,$$
$$\frac{d}{dt}W = -\kappa_w R^{-1} W + \sigma_w UW - \sigma_m MW - \sigma_v VW,$$
$$\frac{d}{dt}M = -\kappa_m R^{-1}(M-1) + \sigma_m W^2 - \sigma_u UV,$$

where M represents the mean flow, V the amplitude of the streamwise vortices, U that of the streaks, and W a supplementary variable accounting for three-dimensional features of the perturbed flow. Constants κ describe the viscous effects and coefficients σ are coupling constants arising from the NS advection term. Their values are such that the kinetic energy is conserved in the sense of Exercise 2.5.2. In addition to its base state $(M = 1, U = V = W = 0)$, the model has several nontrivial states for R

---

[42] F. Waleffe, "On a self-sustaining process in shear flows," Phys. Fluids **9** (1997) 883–900.

sufficiently large, representing possible operating points of the cycle mentioned above. In particular, in the perturbation equations around the base state, the equation for U contains a term in V (from the linearisation of the term $\sigma_u MV$) that accounts for the *lift-up* mechanism at the origin of transient energy growth (Figure 7.27). The boundary of the attraction basin of the base state has been studied in detail for this system and complex transients reminiscent of what is observed in laboratory experiments below the global stability threshold have been shown to exist.[43] More realistic modelling is the subject of ongoing work.

In the following we shall no longer consider the instability mechanisms at the origin of unsteadiness in turbulent flows but rather focus on the statistical aspects of the flows that develop beyond the transitional regime.

7.4 Exercises

7.4.1 *Velocity-vorticity perturbation equations*

Starting from Equations (7.17–7.20) for the perturbation, show that the pressure fulfils a Poisson equation:

$$\nabla^2 p = -2v\partial_y U \qquad (7.52)$$

by computing the divergence of the momentum equations and using the continuity equation. Applying the three-dimensional Laplacian to the equation for v, next eliminate the pressure variable thanks to (7.52) and show that this yields:

$$(\partial_t + U\partial_x)\nabla^2 v - \partial_{yy}U\partial_x v = R^{-1}\nabla^4 v. \qquad (7.53)$$

Finally, by cross differentiation and subtraction of equations for u and w show that the equation for $\Omega_y = \partial_z u - \partial_x w$ reads

$$(\partial_t + U\partial_x)\Omega_y - R^{-1}\nabla^2\Omega_y = -\partial_y U \partial_z v. \qquad (7.54)$$

Notice that (7.53) is formally identical to the Orr–Sommerfeld equation (7.30) and is closed for v, while (7.54), called the *Squire equation*, is not closed for Ω_y but contains a forcing term in v.

Show that the the no-slip condition $u = v = w = 0$ at a solid wall implies the boundary conditions $v = \partial_y v = 0$ and $\Omega_y = 0$.

---

[43] O. Dauchot and N. Vioujard, "Phase space analysis of a dynamical model for subcritical transition to turbulence in plane Couette flow," Eur. Phys. J. B **14** (2000) 377–381.

7.4.2 Derivation of Squire's theorem

The theorem states that for a one-directional plane flow $\mathbf{v}_0 = U(y)\hat{\mathbf{x}}$, Figure 7.1a, the most dangerous perturbations are two-dimensional, i.e. without spanwise velocity component and independent of the transverse coordinate z.

1) Write down the linearised NS equations (7.17–7.20) for a normal mode

$$[\mathbf{v}, p](x, y, z, t) = [\bar{\mathbf{v}}, \bar{p}](y) \exp[i(k_x(x - ct) + ik_z z)].$$

2) Notice the symmetrical role of $k_x u$ and $k_z w$ in the continuity equation and derive the equation governing $k_x u + k_z w$ by appropriate combinations of the equations for u and w.

3) Check that the equations for $k_x u + k_z w$ and v are those of a two-dimensional problem in x and y (7.24–7.26) for some wavevector $\tilde{k}$ and a Reynolds number $\tilde{R}$ to be identified.

4) Conclude the argument by comparing the critical Reynolds numbers corresponding to a three-dimensional mode (k_x, k_y) and the associated two-dimensional mode $(\tilde{k}, 0)$ in the viscous case, and the growth rates of these modes in the inviscid case.

7.4.3 Derivation of Fjørtoft's theorem

Rayleigh's theorem is obtained by working with the imaginary part of equation (7.34). Fjørtoft's improvement is gained by manipulating its real part.

1) Use Rayleigh's theorem to show that the real part of the phase velocity c_r can be replaced by any real speed $\bar{U}$ in the real part of (7.34).

2) Choose the speed $\bar{U} = U_s = U(y_s)$ at an inflection point ($U''(y_s) = 0$) as a reference speed and expand $U(y) - U_s$ and U'' in Taylor series in the neighbourhood of y_s.

3) Insert these expansions in the identity for the real part of (7.34) and, assuming that the velocity profile $U(y)$ is monotonic with a single inflection point, derive a necessary condition for instability by the argument used to obtain Rayleigh's criterion. Express the result by noticing that $U'(y)$ is the opposite of the base flow vorticity $(\boldsymbol{\nabla} \times \mathbf{v})_z = \partial_x v_y - \partial_y v_x$.

7.4.4 Stability of some linear-by-part velocity profiles

Instability modes of profiles in the form (7.36–7.39) are searched here by direct computation. (For unbounded domains, the boundary conditions at infinity are obtained by expressing that the solutions do not diverge exponentially.)

• Derivation of matching conditions (7.38) and (7.39).

1) Express the continuity of pressure from equation (7.24) for the streamwise velocity component u at discontinuity points of U and/or U' and derive condition (7.38).

2) Write down the equation governing the motion of a material point with cross-stream coordinate Y belonging to the interface between the two flows, first in full generality and next for a normal mode (i.e., for $Y = \tilde{Y} \exp(ik(x - ct))$). Noticing that by definition of the cross-stream velocity component one has $dY/dt = v$, obtain $\tilde{Y}$ as a function of $\tilde{\psi}$ and derive the matching condition (7.39).

• Kelvin–Helmholtz instability and mixing layer.

1) Find the dispersion relation of normal modes for a confined mixing layer with velocity profile:

$$U(y) = U_1 \quad \text{for} \quad 0 < y < y_2,$$
$$U(y) = U_2 \quad \text{for} \quad y_1 < y < 0.$$

where $y_1 < 0$ and $y_2 > 0$ are the cross-stream bounds of the fluid vein.

2) Same question for the smoothed velocity profile:

$$U(y) = +1 \quad \text{for} \quad 1 < y < y_2,$$
$$U(y) = y \quad \text{for} \quad -1 < y < 1,$$
$$U(y) = -1 \quad \text{for} \quad y_1 < y < -1.$$

Take $y_1 = -y_2$. Consider first y_2 large and next $y_2 > 1$ but finite.

3) Same question for the monotonic profile defined as

$$U(y) = \alpha y + \tfrac{1}{2}(1 - \alpha)\bigl(|y + 1| - |y - 1|\bigr)$$

on the interval $[-b; b]$, $b > 1$ and discuss the solutions according to the value of α.

a) Show that, for all α, modes with $k \to \infty$ are not unstable.

b) Show that there cannot be unstable modes when $\alpha > 1$.

c) For $\alpha < 1$, show that there exist unstable modes when $k \to 0$ and

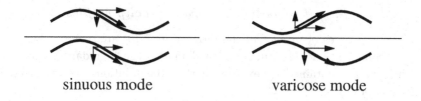

Fig. 7.28 Streamlines of the unstable modes of the plane jet.

$\alpha \approx 1$ (the complete dispersion relation should be determined outside this neighbourhood). Interpret these results in view of Fjørtoft's theorem.

- Stability of the plane jet.

The base velocity profile of a plane jet (with speed $U = 1$) of some inviscid fluid entering a large vessel of the same fluid at rest can be taken in the form

$$U(y) = 0 \quad \text{for} \quad y < -1 \quad \text{and} \quad y > 1,$$
$$U(y) = 1 \quad \text{for} \quad -1 < y < 1.$$

1) Compute the dispersion relation while taking care of the parity of the solutions in the cross-stream coordinate y.

2) Discuss the solutions and explain why the even mode is called 'sinuous' and the odd mode 'varicose' (cf. Figure 7.28).

7.4.5 Absolute/convective instability (CGL equation)

Consider the linearised Ginzburg–Landau equation in the following form:

$$\partial_t A - V \partial_x A = rA + (1 + i\alpha) \partial_{x^2} A$$

where A is the amplitude of the envelope of the primary waves generated by some instability mechanism and r is the corresponding control parameter, diffusion is normalised to 1 and α accounts for the dispersion of the waves (Chapter 6, p. 243). A supplementary term $V \partial_x A$ has been introduced to mimic the effects of an average flow at some speed V on the pattern.

1) Determine the dispersion relation of plane waves solutions in the form $A \propto \exp(i(kx - \omega t))$ and derive their growth rate. When are they unstable according to the definition and what is their (real) phase speed?

2) When $V \neq 0$, the question arises of the convective or absolute nature of the instability. Compute the group velocity of plane waves $v_g = \partial \omega / \partial k$,

then the complex wavevector $k^{(0)}$ for which $v_g = 0$, and find its growth rate $\omega_i^{(0)}$ as a function of r and V.

3) Conclude by applying the Briggs–Bers criterion, and discuss the respective roles of the instability mechanism (through r) and the advection (through V) on the absolute/convective character of the instability.

Chapter 8

Developed Turbulence

In the previous chapters, we focused our attention on the transition to turbulence in closed systems and then in open flows. This problem was essentially studied from the viewpoint of mechanisms and associated thresholds for idealised base flows. However, when one tries to evaluate the values of the control parameters in concrete cases, one finds that they are most of the time far beyond the thresholds determined theoretically. This is in particular the case of most flows of geophysical or engineering interest (flows in tubes, around vehicles or obstacles, combustion, ...) while the opposite situation, laminar flow, is quite exceptional (highly viscous fluids, lubrication, flows in capillaries of porous media, ...).

From the study of instability mechanisms, we can expect at most hints on the *details* of how things evolve *locally* and *instantaneously* in a turbulent flow. But this *microscopic* account, as useful as it can be to the understanding of irregularities and chaos in the flow, must be replaced by a statistical approach able to deal with their *macroscopic* regularities, in much the same way as, in gases, collisions between molecules lose interest when thermodynamics is relevant. As we shall see this analogy is fruitful but must not be pushed too far. We, indeed, have to stress immediately that turbulence is not a property of a fluid (in some new state of matter) but of its *motion* at scales large when compared to molecular dimensions.

As far as applications are concerned, turbulence is first of all characterised by strong *mixing* properties due to intense velocity fluctuations over a wide range of scales (§8.1). One of the objectives of the statistical theory to be built is thus to separate the mean flow from fluctuations (§8.2). Turbulent mixing implies statistical smoothing which is tempting to understand as a diffusion. As current experience tells us, it is more efficient by several orders of magnitude than molecular diffusion. Though it is not well founded theoretically, the notion of *turbulent diffusivity* is then introduced (§8.3) and we illustrate its use in the example of a fully turbulent boundary layer to predict the shape of the mean flow. We conclude the chapter by hinting at why and how to go beyond this elementary approach to deal with situations of practical interest.

In this chapter we are essentially interested in completing the overall perspective centred on the dynamics of complex systems (mostly involving fluids in flow) by giving a few ideas of what happens when a hierarchy of interacting structures is involved, and developed turbulence should just be taken as an example in continuation with what precedes. However, we choose to favour simple approaches of practical value in applications, while avoiding to discuss more theoretical issues about the nature of turbulence.[1] The chapter just touches on a wide subject, many expositions of which can be found in the literature by more qualified authors. Here, let us mention a few books and defer to the corresponding part of the general bibliography for more references. [Tennekes and Lumley (1972)] or [Lesieur (1997)] may first serve as thorough introductions. The monograph [Frisch (2001)], in the perspective of Kolmogorov's work, has a more theoretical flavour. Finally, [Pope (2000)] adds a particularly clear and interesting light on modelling and numerical issues.

8.1 Scales in Developed Turbulence

Let us come back to fundamental characteristics of developed turbulence and the different spatiotemporal scales involved. Their origin has been briefly and schematically illustrated in Figure 1.12, p. 23. Here we re-examine them in a more quantitative way.

8.1.1 *Production scale*

Having noticed that turbulent flows develop at high Reynolds numbers, let us try to be a little more precise and call ℓ_0 and $\tilde{v}_0$ the external characteristics of a given flow. Subscript '0' here indicates the scale of motions at the starting point of the hierarchy evoked above and generated by the nonlinear advection term $\mathbf{v} \cdot \nabla \mathbf{v}$. Tildes are attached to fluctuating quantities, the separation of the *mean flow* from *fluctuations* will be discussed later.

---

[1] A nice popular presentation of these issues is given by G. Falkovich and K.S. Sreenivasan: "Lessons from hydrodynamic turbulence," Physics Today, April 2006, pp. 43-49.

Fig. 8.1 Turbulent flow behind a grid: From production to decay. After a picture in [van Dyke (1982)].

Let $\tilde{v}_0$ measure the order of magnitude of velocity fluctuations and ℓ_0 be the typical size of eddies produced by the instabilities leading to local chaos. The relation between these scales and the apparent external characteristics of the flow may not be immediate. For example in grid turbulence (Fig. 8.1), ℓ_0 may *a priori* depend on the diameters of the bars, their wakes producing the eddies, on the distances between the bars which move the points where different wakes merge to produce a homogeneous turbulent flow, as well as on the typical diameter L of the channel that undoubtedly gives an upper bound to the size of the eddies present in the flow. For its own part, $\tilde{v}_0$ should be related to the transverse velocities generated by the vortices shed by the bars, which in turn scale with the average speed of the flow U. Some uncertainty thus exists about the value of the quantities entering the definition of the Reynolds number $R_0 = \tilde{v}_0 \ell_0 / \nu$. This is what will later encourage us, on p. 320, to define a Reynolds number conventionally noted R_λ and based on a length called the *Taylor micro-scale*, which can be derived from experimental measurements.

Let us notice that the hierarchy of scales gets wider as turbulence gets more developed. The multiplicative process from which it results will be analysed in a logarithmic perspective that will not be sensitive to the starting point $(\ell_0, \tilde{v}_0)$. Even if it differs from the naive nominal Reynolds computed from U and L by some unknown proportionality factor, we shall assume that the R_0 built on the production scale $(\ell_0, \tilde{v}_0)$ and the fluid's kinematic viscosity ν are always very large compared to one (powers of ten!) in situations of developed turbulence. This assumption dismisses an intermediate regime called *soft turbulence* still too close to the transitional regime and where the nature of the concrete mechanisms that lead to local chaos cannot be ignored, a regime we alluded to in the case of turbulent convection in §3.2.5 (see especially Figure 3.14, p. 113).

At the production scales, the viscous relaxation time of velocity fluctuations $\tilde{v}_0$ is large when compared to their turn-over time $\tau_0 = \ell_0/\tilde{v}_0$, since we have $\tau_v = \ell_0^2/\nu$, so that the ratio $\tau_v/\tau_0 = R_0$ is by assumption very large. Viscous dissipation can thus be neglected, leaving us just with the nonlinear advection term $\mathbf{v}\cdot\nabla\mathbf{v}$. As already discussed in §1.3.5, p. 23, when operating on eddies of typical size $k_0 \sim 1/\ell_0$, it tends to produce eddies of typical size $2k_0 \sim \ell_0/2$. This implies an energy transfer toward scales $\ell < \ell_0$ that should be independent of viscosity.

Following a well-established tradition, the energy flux from large to small scales is denoted ϵ. The fundamental assumption that allows us to estimate its value is of dimensional nature (see the exercises). It consists in saying that during a turn-over time τ_0 of eddies with size ℓ_0, a fraction (fixed on average) of the kinetic energy *per* unit mass present in these eddies $\sim \tilde{v}_0^2$ is transferred to smaller ones:

$$\epsilon \sim \tilde{v}_0^2/\tau_0 = \tilde{v}_0^3/\ell_0\,. \tag{8.1}$$

8.1.2 Inertial scales and the Kolmogorov spectrum

Since the advection term produces scales of smaller and smaller sizes, of order $\ell_0, \ell_0/2, \ell_0/4\ldots$ let us examine its effects on an eddy of size $\ell \ll \ell_0$ and velocity $\tilde{v}_\ell$, with turn-over time $\tau_\ell = \ell/\tilde{v}_\ell$. The corresponding Reynolds number reads $R_\ell = \tilde{v}_\ell \ell/\nu$ and, even if $\ell \ll \ell_0$, as long as $R_\ell \gg 1$, the corresponding motions carry the energy and transfer part of it, but have no time to dissipate it into heat, hence their name of *inertial* scales. As the scale division process is iterated, correlation among eddies is expected to decrease, so that it can easily be accepted that fluctuations at inertial scales are *homogeneous* and *isotropic*, that is to say invariant by translation and rotation, see Figure 8.2.

At steady state the energy cannot accumulate in some given scale. Accordingly, the amount that is extracted *per* unit time from larger scales has to be transferred to smaller ones by the nonlinear advection term. Assuming that the instantaneous transfer does not fluctuate and is equal to ϵ all along the cascade, i.e. independent of ℓ, one can estimate the velocity

Fig. 8.2 Sufficiently far from the production region, here a grid, one can consider that turbulence is locally homogeneous and isotropic when the Reynolds number is large enough. The *homogeneity* and *isotropy* concepts have to do with the statistical invariance of the flow with respects to translation and rotation, as suggested by the arrows supposed to move the circled region. After a picture in [van Dyke (1982)].

fluctuation $\tilde{v}_\ell$ using the argument leading to (8.1) for scale ℓ_0, which gives:
$$\epsilon \sim \tilde{v}_\ell^2/\tau_\ell = \tilde{v}_\ell^3/\ell \, ,$$
from which $\tilde{v}_\ell$ can be derived and thus the Reynolds number at the corresponding scale:
$$\tilde{v}_\ell \sim \tilde{v}_0 \left(\ell/\ell_0\right)^{1/3} = (\epsilon \ell)^{1/3} \tag{8.2}$$
and
$$R_\ell = \frac{\tilde{v}\ell}{\nu} = \frac{\tilde{v}_0(\ell/\ell_0)^{1/3}\ell}{\nu} = \frac{\tilde{v}_0 \ell_0}{\nu} \frac{\ell}{\ell_0} \left[\frac{\ell}{\ell_0}\right]^{1/3} = R_0 \left(\ell/\ell_0\right)^{4/3} \tag{8.3}$$

This energy transfer through the *inertial cascade* can continue as long as R_ℓ stays much larger than one, so that the neglect of the viscous dissipation remains legitimate, see Figure 8.3.

Turbulence in the inertial regime is usually characterised by its energy spectrum $E(k)$ defined from $K = \int_0^\infty E(k)\, dk$, where quantity K is the total kinetic energy *per* unit mass contained in the flow. Splitting the k-space in concentric shells, one can write
$$K = \sum_p K_p \, ,$$
where 'p' labels scale ℓ_p with wavevector $k_p = 2\pi/\ell_p$. The energy spectrum is homogeneous to an energy *per* unit wavevector (and *per* unit mass). On dimensional grounds this gives:
$$K_p = k_p E(k_p) \sim \tilde{v}_{\ell_p}^2 \, , \tag{8.4}$$

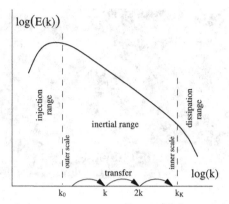

Fig. 8.3 Inertial cascade in wavevector space ($k = 2\pi/\ell$) from production to dissipation.

so that using (8.2) one gets:
$$K_p \sim \tilde{v}_0^2 (\ell_p/\ell_0)^{2/3}.$$
Coming back to k-space, omitting index p, and substituting this expression in relation (8.4) one obtains:
$$E(k) = C\epsilon^{2/3} k^{-5/3}. \tag{8.5}$$
This is the famous *Kolmogorov spectrum*. Constant C ($\simeq 1.5$ from empirical adjustment with experiments) is supposed to be universal.

In experiments, what is usually recorded is the time series of fluctuations at a given point rather than a chart of the whole fluctuation field at a given time. In order to pass from wavevectors to frequencies, it is sufficient to observe that the time dependence of a quantity associated with a small scale contains a trivial part due to the transport by the large scale motion. This is the essence of Taylor's *frozen turbulence hypothesis*. Especially in wind or water tunnels, with average speed V at the location of measurement, one gets $\omega = Vk$ and thus $d\omega = Vdk$, so that, in k as well as in ω, a power law with exponent $-5/3$ is expected. In fact the value of this exponent derives from an argument which is not rigourous (see Note 1) and has to be corrected for nontrivial effects of fluctuations to be alluded to later (§8.4) but Figure 8.4 shows that the guess is not too bad.

8.1.3 Dissipation scales

The turn-over time of eddies τ_ℓ can be obtained from $\epsilon \sim \tilde{v}_\ell^2/\tau_\ell \sim (\ell/\tau_\ell)^2/\tau_\ell \sim \ell^2/\tau_\ell^3$ which yields $\tau_\ell \sim (\ell^2/\epsilon)^{1/3}$, whereas the viscous diffusion time is still given by $\tau_{\rm v} \sim \ell^2/\nu$. As long as the first one is much

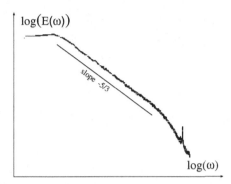

Fig. 8.4 Kinetic energy $E(\omega)$ contained in the longitudinal velocity fluctuations in the ONERA-S1 wind tunnel in Modane, France. Courtesy Y. Gagne (Grenoble University, 1987).

shorter than the second one, viscosity has no sufficient time to smooth out fluctuations and inertial effects dominate. This is the regime considered up to now. However, as the cascade proceeds, τ_ℓ decreases as $\ell^{2/3}$ while τ_v, initially larger, decreases much more rapidly as ℓ^2. Viscosity ceases to be negligible when these two times are of the same order of magnitude. This leads directly to the definition of the *Kolmogorov dissipation scale* ℓ_K:

$$\ell_K^2/\epsilon = \left[\ell_K^2/\nu\right]^3$$

or

$$\ell_K = \left(\nu^3/\epsilon\right)^{1/4} = \left(\nu^3 \ell_0 / \tilde{v}_0^3\right)^{1/4}, \tag{8.6}$$

$\ell_K \sim 1/k_K$ in Figure 8.3. It is easily checked that the Reynolds number corresponding to this scale is of the order of one, so that the different regimes match correctly: imposing $R = R_K = 1$ in (8.3) for the Reynolds number in the inertial range gives the same result that can be rewritten as

$$\ell_K/\ell_0 = R_0^{-3/4}. \tag{8.7}$$

This leads us to point out that the difference between similar turbulent flows manifests itself only at the smallest scales (Exercise 8.5.3). The film *Turbulence* [Stewart (1972)] illustrates this feature using cinema *special effects* as an example: the eye is mostly sensitive to large structures and is easily mistaken about the order of magnitude of the Reynolds number, which allows specialists to blast models rather than full-scale film sets.

Structures with sizes $\ell \ll \ell_K$, smoothed by dissipation, are completely enslaved to larger scales. In developed turbulence, eddies sufficiently independent to act as degrees of freedom must be at least as large as ℓ_K. The

number of such eddies that can be piled up in a volume of order ℓ_0^3 is thus:
$$N_{\text{turb}} \propto (\ell_0/\ell_K)^3 \sim R^{9/4}, \tag{8.8}$$
which is Landau's estimate of the number of degrees of freedom in developed turbulence.

8.1.4 Remarks

8.1.4.1 Taylor micro-scale

Starting from the dissipation rate ϵ and an estimate of the intensity of large scale velocity fluctuations $\tilde{v}_0$, one can define another characteristic length λ called the *Taylor micro-scale*. The order of magnitude of velocity gradients is indeed given by $\nabla \tilde{v} \sim \tilde{v}_0/\lambda$. The dissipation is homogeneous to $[\nu][\nabla \tilde{v}]^2$, hence $\epsilon = \nu \tilde{v}_0^2/\lambda^2$, up to numerical factors, so that:
$$\lambda/\ell_0 = R_0^{-1/2},$$
by substitution. The Reynolds number R_λ is then defined as $R_\lambda = \tilde{v}_0 \lambda/\nu$, so that:
$$R_\lambda = R_0^{1/2}.$$
On the other hand, comparing λ with ℓ_K one gets:
$$\lambda/\ell_K \propto R_0^{1/4} \propto R_\lambda^{1/2}$$

Taking the square root to pass from R_0 to R_λ leads to less impressive numbers than estimates from primitive quantities. In practice, R_λ is also a less subjective quantity since λ and $\tilde{v}_0$ can both be derived from experiments by separate measurements of the root-mean-square values of the velocity fluctuations ($\tilde{v}_0$) and of the velocity gradient fluctuations ($\tilde{v}_0/\lambda$), which makes it appealing to characterise the turbulence level empirically.

8.1.4.2 Decaying turbulence

To conclude this section, let us insist on the fact that, without an energy supply sustaining it, turbulence ultimately decays since it develops in a dissipative medium. In the absence of permanent forcing, a flow starting from a given set of turbulent initial conditions always returns to the rest state at the end of a long *turbulent transient* (Exercise 8.5.4). For open flows such as those in Figure 8.1 and 8.2, this happens naturally as the fluid is followed downstream. Close to the grid, turbulence is intense but it decays while being carried away since no further perturbation is introduced in the flow. Since small structures are more heavily damped than large ones, they disappear shortly and only large eddies remain far downstream.

8.2 Mean Flow and Fluctuations

8.2.1 *Statistical approach*

The turbulent flow now being conceived as a stochastic process, it becomes necessary to specify its statistical characteristics. The nature of the corresponding mathematical theory rests on the definition of probability distributions representing *statistical ensembles*, that is to say a supposedly infinite collection of systems submitted to the same constraints, prepared in the same way, and among which a given observed state is drawn *at random*. In this context, one defines *ensemble averages*:

$$\langle Y \rangle = \sum_i Y_i \varpi_i, \tag{8.9}$$

where Y_i denotes the value of observable Y in state i with probability ϖ_i, which is determined as the fraction of systems in the ensemble that are in state i, the specific flow configuration realised at the considered time ($\varpi_i = \lim_{N \to \infty} N_i/N$). Higher order statistical moments are defined in a similar way. Here the nature of the observable Y is not specified, this can be for example the value of a given velocity component at some particular location. What matters is our ability to prepare the appropriate ensemble since a large number of similar experiments should be realised in principle. This might be less difficult in numerical experiments than in the laboratory despite the fact that N is supposed to tend to infinity. Anyway, the statistical approach using ensembles remains mostly a (good) theoretical concept to which it is advisable to oppose an empirical viewpoint resting on measurements taken on a single specific realisation that has developed in the course of time. *Time averages* are defined as:

$$\overline{Y}(t) = \lim_{T \to \infty} \frac{1}{T} \int_t^{t+T} Y(t') \, dt', \tag{8.10}$$

where Y is the considered observable, t the current time, and T a finite but long duration that in principle should be made arbitrarily large. According to this viewpoint, it is assumed that during a typical experiment the flow explores all the accessible configurations, weighted by the fraction of time it spends in a given state, which finds its foundation in the instability of individual trajectories for a non-predictable chaotic system, as already discussed in §5.1.

The regime is *stationary* when the so-defined average quantity $\overline{Y}(t)$ converges toward a value independent of t as T increases. When the flow depends on time, $\overline{Y}$ defines a sliding average at fixed T. This definition remains satisfactory provided that T can be kept sufficiently small when compared to the typical evolution time of the flow while being long enough with respect to time scales of the turbulent fluctuations. The theoretical prediction of the flow's statistical properties takes for granted the equivalence of the two types of calculation:

$$\text{ensemble average} \equiv \text{time average}$$

This is the so-called *ergodic* hypothesis. In what follows we shall consider ensemble averages (8.9) while not forgetting that they will be evaluated using time averages (8.10), which is reasonable in the stationary case but may raise difficulties otherwise.

8.2.2 Reynolds averaged Navier–Stokes equation

Let us apply the ensemble averaging procedure defined above to the continuity and Navier–Stokes equations for an incompressible flow. For the sake of compactness, here and as often as possible in the rest of the chapter, equations will be written using simplified notations: $\partial_{x_j} \equiv \partial_j$, $j = 1, 2, 3$ with $x_1 = x$, $x_2 = y$, $x_3 = z$, and Einstein's convention of implicit summation over repeated subscripts, so that $\partial_j v_j \equiv \sum_{j=1}^{3} \partial_j v_j$, and so on. However, when the specification of coordinate axes will be needed, the velocity components will recover the names they had in the previous chapter, i.e. u for x_1 (streamwise), v for x_2 (cross-stream), and w for x_3 (spanwise).

The primitive equations then read:

$$\partial_j v_j = 0, \tag{8.11}$$

$$\rho(\partial_t + v_j \partial_j) v_i = \partial_j \sigma_{ij}, \tag{8.12}$$

where σ is the stress tensor. Viscous effects are parameterized by the kinematic viscosity $\nu = \mu/\rho$ where the shear viscosity μ and the density ρ are supposed to be constant. For a Newtonian fluid we thus simply get:

$$\sigma_{ij} = -p\,\delta_{ij} + 2\mu s_{ij} \quad \text{with} \quad s_{ij} = \tfrac{1}{2}(\partial_i v_j + \partial_j v_i). \tag{8.13}$$

Following Reynolds, we isolate the fluctuations from the mean flow by requiring that they average to zero. For convenience and in order to make the distinction more obvious, we denote all fluctuating quantities with tilded lower-case letters and mean values with upper-case letters:

$$\mathbf{v} = \mathbf{V} + \tilde{\mathbf{v}}, \qquad p = P + \tilde{p}. \tag{8.14}$$

By definition, we have:

$$\langle \mathbf{v} \rangle \equiv \mathbf{V}, \quad \langle p \rangle \equiv P \quad \text{and} \quad \langle \tilde{\mathbf{v}} \rangle = 0, \quad \langle \tilde{p} \rangle = 0. \tag{8.15}$$

Averaging defined by (8.9) is a linear operation that commutes with differentiation, yielding:
$$\langle \partial_t \mathbf{v} \rangle = \partial_t \langle \mathbf{v} \rangle = \partial_t \mathbf{V}$$
($\equiv 0$ in the stationary case) and
$$\langle \nabla \mathbf{v} \rangle = \nabla \langle \mathbf{v} \rangle = \nabla \mathbf{V}.$$
Averaging the continuity equation (8.11) leads to:
$$0 = \langle \partial_j v_j \rangle = \langle \partial_j (V_j + \tilde{v}_j) \rangle = \langle \partial_j V_j \rangle + \langle \partial_j \tilde{v}_j \rangle = \partial_j V_j + \partial_j \langle \tilde{v}_j \rangle$$
that is to say:
$$\partial_j V_j = 0 \qquad (8.16)$$
and, by subtraction of (8.11):
$$\partial_j \tilde{v}_j = 0. \qquad (8.17)$$

The introduction of (8.14) in the advection term $(V_j + \tilde{v}_j)\partial_j(V_i + \tilde{v}_i)$ gives four terms, $V_j\partial_j V_i$, $\tilde{v}_j\partial_j V_i$, $V_j\partial_j \tilde{v}_i$, and $\tilde{v}_j\partial_j \tilde{v}_i$. Only the first and the last terms survive to the averaging. Like for primitive quantities, we can define tensors associated to the average flow and the fluctuations:
$$\Sigma_{ij} = -P\delta_{ij} + 2\mu S_{ij} \quad \text{with} \quad S_{ij} = \tfrac{1}{2}\left(\partial_j V_i + \partial_i V_j\right) \qquad (8.18)$$
and
$$\tilde{\sigma}_{ij} = -\tilde{p}\delta_{ij} + 2\mu \tilde{s}_{ij} \quad \text{with} \quad \tilde{s}_{ij} = \tfrac{1}{2}\left(\partial_j \tilde{v}_i + \partial_i \tilde{v}_j\right). \qquad (8.19)$$
The averaged Navier–Stokes equation then reads:
$$\rho(\partial_t V_i + V_j \partial_j V_i + \langle \tilde{v}_j \partial_j \tilde{v}_i \rangle) = \partial_j \Sigma_{ij}.$$
The third term on the left hand side can be transformed as:
$$\langle \tilde{v}_j \partial_j \tilde{v}_i \rangle = \langle \partial_j(\tilde{v}_j \tilde{v}_i) - \tilde{v}_i \partial_j \tilde{v}_j \rangle = \partial_j \langle \tilde{v}_j \tilde{v}_i \rangle - \langle \tilde{v}_i(\partial_j \tilde{v}_j) \rangle = \partial_j \langle \tilde{v}_j \tilde{v}_i \rangle,$$
using continuity condition (8.17) to cancel the second term of the third member of this series of equalities.

In component form, the *Reynolds stress tensor* is defined as[2]
$$\tau_{ij} = -\rho \langle \tilde{v}_i \tilde{v}_j \rangle. \qquad (8.20)$$
By identification with the mean stress tensor one then gets the mean flow equation in the form:
$$\rho(\partial_t + V_j \partial_j)V_i = \partial_j \left(-P\delta_{ij} + 2\mu S_{ij} + \tau_{ij}\right). \qquad (8.21)$$

---

[2]The density ρ is still present in this definition so that, dimensionally, τ_{ij} is indeed a force *per* unit surface.

It is immediately remarked that the set (8.16, 8.21) is not closed since the r.h.s. of (8.21) cannot be expressed entirely as a function of P and $\mathbf{V}$ but refers to the Reynolds stress, i.e. the average of a fluctuating term that remains to be evaluated. This would be achieved by writing down the equations for the fluctuations by subtracting (8.21) from (8.12), which gives:

$$\rho(\partial_t \tilde{v}_i + V_j \partial_j \tilde{v}_i + \tilde{v}_j \partial_j V_i + \tilde{v}_j \partial_j \tilde{v}_i) = \partial_j(-\tilde{p} + 2\mu\,\tilde{s}_{ij} - \tau_{ij}), \qquad (8.22)$$

then by deriving equations for products $\tilde{v}_i \tilde{v}_j$ from $\partial_t(\tilde{v}_i \tilde{v}_j) = \tilde{v}_i \partial_t \tilde{v}_j + \tilde{v}_j \partial_t \tilde{v}_i$, adding equations such as (8.22) after multiplication by the appropriate fluctuating velocity components, then averaging and solving the so-obtained system.

These operations produce triple products in $\tilde{v}_i$, $\tilde{v}_j$, $\tilde{v}_k$, leaving us with triple correlation terms after averaging, and so on. We have therefore to face the well known *closure* problem of turbulence theory: new terms appear in equations at a given order that can be determined only by continuing the computation at higher orders. In a number of applications the problem is settled by imposing a *closure relation* among moments at a given order, that is to say by postulating relations that give the new unknown quantities in terms of already determined quantities.

Instead of considering the equations governing every component of the Reynolds stress tensor we shall limit ourselves to the derivation of the equation for its trace which, up to a factor $1/2$, is nothing but the average fluctuating energy

$$\tilde{e} = \tfrac{1}{2}\rho \langle \tilde{\mathbf{v}}^2 \rangle = \tfrac{1}{2}\rho \langle \tilde{v}_i \tilde{v}_i \rangle = \tfrac{1}{2}\tau_{ii}\,.$$

8.2.3 Energy exchanges in a turbulent flow

The distinction between mean flow and turbulent fluctuations leads us to examine how these two components of the flow exchange energy. Directly from the definitions, we get:

$$E_{\text{tot}} = \tfrac{1}{2}\rho\langle \mathbf{v}^2 \rangle = \tfrac{1}{2}\rho \langle (\mathbf{V}+\tilde{\mathbf{v}})^2 \rangle = \tfrac{1}{2}\rho \left[\langle \mathbf{V}^2\rangle + 2\langle \mathbf{V}\cdot\tilde{\mathbf{v}}\rangle + \langle \tilde{\mathbf{v}}^2\rangle\right].$$

Noticing that $\langle \mathbf{V}^2 \rangle = \mathbf{V}^2$ and $\langle \mathbf{V}\cdot\tilde{\mathbf{v}}\rangle = \mathbf{V}\cdot\langle\tilde{\mathbf{v}}\rangle = 0$, we simply get:

$$E_{\text{tot}} = \tfrac{1}{2}\rho \mathbf{V}^2 + \tfrac{1}{2}\rho\langle\tilde{\mathbf{v}}^2\rangle = E + \tilde{e}\,.$$

In order to compute the kinetic energy of the mean flow, Eq. (8.21) for component V_i is multiplied by V_i itself, and next the sum over the subscript

i is performed. After a few manipulations and using definition (8.20) for the Reynolds stress, we obtain:

$$(\partial_t + V_j\partial_j)E = -\partial_j\left(PV_j - 2\mu S_{ij}V_i + \tau_{ij}V_i\right) - 2\mu S_{ij}S_{ij} - \tau_{ij}S_{ij}. \quad (8.23)$$

This equation is valid locally. The first term on the r.h.s., $-\partial_j(\ldots)$, corresponds to a flux term that, in a global balance equation, can be integrated out as a surface term. Let us consider the other terms. The next one, $-2\mu S_{ij}S_{ij}$, obviously accounts for the energy dissipated by the mean flow through viscous friction. It is most often completely negligible since the large scale mean shear is small when compared to the shear from the small scale eddies. The most interesting term is the last one, $-\tau_{ij}S_{ij}$, which represents the work done by the mean flow on the fluctuations, and thus the energy transferred to the smaller scales.

Let us now perform the complementary computation for the fluctuating kinetic energy. From the equation for the total energy, after averaging and subtraction of the mean flow kinetic energy equation just determined we get:

$$(\partial_t + V_j\partial_j)\tilde{e} = -\partial_j\left(\langle\tilde{v}_j\tilde{p}\rangle - 2\mu\langle\tilde{s}_{ij}\tilde{v}_i\rangle + \tfrac{1}{2}\rho\langle\tilde{\mathbf{v}}^2\tilde{v}_j\rangle\right) - 2\mu\langle\tilde{s}_{ij}\tilde{s}_{ij}\rangle + \tau_{ij}S_{ij}. \quad (8.24)$$

Discarding again the flux gradient term and considering what remains, one easily identifies the fluctuating viscous dissipation $-2\mu\langle\tilde{s}_{ij}\tilde{s}_{ij}\rangle$, which is a negative definite contribution and the turbulent energy production $+\tau_{ij}S_{ij}$, directly subtracted to the energy of the mean flow.

A stationary regime can be obtained only if these two terms balance each other, otherwise, the evolution is transient (decaying turbulence). We now turn to the determination of the mean flow.

8.3 Mean Flow and Effective Diffusion

8.3.1 *Mixing length and eddy viscosity*

As we have already pointed out, turbulence is characterised by diffusivity coefficients that are considerably augmented with respect to their microscopic counterparts, viscosity in Stokes' law, molecular diffusivity in Fick's law. It is however advisable to understand how the molecular properties arise in kinetic theory of gases.

On general grounds, the macroscopic diffusion equation for the density of some microscopic physical quantity Y reads

$$\partial_t Y = D_Y \partial_{x^2} Y,$$

where D_Y is the relevant diffusivity, homogeneous to $[L]^2/[T]$. It describes the smoothing of inhomogeneity by redistribution of that quantity as it is randomly transported by the microscopic agents (molecules). The diffusivity can be estimated through a dimensional argument in which microscopic times and lengths involved in their motions are introduced. The kinematic viscosity related to the redistribution of linear momentum is for example given by:

$$\nu \sim \frac{\xi^2}{\tau_c} = \xi\bar{c},$$

in which ξ is the mean free path, τ_c the average time between collisions, and $\bar{c} = \xi/\tau_c$ the average speed of molecules at the considered temperature. A detailed computation would give $\nu = \xi\bar{c}/3$ since on average only one third of the molecules contribute to the transfer of momentum in the direction of the gradient of velocity. Furthermore, in a gas all molecular diffusivity coefficients have the same order of magnitude since the physical quantities are transported by the same agents, the molecules themselves. Hence, the Prandtl number, ratio of the kinematic viscosity (diffusivity of momentum) to the thermal diffusivity (for energy) is of order one. Things are different in viscous condensed media where heat diffusion is fast since it also involves vibration modes of a local pseudo-lattice. An analogous situation holds in mixtures where the diffusivity of a solute can be very slow when compared to viscous damping when the solute's molecules are much bigger than the solvent's molecules (Schmidt number $S = \nu/D \gg 1$, see also Exercise 3.3.3 where the Lewis number $L = D/\kappa$ was defined and Note 17, p. 192, about mixing).

The idea underlying the introduction of an *eddy viscosity* rests on a heuristic argument that simply replaces the thermal velocity $\bar{c}$ by the typical *mixing velocity* $\tilde{v}_{\text{mix}}$, e.g. the root-mean-square of $\tilde{v}$ at some relevant place, and the mean free path ξ by a corresponding length ℓ_{mix}, called the *mixing length* (Prandtl):

$$\nu_t = \tilde{v}_{\text{mix}} \ell_{\text{mix}}.$$

This concept is useful but its use is delicate because its justification is purely dimensional: the specific values of its two ingredients are not specified *a priori*, in contrast with the case of gases where the thermal speed and the mean free path are unambiguously defined and where the latter remains usually extremely small when compared to the scale of the applied gradients. We already know in particular that, in turbulence, many eddies are active over a whole range of scales in turbulence, and that there is no real gap between the scales typical of the mean flow and those typical of turbulent fluctuations. This caveat being stated, below we elaborate on the concept of eddy viscosity, noticing that the argument can sometimes be developed in terms of time scales rather than space scales (e.g. in Exercise 8.5.5).

Up to now we have done as if the scales (speed, length) were constant but most flows of practical interest develop downstream, boundary layers, jets, wakes, mixing layer. In the case of a laminar flow controlled by the molecular viscosity, the width of the sheared region δ_m in the cross-stream direction y depends on the streamwise coordinate x in a way correctly predicted by the dimensional argument in which, the advection $U\partial_x U$ is balanced by the viscous diffusion $\nu \partial_{y^2} U$ in the Navier–Stokes equation. For plane flows, this yields $U^2/X = \nu U/\delta_\mathrm{m}^2$, where X is the distance from the origin of the flow (leading edge of the plate for a boundary layer). Physically X/U can be understood as the advection time from this origin to the point of interest and δ_m as the width of the region affected by viscous damping during that time.

When trying to apply the dimensional argument to a turbulent flow in the x direction with a mean flow profile $U(y)$ at position X, the estimate of the advection time does not change (X/U) while the momentum fluctuations are no longer diffusively damped but rather transported by eddies with characteristic speed $\tilde{v}_\mathrm{mix}$. This speed is not the equivalent of a well-defined thermal speed in a near-equilibrium gas, but on the contrary a scale-dependent quantity. A crude guess, to be refined in the Sec. 8.3.2 below, would be to assume $\tilde{v}_\mathrm{mix}$ constant and to evaluate the thickness of the layer affected by turbulence dimensionally as $\delta_\mathrm{t} = \tilde{v}_\mathrm{mix} \times (X/U)$. Unfortunately, in flows developing downstream, this assumption is not tenable and $\tilde{v}_\mathrm{mix}$ must vary (slowly) with X. The widening of the turbulent layer cannot be estimated without pushing the argument further or taking into account experimental observations (see Sec. 8.4, later on). The behaviour of $\delta_\mathrm{t}/X = \tilde{v}_\mathrm{mix}/U$ indeed depends on the class of flow considered. In boundary layers $\tilde{v}_\mathrm{mix}/U \to 0$ whereas for jets and mixing layers $\tilde{v}_\mathrm{mix}/U \to$ Constant.

8.3.2 Determination of the mean flow

In what follows we restrict ourselves to the case of a fully developed turbulent boundary layer along a plane wall in direction x at stationary state $(\partial_t U \equiv 0)$, in the absence of pressure gradient $(\partial_x P \equiv 0)$. We assume that we are far enough from the leading edge, so that its thickness δ_t, defined as the distance at which the speed is roughly that of the external flow, is very large, and the fluctuations have sufficient room to develop over a wide range of scales. We also neglect the thickening of the layer so that the mean flow $\mathbf{V}$ is reduced to its streamwise component $U\hat{\mathbf{x}}$.

From now on, we consider only quantities *per* unit mass, which means that (8.21–8.24) have been divided by ρ,[3] with the replacements $\mu \mapsto \nu$, $(P/\rho, \tilde{p}/\rho) \mapsto (P, \tilde{p})$, and that stresses or pressures are homogenous to velocities squared, especially the Reynolds stress defined by (8.20), $\tau_{ij} \mapsto -\langle \tilde{v}_i \tilde{v}_j \rangle$.

Close to the wall, the no-slip condition imposes that all the velocity components tend to zero. Accordingly the flow can be characterised by the mean shear at the wall $\partial_y U|_{y=0}$, or rather by the friction at the wall, a stress *per* unit mass denoted τ^*:

$$\tau^* = \nu \partial_y U|_{y=0},$$

since the Reynolds stress does not contribute to the stress *at* the wall due to the cancellation of the transverse fluctuating velocity component there. Experimentally, it is not possible to evaluate τ^* using this relation since the mean flow profile is not measured with sufficient precision in the limit $y \to 0$ (a derivative has to be taken further, which does not improve the situation). However at very large Reynolds numbers this quantity is transferred to the bulk without losses, so that it can be evaluated from the measurement of the mean flow profile and the Reynolds stresses farther from the wall. Anyway, this stress is dimensionally homogeneous to the square of a speed, called the *friction velocity* and here denoted $\tilde{v}^*$, so that we set:

$$\tau^* = (\tilde{v}^*)^2.$$

Using this speed and the kinematic viscosity ν one can construct a length unit y^* through:

$$y^* = \nu/\tilde{v}^*.$$

The *wall units* $\tilde{v}^*$ and y^* then serve to rewrite the problem close to the wall in dimensionless variables, the so-called *plus-variables*:

$$y^+ = y/y^* \quad \text{and} \quad U^+ = U/\tilde{v}^*. \tag{8.25}$$

In the immediate vicinity of the wall, $y^+ \sim 1$, in a region called the *viscous sublayer*, the mean velocity gradient is very large, so that the viscous contribution of the mean strain tensor to the mean stress tensor is not

---

[3]or equivalently, just set $\rho \equiv 1$.

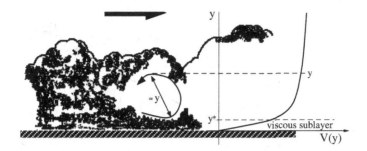

Fig. 8.5 Fully developed boundary layer. Outside the viscous sublayer with thickness of the order of y^*, the most natural length scale of the eddies that disperse momentum efficiently is the distance y to the wall for dimensional reasons. The logarithmic mean profile follows.

negligible (see later). But it is empirically observed that this gradient decreases very rapidly and that one can neglect it somewhat beyond the viscous sublayer. We now focus on the determination of the mean velocity profile in that region where the stress tensor is reduced to its Reynolds stress part τ_{ij}.

Modeling the Reynolds stress using an *eddy viscosity*, one obtains:

$$\tau_{xy} = \nu_t(y) \partial_y U \,.$$

The assumptions that have been made simplify the mean flow equation (8.21) considerably since one is left with:

$$\partial_y \tau_{xy} = 0, \quad \text{hence:} \quad \nu_t(y) \partial_y U = \text{Constant},$$

where the constant on the right hand side cannot be something other than the wall stress $\tau^* = (\tilde{v}^*)^2$ introduced earlier.

The variation of ν_t with y remains to be fixed. One then assumes that most of the turbulent dispersion at distance y from the wall is due to the largest possible eddies that can transfer momentum to the wall from that position (Fig. 8.5). Quantities $\tilde{v}^*$ and y are the only ones we have at our disposal to construct ν_t, so that:

$$\nu_t(y) = \chi \, y \, \tilde{v}^* \,, \tag{8.26}$$

where χ is a dimensionless proportionality coefficient called the *Kármán constant*. Integrating $(\chi \, y \, \tilde{v}^*) \, \partial_y U = (\tilde{v}^*)^2$, that simplifies as:

$$\partial_y U = \frac{1}{\chi} \frac{\tilde{v}^*}{y} \,, \tag{8.27}$$

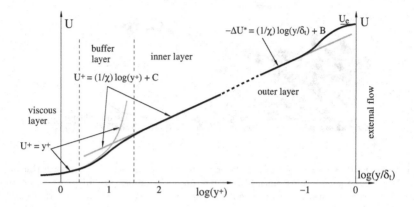

Fig. 8.6 Mean flow profile in a fully developed turbulent boundary layer as it comes out of the mixing length theory.

we obtain:
$$U(y) = \frac{\tilde{v}^*}{\chi}\ln(y) + \text{Constant}. \qquad (8.28)$$

At this stage it is useful to turn to wall units, which gives us the universal law:
$$U^+(y^+) = \frac{1}{\chi}\ln(y^+) + C. \qquad (8.29)$$

Kármán's logarithmic profile gives a satisfactory description of the bulk of the fully developed boundary layer, which justifies the assumptions made to obtain it *a posteriori*. Empirical fits against experiment give us:
$$\chi \simeq 0.41 \quad \text{and} \quad C \simeq 5.$$

These constants are also supposed to be universal.

The logarithmic behaviour predicted by (8.29) cannot be extrapolated indefinitely, neither for $y \to 0$ nor for $y \to \infty$ since the logarithm diverges in both limits. It is only valid in the internal part of the layer called the *inertial* layer, and sufficiently far beyond the viscous sublayer where we must come back to the full expression of the stress in (8.21) but this time, neglect the contribution of the Reynolds stress. We then simply get:
$$\nu \partial_y U = \tau^*,$$

which we integrate to obtain:
$$U(y) = (\tau^*/\nu)y = [(\tilde{v}^*)^2/\nu]\,y = \tilde{v}^*(\tilde{v}^*y/\nu) = \tilde{v}^*(y/y^*)$$

(there is no integration constant since $U|_{y=0} = 0$). Turning to wall units this reads

$$U^+ = y^+.$$

Experiments show that the log law is acceptable for $y^+ > 40$ and the linear law for $y^+ < 3$. The matching of these two variation laws is achieved in the region $3 < y^+ < 40$ called the *buffer layer*.

The logarithmic divergence of the profile for y large is equally unacceptable. The bulk log-layer flow has now to be matched with the exterior flow $U = U_\infty$ for $y > \delta_t$. In the region where this matching takes place, the velocity scale is still $\tilde{v}^*$ (we have nothing else) but the length scale is now the thickness δ_t of the layer. Dimensionally, we must have

$$\partial_y U = \frac{\tilde{v}^*}{\delta_t} f(y/\delta_t)$$

where f is an unknown function of the dimensionless variable $\bar{y} = y/\delta_t$.

This expression of $\partial_y U$ in the external region must be matched with that given by (8.27), valid in the interior region. This implies $f(\bar{y}) \sim 1/\chi\bar{y}$ for $\bar{y} \ll 1$. The speed of the exterior flow being denoted as U_∞, the velocity in the external region of the layer is thus obtained from the integration of:

$$\partial_{\bar{y}} U = \frac{\tilde{v}^*}{\chi\bar{y}},$$

that is:

$$U(\bar{y}) = U_\infty + \frac{\tilde{v}^*}{\chi} \log(\bar{y})$$

(so that $U = U_\infty$ for $\bar{y} = 1$).

Since we are interested in the domain $y \leq \delta_t$, it is advisable to consider the *velocity defect* $[U_\infty - U(y)]$ normalized by $\tilde{v}^*$ and to measure y using its natural unit δ_t, hence:

$$\Delta U^* = \frac{U_\infty - U(y)}{\tilde{v}^*} = -\frac{1}{\chi} \log(\bar{y}).$$

The calculation can be adapted to other situations, e.g. the determination of the pressure head necessary to drive a fully turbulent flow in a plane channel. This can be done along the same lines by considering two turbulent boundary layers placed face to face, each with its viscous sublayer, buffer, inertial, and external layers, which implies sufficiently large Reynolds numbers. The essential assumption is the existence of a single velocity scale v^*. The argument breaks down when several scales are relevant.

8.4 Beyond the Elementary Approach

8.4.1 *The structure of turbulent flows*

In the presentation of developed turbulence adopted up to now, we have considered statistical characteristics of the flow at the lowest possible order, relying on the mean-flow/fluctuation decomposition (in fact the root-mean-square of fluctuations). Things become much more complicated beyond this elementary level. In fact a large part of the dynamics of turbulent flows has been skipped. Whereas dimensional arguments give the essential of the Kolmogorov spectrum (8.5) that accounts for two-point correlation of the velocity field,[4] the analysis of higher-order correlations is more delicate since the closure problem cannot be escaped. A limitation comes from Landau's objection to the Kolmogorov theory, according to which the dissipation rate *per* unit mass ϵ cannot be constant but rather fluctuates in time and space, which is called *inertial range intermittency*. The excitation of eddies at a given scale is indeed sporadic and the energy is distributed at random (local chaos) between the different daughter-eddies (size $\ell/2$) generated by nonlinearity from the mother-eddy (size ℓ). In fact triads of wavevectors are implied, such that $\mathbf{k}_1 + \mathbf{k}_2 + \mathbf{k}_3 = 0$. Elementary arguments developed previously assume, e.g. $|\mathbf{k}_1| \sim |\mathbf{k}_2| \sim k = 2\pi/\ell$ and $|\mathbf{k}_3| \sim 2k$, but the combination also produces wavevectors such that $|\mathbf{k}_3| \ll k$: a transfer also takes place toward larger scales though the transfer toward smaller scales is statistically dominant, so that the simplest image of the energy cascade prevails.

The fine statistical characterisation of turbulence rests on the study of velocity increments as functions of the distance, the so-called *structure functions*:

$$\Delta_i v_j(r) = v_j(\mathbf{x} + r\hat{\mathbf{x}}_i) - v_j(\mathbf{x}),$$

r being the distance measured in direction i indicated by the unit vector $\hat{\mathbf{x}}_i$, parallel $(i = j)$ or perpendicular $(i \neq j)$ to the velocity component j considered. The quantity of primary concern is of course the second moment directly related to the turbulent energy, but other quantities defined in terms of gradients of the velocity field are also interesting:

$$M_n^{[ij]} = \frac{\langle (\partial_i \tilde{v}_j)^n \rangle}{\langle (\partial_i \tilde{v}_j)^2 \rangle^{n/2}},$$

and most especially the *skewness* coefficient $S = M_3^{[ii]}$ and the *flatness* coefficient $F = M_4^{[ii]}$ of longitudinal gradients $(i = j)$ that would be equal

---

[4]See §5.2.1 for the relation between Fourier spectra and correlation functions.

to 0 and 3, respectively, if turbulence were a plain centred Gaussian random process.[5]

Let us come back to the moments of the probability distribution of the increments $\Delta_i v_j(r)$

$$D_n = \langle \Delta_i v_j^n \rangle.$$

and especially the *longitudinal structure functions* which corresponds to $i = j$. The argument that leads to (8.2) would predict

$$D_n = C_n(\epsilon r)^{n/3},$$

which is a particular case of a more general *scaling assumption*:

$$D_n = C_n(\epsilon r)^{\zeta_n},$$

which defines the scaling exponent ζ_n. For $n = 3$, the value $\zeta_3 = 1$ is exact, as shown by Kolmogorov who also found $C_3 = 4/5$ (the '4/5 law', one of the few exact results in developed turbulence). For $n = 2$, the argument predicts $\zeta_2 = 2/3$, which yields the $k^{-5/3}$ energy spectrum. As seen from Figure 8.4 this not too badly verified though quantitative experimental determination rather gives $\zeta_2 \approx 0.7$. However, when searching to adjust moments D_n to power laws r^{ζ_n}, one observes that, for $n \neq 3$, experiments deviate increasingly and nonlinearly from Kolmogorov's simple prediction $\zeta_n^{(K)} = n/3$ due to the intermittency phenomenon mentioned earlier. The scaling is said to be *anomalous*. Figure 8.7 sketches this result in terms of the relative difference $\Delta \zeta_n = (\zeta_n - \zeta_n^{(K)})/\zeta_n^{(K)}$ as a function of the order n, where values of ζ_n are compiled from laboratory and numerical experiments. Theoretical considerations relative to this problem are reviewed in [Frisch (2001)].[6]

Another equally important feature of turbulent flow is the presence of coherent structures, which are long-lived but unsteady eddies containing a large part of the turbulent energy. Their existence was mentioned on p. 286 when we alluded to the post-transitional regimes in free shear flows. Figure 8.8 (left) illustrates the case of the mixing layer which develops when two streams join at the trailing edge of a splitter plate. The illustration here is borrowed from the work of Brown and Roshko:[7] the two parallel streams

---

[5] The *kurtosis* $K = F - 3$ of a Gaussian process is zero. When $K < 0$ the probability distribution function (PDF) is sharper than the Gaussian curve, otherwise it is flatter.

[6] S.Y. Chen *et al.*: "Anomalous scaling of low-order structure functions of turbulent velocity," J. Fluid Mech. **533** (2005) 183–192. In this reference, the computation of moments was extended to values below 2 down to −1 and $|\Delta_i v_j(r)|$ was taken to be able to compute ζ_n with n fractional or negative.

[7] G.L. Brown and A. Roshko: "On density effects and large structures in turbulent mixing layers," J. Fluid Mech. **64** (1974) 775–816.

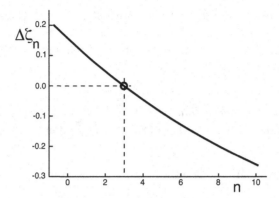

Fig. 8.7 Anomalous scaling in developed turbulence: Relative difference $\Delta\zeta_n$ as a function of the order n of the moments used to compute the structure functions. Kolomogorov's exact result at $n = 3$, $\Delta\zeta_n = 0$ is marked with a circle. Adapted from data in Note 6.

(top: helium, $U_1 = 5$ m/s; bottom: nitrogen, $U_2 = 1.9$ m/s; pressure: 4 atm.) merge to form a wedge of turbulent flow that expands linearly from the trailing edge of the splitter plate. Visualisation is made possible by the density difference between the gases, inducing a large difference in refraction index. Though the wedge is statistically independent of time, the eddies evolve chaotically without any fixed repeatable pattern of spanwise rollers, here seen from the side, on top of which streamwise unsteady vortices form. This kind of structure is conspicuous in other free shear flows, for example in turbulent jets and wakes of bluff bodies, the flow over a backward facing step, etc.

Coherent structures are also present in the wall-bounded flows, the other broad category of flows that we encounter repeatedly. Their existence has been discovered more recently and they are still the subject of active research.[8] An example from numerical simulations[9] of the plane Couette

---

[8] See e.g. the 2008 meeting *Wall Bounded Shear Flows: Transition and Turbulence* at the Isaac Newton Institute, Cambridge, UK; videos and presentations can be found at http://www.newton.ac.uk/webseminars/pg+ws/2008/hrt/hrtw01/. See also the contributions to "Scaling and structure in high Reynolds number wall-bounded flows," B. McKeon, ed., Phil. Trans. R. Soc. A **365** (2007), March 15 theme issue.

[9] Simulations were performed using Gibson's open source code CHANNELFLOW (http://www.channelflow.org/) at $\tilde{R} = Uh/\nu = 750$, where $\tilde{R}$ is defined according to the tradition, as discussed on p. 300. The domain was $L_x \times L_z = 64 \times 32$ with periodic boundary conditions. $N_y = 35$ Chebyshev polynomials were used to account for the wall-normal dependence. The advection term of the Navier–Stokes equation was evaluated according to a pseudo-spectral scheme (§B.2.2) with $N_x = 384$ and $N_z = 384$ collocation

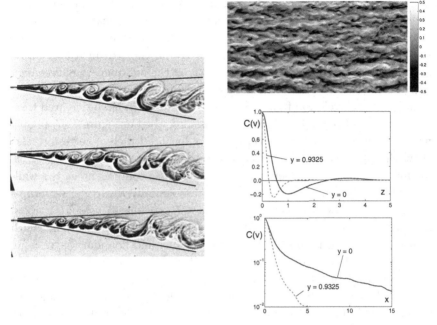

Fig. 8.8 Left: Coherent structures in a mixing layer. Three images taken at random times show the presence of the structures but the absence stable fixed pattern. Suggested by the lines added to the pictures enveloping of the turbulent region, the mixing region expands linearly with the distance to its origin. After pictures in Note 7. Right: Very large-scale structures observed in simulations of turbulent plane Couette flow, see text and Note 9. Top: streamwise velocity component u in the mid-plane ($y = 0$). Bottom: auto-correlation of the wall-normal velocity component v in the mid-plane ($y = 0$, full line) and close to the moving wall ($y = 0.9325$, dashed line).

flow introduced in §7.3.4.3 is displayed in Figure 8.8 (right). Simulation conditions are well above the transitional range discussed in that section. The top image features an instantaneous view of the flow pattern in the mid-plane, a place where the streamwise component u of the flow vanishes on average, owing to symmetry. Long-lived streamwise modulations are observed with maximum amplitude ± 0.5, about a half of the maximum velocity in the fluid ($U = \pm 1$ in dimensionless form). The next two panels display auto-correlation functions of the wall-normal velocity component v.

points. The picture in Figure 8.8 (top, right) was taken at $t = 2450\,h/U$. The correlation functions were obtained by averaging over 30 images taken every $\Delta t = 50\,h/U$. These simulations closely reproduce the case considered by J. Komminaho et al. ("Very large structures in plane turbulent Couette flow," J. Fluid Mech. **320** (1996) 259–285) who were among the first to study these structures quantitatively in detail.

In the spanwise direction z, the middle panel indicates the presence of a recirculation which, in units of the half-gap h, is statistically coherent on a scale ~ 0.5 close to the wall and ~ 1 in the mid-plane. As seen from the lower panel, this perturbation just decays in the streamwise direction x, rapidly close to the wall but more slowly in the mid-plane, in agreement with the impression left by the snapshot placed above them. Whereas the profile of the correlation function close to the wall can be anticipated from the existence of the self-sustaining process described in Figure 7.27, p. 305, which is expected to take place, and stay localised there, the results also show the presence of large-scale structures occupying the full gap, with distinct characteristics. Completed by a spanwise component w, the wall-normal velocity component v forms unsteady long-lived vortical structures along the streamwise direction, inducing the streamwise velocity component u by *lift-up* operating, this time, on the turbulent mean profile. Such structures are also present in turbulent channel and boundary-layer flows where they scale with the largest possible length, the channel width or the boundary layer thickness.

In wall-bounded flows like in free shear flows, mechanisms analogous to those rendering the laminar flow unstable seem to root the existence of coherent but fluctuating structures absent from the statistical description based on the plain mean-flow/fluctuation decomposition. This motivates the modelling effort to be described in the next section and the numerical strategy to be developed in §8.4.3 to cope with these structures.

8.4.2 Turbulence modelling

The approach followed in §8.3.2 to determine the turbulent boundary layer mean profile rests on the intuitive concept of *eddy viscosity*. Trying to close the hierarchy of equations right at lowest level using appropriate extensions to this model seems both natural and desirable in order to deal also with other more complicated situations. An excellent introduction to that problem is [Pope (2000)] and a thorough discussion of models used in numerical fluid dynamics is [Wilcox (2000)].

According to Boussinesq (1897), Reynolds stresses can be modelled by means of an eddy viscosity assumption expressed as:

$$-\langle \tilde{v}_i \tilde{v}_j \rangle + \tfrac{2}{3} k \delta_{ij} = 2\nu_t S_{ij}, \tag{8.30}$$

where $S_{ij} = \tfrac{1}{2}(\partial_i V_j + \partial_j V_i)$ is the mean strain tensor and where, according to common practice, k here denotes the fluctuating kinetic energy per unit mass.[10] Reynolds averaged equations then read:

$$(\partial_t + V_j\partial_j)V_i = -\partial_i \left(P + \tfrac{2}{3}k\right) + \partial_j \left[2(\nu + \nu_t)S_{ij}\right]. \quad (8.31)$$

Depending on the approach, the formulation can be complete or not, i.e. according to whether the expression of the eddy viscosity ν_t is totally (or partially) fixed from the outside (incomplete model), or derived from the mean flow properties through specific equations (complete model).

The simplest possible assumption consists in prescribing ν_t from the local characteristics of the external one-directional flow along x at position X: $\nu_t(X) = U(X)\ell_0(X)/R_t$, where ℓ_0 is some outer scale and R_t a turbulent Reynolds number. But this assumption has a narrow range of applicability since it does not take the least account of the specificity of turbulence.

One step further, the mixing length theory used to treat the case of the boundary layer is a less trivial example of incomplete model founded on a clear physical assumption. Coming back to (8.26), p. 329, it was assumed that $\nu_t = \tilde{v}_{\mathrm{mix}} \ell_{\mathrm{mix}}$ where $\tilde{v}_{\mathrm{mix}} = \tilde{v}$ and ℓ_{mix} are two quantities relative to the turbulent mixing. More generally, Prandtl suggested taking $\tilde{v}_{\mathrm{mix}}/\ell_{\mathrm{mix}} \propto |\partial_y U|$ for a mean flow U along x sheared along y, which is dimensionally correct though this proposal has no real physical grounds. One thus gets:

$$\nu_t = \ell_{\mathrm{mix}}^2 |\partial_y U|,$$

that has been extended to general flows by Smagorinsky[11] in the form

$$\nu_t = \ell_{\mathrm{mix}}^2 (2S_{ij}S_{ij})^{1/2}. \quad (8.32)$$

The two formulations are easily checked to be equivalent when used in Prandtl's conditions: evaluating the second form with a mean strain tensor S_{ij} reduced to[12] $S_{12} = S_{21} = \tfrac{1}{2}\partial_2 V_1$, one gets: $\nu_t = [2(S_{12}S_{12} + S_{21}S_{21})]^{1/2} = \left[4(\tfrac{1}{2}\partial_2 V_1)^2\right]^{1/2} = |\partial_2 V_1|$.

All this however does not make a complete model since an expression for ℓ_{mix} is still requested. It is thus advisable to turn to complete models, still resting on the eddy viscosity assumption, but free of external specification of v_{mix} and/or ℓ_{mix} on a case-by-case basis. These models

---

[10] k is the established notation for this quantity previously denoted $\tilde{e}$ in §8.2.3. The risk of confusion with a wavevector is negligible. The presence of term $\tfrac{2}{3}k\delta_{ij}$ in expression (8.30) guarantees that the trace of the tensor is zero, owing to the mean flow incompressibility condition, $\partial_i V_i = 0$.

[11] J. Smagorinsky, "General circulation experiments with the primitive equations: I. The basic equations," Mon. Weather Rev. **91** (1963) 99–164.

[12] We recall that '1' ≡ 'x', '2' ≡ 'y', and $V_1 \equiv U$.

can be developed at several levels depending on the number of additional dynamical equations closing the Reynolds averaged equations. From this viewpoint, mixing length models are zero-equation models (or 'algebraic' models). One-equation models determine the velocity scale $\tilde{v}_{\text{mix}}$ from the turbulent kinetic energy k by setting $\tilde{v}_{\text{mix}} \propto \sqrt{k}$. The latter quantity is then obtained from Eq. (8.24), with appropriately modelled unknown terms. This equation is rewritten here in the generic form:

$$(\partial_t + V_j \partial_j)k = \partial_j \mathcal{F}_j^{(k)} + \mathcal{P}^{(k)} - \mathcal{D}^{(k)}, \tag{8.33}$$

where $\mathcal{F}^{(k)}$, $\mathcal{P}^{(k)}$ and $\mathcal{D}^{(k)}$ represent flux, production, and dissipation terms, respectively. The actual flux term is replaced by a diffusion term:

$$\mathcal{F}^{(k)} = -\kappa_t^{(k)} \nabla k,$$

where the turbulent diffusivity of energy κ_t is related to the eddy viscosity (turbulent diffusivity of momentum) ν_t by means of an effective Prandtl number traditionally noted σ_k, i.e. $\kappa_t = \nu_t / \sigma_k$. In full generality, the production term reads $\mathcal{P} = -\langle \tilde{v}_i \tilde{v}_j \rangle S_{ij}$. The eddy viscosity assumption turns it into:

$$\mathcal{P}^{(k)} = 2\nu_t S_{ij} S_{ij},$$

which is positive definite. The dissipation term, corresponding to the energy ϵ transferred from the large scales is modelled by (8.1), which yields:

$$\mathcal{D}^{(k)} = C_D \frac{k^{3/2}}{\ell_{\text{mix}}},$$

where C_D is one of the constants in the model. The presence of ℓ_{mix} in that equation, as well as in the relation that defines the eddy viscosity, $\nu_t \propto k^{1/2} \ell_{\text{mix}}$, still makes this type of modelling incomplete. This limitation is raised by adding a supplementary equation.

The best known two-equation model is the k-ϵ model that collects all that precedes but, instead of adding an equation governing ℓ_{mix}, uses the fact that the missing length can be constructed from k and ϵ as $\ell_{\text{mix}} \propto k^{3/2}/\epsilon$, which is just Eq. (8.1) solved for ℓ given ϵ. The eddy viscosity is then taken as:

$$\nu_t = C_\mu \frac{k^2}{\epsilon},$$

where C_μ is a constant. The equation for ϵ that closes the system is copied from (8.33):

$$(\partial_t + V_j \partial_j)\epsilon = \partial_j \mathcal{F}_j^{(\epsilon)} + \mathcal{P}^{(\epsilon)} - \mathcal{D}^{(\epsilon)}, \tag{8.34}$$

in which the r.h.s. terms are modelled as before, in particular:

$$\mathcal{F}^{(\epsilon)} = -\kappa_t^{(\epsilon)} \nabla \epsilon,$$

with $\kappa_t^{(\epsilon)} = \nu_t/\sigma_\epsilon$, where σ_ϵ is the Prandtl number relative to the "diffusion" of ϵ. The production and dissipation terms are further taken in the form

$$\mathcal{P}^{(\epsilon)} = C_{\epsilon 1} \frac{\mathcal{P}^{(k)} \epsilon}{k}, \qquad \mathcal{D}^{(\epsilon)} = C_{\epsilon 2} \frac{\epsilon^2}{k},$$

which can be justified by noticing that quantity k/ϵ is homogeneous to a time τ and that a dimensionally consistent way to pass from (8.33) to (8.34) is by dividing (8.33) by this time and introducing phenomenological dimensionless constant to account for the differences between the processes governing the dynamics of k and ϵ. Constants appearing in the model, C_μ, $C_{\epsilon 1}$, $C_{\epsilon 2}$, σ_k and σ_ϵ are generally adjusted by demanding that the predictions in a few test cases (decaying turbulence, log-law in boundary layers, etc.) correspond to empirical data. To go further, consult [Pope (2000)], Chapter 10.

8.4.3 *Large eddy simulations*

The variables introduced above refer to ensemble averages. However, quantitative information collected from the study of specific realisations (structure and temporal history) are at least as important as the properties of statistical ensembles ideally corresponding to these experiments. There is therefore a pressing need for numerical simulations of Navier–Stokes equation at large Reynolds numbers, in industrial or geophysical configurations for example. The evolution of small scales is of little interest, except that they have to be present to ensure energy dissipation. On the contrary, large scale motions that carry the energy and achieve mixing are of more interest. This suggests performing a *filtering* of the equations that eliminates small scales, as implemented in *large eddy simulations* (LES). The need for a filtering of primitive equations is clearly understood from Landau's estimate (8.8) of the number of degrees of freedom relevant to developed turbulence, which might open a pessimistic perspective: despite the present observed increase of computational power, Reynolds numbers accessible to direct numerical simulation of Navier–Stokes equation will, for long, stay modest (see Exercise 8.5.1). Considering spectral approaches to the simulation problem (§B.2.2) will allow us to better understand why and how one can develop an alternative strategy.

In practice large scale eddies are reconstructed from a set of small wavevectors of the order of the inverse of ℓ_0, the scale in which energy is injected in the flow by the main instability mechanism. Few steps in the cascade are necessary to reach complete disorder. These steps depend little on the precise value of the Reynolds number. Let us call[13] $k_{\rm LE}$ the maximum wavevector necessary to account for large eddies (hence subscript "LE"). One thus gets $k_{\rm LE} \propto k_0$, with a rather small proportionality factor, typically 2^n for n steps, with $n = 3$ or 4. In order of magnitude, the number of such modes is thus given as the volume of a sphere with radius $k_{\rm LE}$ in Fourier space: $N_{\rm LE} \propto k_{\rm LE}^3$. In contrast, the value of $k_{\rm K}$, the wavevector associated to the Kolmogorov scale (8.6) marking the end of the inertial cascade, is strongly dependent on R: $k_{\rm K} \propto R^{3/4}$. The corresponding number of active modes, given by the Landau estimate $N_{\rm diss} \propto k_{\rm K}^3 \sim R^{9/4}$, is rapidly diverging. Let us denote $k_{\rm max}$ the largest wavevector included in the simulation (typically, $k_{\rm max} = \pi/\Delta x$, where Δx measures the space resolution). In principle the injected energy will be correctly dissipated at the given value of R if $k_{\rm max} \geq k_{\rm K}$ and thus $N_{\rm max} \propto k_{\rm max}^3 \geq N_{\rm diss}$, which is impossible for the values of R of interest in applications.

In fact, the large scales will be correctly described provided that the energy is correctly transferred through the scale derived from $k_{\rm LE}$. Accordingly, simulating the whole range of scales between $k_{\rm LE}$ and $k_{\rm K}$ in the inertial cascade is necessary only if one does not know how to parameterize this transfer through $k \sim k_{\rm LE}$. The filtering is thus an operation by which numerous "useless" modes ($N_{\rm diss} - N_{\rm LE} \gg N_{\rm LE}$) are eliminated to the benefit of few modes below $k_{\rm LE}$ "useful" to the description of large scale motions. A detailed operational presentation of the method is out of question within the limited scope of these notes – for this one may consult [Pope (2000)], Chapter 13 – but just give a qualitative idea of it, resting on, and extending, the discussions in the previous subsection.

The ensemble averaging leads to a splitting of the hydrodynamic fields into two components (8.14) that fulfil (8.15). Here the flow is similarly decomposed into *filtered* and *residual* quantities, $(\mathbf{v}, p)^{\rm f}$ and $(\mathbf{v}, p)^{\rm r}$ with superscripts 'f' and 'r', respectively. The filtered velocity field in physical space is obtained from the *convolution* of the original field with a filter function G_Δ:

$$\mathbf{v}^{\rm f}(\mathbf{x}, t) = \int G_\Delta(\mathbf{y}, \mathbf{x}) \mathbf{v}(\mathbf{x} - \mathbf{y}, t) \, {\rm d}^3 \mathbf{y}, \qquad (8.35)$$

---

[13] From now on, k no longer refers to the fluctuating kinetic energy but again serves to label wavevectors.

where $G_\Delta(\mathbf{y}, \mathbf{x})$ is some appropriately normalised integral kernel ($\int G_\Delta(\mathbf{y}, \mathbf{x}) \, d^3\mathbf{y} = 1$) localised in space over a region of typical size Δ. The filter is *homogeneous* if it does not depend on position $\mathbf{x}$, which is usually the case except close to a wall). Furthermore it is *isotropic* if it depends only on the length $|\mathbf{y}|$ and not the orientation of $\mathbf{y}$. The simplest homogeneous isotropic filter corresponds to the averaging over a sphere with diameter Δ (the averaging over a cube of side Δ would not be isotropic since the result would depend on the local flow direction with respect to the coordinate axes). Smoother filters, e.g. Gaussian, are usually considered.

Inspired by the initial remark about modes in Fourier space, one can also work in spectral space and perform the filtering in that space where the convolution product of two functions in physical space is just the ordinary product of their Fourier transforms. The Fourier transform of the filter function in physical space is called the *transfer function*. The transfer function of a Gaussian function is another Gaussian function, whereas the transfer function of a square window displays damped oscillatory wings in Fourier space. Conversely, the filter function of the low-pass square filter in spectral space also displays wings in physical space. The simplest spectral filtering is indeed achieved by such low-pass filtering that amounts to projecting the dynamics onto the subspace of Fourier modes such that $k \leq k_\Delta$, where k_Δ is the *cut-off* wavevector (i.e. truncating the Fourier decomposition of the fields beyond k_Δ).

The residual velocity field is by definition the difference between the primitive field and the filtered field:

$$\mathbf{v}^r(\mathbf{x}, t) = \mathbf{v}(\mathbf{x}, t) - \mathbf{v}^f(\mathbf{x}, t).$$

Filtering is less simple than averaging. First it is to be noted that except when it comes to a projection over a set of modes, and contrary to averaging (see (8.15), p. 322), filtering is not an *idempotent* operation. Applying the filter to an already filtered quantity does not leave it unchanged, since filtering a residual does not give zero. One gets:

$$\mathbf{v} \equiv \mathbf{v}^f + \mathbf{v}^r \quad \Rightarrow \quad \mathbf{v}^f = \left[\mathbf{v}^f\right]^f + \left[\mathbf{v}^r\right]^f \quad \text{with} \quad \left[\mathbf{v}^r\right]^f \neq 0 \quad \text{(in general)}.$$

In the same way, it is easy to show that time differentiation commutes with filtering but that space differentiation commutes only if the filter is homogeneous, which is shown from a direct computation starting with (8.35):

$$\partial_i v_j^f = [\partial_i v_j]^f + \int v_j(\mathbf{x} - \mathbf{y}) \partial_i G_\Delta(\mathbf{y}, \mathbf{x}) \, d\mathbf{y}.$$

Except in special cases (e.g. close to a wall) one can work with a homogeneous filter and try to get, in this simple case, the equations governing the filtered fields to be explicitly resolved in a simulation. In the following, "filtered" and "resolved" will thus be synonymous terms since the aim is to develop a computation from which the residues are absent and their presence just modelled. Filtering the continuity equation immediately gives:

$$[\partial_j v_j]^f = \partial_j v_j^f = 0,$$

and, by subtraction:

$$\partial_j v_j^r = 0$$

for the residue. Difficulties with the momentum equations are analogous to those that led to the introduction of the Reynolds stresses, p. 323. In full generality, these equations read:

$$\rho \left(\partial_t v_i^f + v_j^f \partial_j v_i^f \right) = \partial_j \left(\sigma_{ij}^f + \tau_{ij}^f \right), \qquad (8.36)$$

where σ_{ij}^f is the stress tensor defined in the usual way with the filtered quantities:

$$\sigma_{ij}^f = -p^f \delta_{ij} + \eta s_{ij}^f \quad \text{with} \quad s_{ij}^f = \tfrac{1}{2} \left(\partial_i v_j^f + \partial_j v_i^f \right).$$

The supplementary term τ_{ij}^f plays the role of the Reynolds stress in the ensemble averaging approach. Its presence is due to the fact that the term $[v_i v_j]^f$ that appears upon filtering the advection term is not equal to the product $v_i^f v_j^f$ of the filtered components of the velocity. Introducing the decomposition $\mathbf{v} = \mathbf{v}^f + \mathbf{v}^r$, one readily gets:

$$[v_j v_i]^f = [v_j^f v_i^f]^f + [v_j^f v_i^r]^f + [v_j^r v_i^f]^f + [v_j^r v_i^r]^f. \qquad (8.37)$$

The first term on the r.h.s. cannot be expressed as a product of resolved quantities but the difference $[v_j^f v_i^f]^f - v_j^f v_i^f$, called the *Leonard stress*, is usually quantitatively small. The two next terms involve resolved and residual scales, and the last term, called *subgridscale* (though at this stage no computational grid has been introduced) involves non-resolved scales only. A closure problem arises, though of a nature different from that in ensemble averaging. It rests on the modelling of the additional tensor τ_{ij}^f in (8.36), which is further decomposed into an isotropic part that contributes to modify the filtered pressure and an anisotropic residue. This anisotropic part is most often modelled using an eddy viscosity assumption that closes the system at this level. A Smagorinsky formula (8.32) may be chosen, where the strain tensor is of course defined in terms of the filtered fields.

In spite of an apparent parallelism, this approach is rather different from the one developed in the previous subsection. As a matter of fact, here the filtered fields are fluctuating quantities while before they were averaged non-fluctuating quantities. The presence of time in equations (8.21–8.24), and later in the k-ϵ model (8.31–8.34) has indeed a different meaning related to the specific nature of the ensemble statistics.

More or less sophisticated models – here we have just suggested the simplest formulation – used in LES present themselves as systems of equations for a "turbulent fluid" with specific rheological properties that would depend on the local state of the flow. They are still largely under development and demand to be validated by comparisons with direct simulations (moderate R) and laboratory experiments (high R), since they settle the actual complexity of turbulent exchanges, cf. (8.37), *via* a closure assumption.

The strategies used to solve the problem of interaction between resolved and residual motions might inspire the scientific approach to the nonlinear dynamics of complex systems such as the climate system to be considered as an example at the end of the next chapter, after a brief summary of the topics touched on up to now.

8.5 Exercises

8.5.1 *Scales in turbulence and numerical simulations*

Direct numerical simulations of Navier–Stokes equation in a situation of developed turbulence are considered, in view of an estimation of the amount of computer resources required for reliable results as a function of the nominal Reynolds number. We assume that the turbulent flow takes place in a cube of size L, that the typical velocity is U, and that the Reynolds number can vary with the kinematic viscosity ν at given L, U.

1) Recall Kolmogorov's estimate of the dissipation scale $\ell_{\rm K}$ relative to L at given R.

2) From a numerical point of view, but without reference to a specific numerical scheme (as discussed in Appendix B), Δx and Δt denoting the space and time integration step, respectively, it is assumed that the small scales are correctly resolved provided that $\ell_{\rm K} \sim 2\Delta x$. Determine the variation with R of the number of grid points needed to resolve the smallest scales.

3) Numerical stability considerations come and limit the time step Δt. In a system dominated by advection (term $\mathbf{v} \cdot \boldsymbol{\nabla}(\ldots)$) the stability is controlled by a *CFL (Courant–Friedrich–Lewy) stability criterion* stipulating that the propagation speed of local (numerical) information $\Delta x/\Delta t$ must remain

larger than the speed of transport of physical information by the velocity field. Numerical stability thus requires $\Delta t < \Delta x/U$. Derive the variation of the number of time steps necessary to perform a simulation over a time interval T with space step Δx, with $T \sim L/U$ (turn-over time).

4) Evaluate the amount of computational power as a function of R in a typical full 3D simulation, and the relative increase of power implied by a doubling of R.

8.5.2 Efficiency of turbulent mixing.

1) A radiator is installed to heat up a room with typical size L (volume L^3). Determine the time necessary to reach thermal equilibrium by assuming pure thermal diffusion using a dimensional argument. Take $L = 5$ m, and $\nu = 15 \times 10^{-6}$ m^2/s. For a gas, the Prandtl number is $\nu/\kappa \simeq 0.7$.

2) Consider now turbulent heating with largest eddies of the size of the room generating smaller eddies driving the process. Still dimensionally, estimate the order of magnitude of the air speed above the radiator by assuming that the kinetic energy is converted from the potential gravitational energy (buoyancy) over a height h above the radiator. Take $g = 10$ m/s^2, a temperature increase $\delta\theta = 10°$ above the ambient temperature, $h = 0.1$ m. The thermal expansion coefficient is $\alpha = 1/300$ K^{-1}.

3) A fraction of this energy is available to stir the air in the room. Assuming that the effective stirring speed is only 20% of the speed computed above, determine the characteristic stirring time at the scale of the room.

8.5.3 Developed turbulence in a cumulus cloud

1) Consider the motions inside a typical small cumulus with diameter of the order of $\ell_0 \sim 300$ m. Assuming that the speed of eddies at this scale is $\tilde{v}_0 \sim 3$ m/s, compute the corresponding Reynolds number ($\nu = 15 \times 10^{-6}$ m^2/s).

2) Estimate the energy dissipation rate *per* unit mass and the total power (in kW) dissipated in the cloud ($\rho = 1.25$ kg/m^3). Evaluate the internal scale ℓ_K at which energy is effectively dissipated by viscous friction.

3) An observation plane goes through the cloud at speed 180 km/h, carrying a hot-wire velocity probe. This apparatus is made of a sensor of length $\delta l = 0.5$ mm. Assuming that it can resolve velocity fluctuations at scales of the order $3\delta l$, what is the maximum frequency of turbulent signals that can recorded by the device? Like in grid turbulence, one can suppose that the

relative velocity of the probe with respect of the fluid is the essential factor in converting spatial fluctuations into temporal fluctuations according to Taylor's *frozen turbulence hypothesis*. Compare the Kolmogorov scale ℓ_K to the scale of the correctly sampled fluctuations. Estimate the turn-over time of large eddies and justify the hypothesis.

8.5.4 Decaying turbulence

Consider a container with volume L^3 containing an initially well stirred fluid that is left to settle. With $\ell = 1$ m, and $\nu = 15 \times 10^{-6}$ m^2/s, and assuming initial velocity fluctuations of order $\tilde{v}_0 = 1$ m/s ($\tilde{v}$ denotes here the root-mean-square velocity in one direction of space), compute the Reynolds number at the beginning of the experiment.

1) Assume that the characteristic scale of turbulent motions remains ℓ and derive the evolution equation for the kinetic energy $\frac{3}{2}\tilde{v}^2$, the evolution law of the velocity and the decay law of the Reynolds number that follows.

2) Compute the time necessary to reach $R \sim 10$ and derive the corresponding fluctuation velocity. At this stage, the viscous dissipation of the large scales is no longer negligible and one must abandon the inertial expression of ϵ to, replace it with the viscous form $\epsilon_{\text{visc.}} = c\nu\tilde{v}^2/\ell^2$. Determine the value of the constant C that assures the continuity of the matching of the two laws at $R = 10$ and the law governing the final stage of settling.

8.5.5 Turbulence in the atmospheric boundary layer

The atmospheric boundary layer is submitted to the Earth's rotation. The angular velocity of a frame linked to the surface is given by $\Omega \sin(\lambda)$, where λ is the latitude. (Coriolis acceleration is given by $2\Omega \times \mathbf{v}_r$, where $\mathbf{v}_r$ is the relative velocity; $f = 2\Omega \sin(\lambda)$ is called the *Coriolis parameter*.)

1) Compute the time scale corresponding to this motion at latitude 45° and next, by a dimensional argument, the thickness of the atmospheric layer that feels the friction on the ground in the laminar case. Take $\nu = 15 \times 10^{-6}$ m^2/s.

2) In fact the atmosphere is turbulent. Estimate the thickness ℓ_{mix} of the boundary layer affected by turbulent fluctuations of the order of 3% of wind velocity (typically 10 m/s) over the same time scale. Determine the corresponding eddy viscosity given by $\nu_t = \tilde{v}_{\text{mix}} \ell_{\text{mix}}$.

8.5.6 Two-dimensional turbulence

The spectrum of velocity fluctuations in the two-dimensional case is different from that in three-dimensional turbulence. In the latter case, the energy cascade toward small scales generates the Kolmogorov $k^{-5/3}$ spectrum. In two dimensions it is admitted that the turbulent processes are controlled by the transport and dissipation of vorticity. In two dimensions the vorticity has a single component $\omega = \partial_x v_y - \partial_y v_x$ (along the normal to the plane in which the motions are supposed to take place). The enstrophy is this then defined as $W = \frac{1}{2}\omega^2$.

1) Consider an eddy of arbitrary size ℓ and velocity v_ℓ, estimate its vorticity ω_ℓ and the associated enstrophy.

2) Let ω_I be the vorticity present in the injection scales ℓ_I and assume that, *per* unit time, eddies loose a fixed fraction of the enstrophy present at these scales. Paralleling the argument for energy injection in three dimensions, estimate the rate β of injection of enstrophy *per* unit time.

3) During the cascade, the size of an eddy is reduced but its vorticity is conserved. Evaluate the typical velocity v_ℓ function of ω_I and ℓ.

4) Dimensionally, the energy spectrum $E(k)$ is an energy *per* unit wavevector (and *per* unit mass). Show that the inertial enstrophy cascade leads to the Kraichnan spectrum $E(k) \propto \beta^{2/3} k^{-3}$.

Chapter 9

Summary and Perspectives

Through all these pages, our main aim has been to present an introduction to the study of complex systems from the viewpoint of nonlinear dynamics. An abstract approach generalising the classical treatment of oscillators has been developed and applied to several concrete situations encountered in hydrodynamics. The validity of this theory eventually rests first on the recognition of macroscopic spatiotemporal *coherence* induced by instability mechanisms that optimally develop with specific scales, as opposed to the incoherence of motions involved in plain relaxation to thermodynamic equilibrium at microscopic scales. In turn this coherence reduces the multiplicity of possible dynamical behaviours, up to a point where the context of low-dimensional *dynamical systems* become an appropriate setting. Complexity then mainly means *chaos*, temporal or spatiotemporal. Such instabilities are generic in macroscopic systems driven far from thermodynamic equilibrium by external stresses. The tendency to return to uniform local configurations implies a resistance to change expressing the *dissipative* character of the global dynamics, which is not incompatible with slow large-scale unsteady evolution. In fluid mechanics, this has been shown to result in *emergent* behaviour, from well-ordered cells or waves, to randomly disorganised, turbulent eddies.

After a brief summary of the results obtained so far (§9.1, 9.2) we examine whether the approach that has been followed can be transposed to other situations where "aggregates" of different nature are in interaction over some range of spatiotemporal scales. Leaving aside applications of technical interest (e.g. flows around obstacles with complex geometry), we focus the discussion on a topic where problems already arise at the modelling stage and next when one wants to draw reliable conclusions from this modelling. Macroeconomics evolution or biodiversity dynamics in ecology belong to this class of problems. Of equal interest to the future of our society is the problem of *climate change*, which we choose to consider in §9.3, mostly since it is still rather close to physics and mechanics (though it might not be completely free of geopolitical afterthoughts).

9.1 Dynamics, Stability and Chaos

The concept of *dynamics* governing a time evolution has been modelled as an *initial value problem* for a differential system written in the form:

$$\tfrac{\mathrm{d}}{\mathrm{d}t}\mathbf{X} = \boldsymbol{\mathcal{F}}(\mathbf{X}), \qquad \mathbf{X}(t_0) = \mathbf{X}_0. \tag{9.1}$$

Next, some operating point called the *base state* and denoted here $\mathbf{X}_\mathrm{b}$ is supposed to be given. The notion of *stability*, that occupies a central position, is easier to approach in a linear context where the departure from the base state $\mathbf{X}' = \mathbf{X} - \mathbf{X}_\mathrm{b}$ is infinitesimal, which allows one to truncate the perturbation expansions beyond the first significant order, leading to problems in the form

$$\tfrac{\mathrm{d}}{\mathrm{d}t}\mathbf{X}' = \boldsymbol{\mathcal{L}}\mathbf{X}', \tag{9.2}$$

Here, $\boldsymbol{\mathcal{L}}$ is a linear operator to which one can apply the familiar tools of linear algebra, and especially *normal mode* analysis. Normal modes are taken in the form:

$$\mathbf{X}' = \hat{\mathbf{X}} \exp(st),$$

where s is the *growth rate*. This leads to an *eigenvalue problem*:

$$(\boldsymbol{\mathcal{L}} - s\boldsymbol{\mathcal{I}})\hat{\mathbf{X}} = 0 \tag{9.3}$$

yielding a series of eigenvalues $\in \mathbb{C}$ called its *spectrum*:

$$s_n = \sigma_n - i\omega_n, \qquad n = 1, 2, \ldots$$

which are further ordered by decreasing values of the σ's: $\sigma_1 \geq \sigma_2 \geq \ldots$

The principal result of this analysis is a representation of the dynamics in a infinitesimal neighbourhood of the base state in terms of a *superposition* of normal modes $\hat{\mathbf{X}}_n$, each with an amplitude $A_n(t)$, function of time and $A_n(t_0) = A_{n,0}$. In the simplest cases, i.e. non-degenerate, this superposition can be written as:

$$\mathbf{X}'(t) = \sum_n A_{n,0} \exp\left[s(t - t_0)\right]\hat{\mathbf{X}}_n.$$

The stability of the base state $\mathbf{X}_\mathrm{b}$ can thus be discussed as a function of the sign of the real parts σ_n of the complex growth-rates s_n of the eigenmodes.

From the linear viewpoint a given mode n is thus *stable* when $\sigma_n < 0$, neutral or *marginal* when $\sigma_n = 0$, and *unstable* when $\sigma_n > 0$. A *sufficient condition* for the instability of $\mathbf{X}_b$ is thus $\sigma_1 > 0$. The imaginary part ω_n of s_n next allows one to distinguish *stationary* modes with $\omega_n = 0$ from *oscillatory* modes with $\omega_n \neq 0$.

In full generality, $\mathcal{F}$, $\mathbf{X}_b$, $\mathcal{L}$ and its spectrum, are functions of control parameters accounting for the different ways of acting on the system, which we symbolise by a single variable r in the present formal setting. A *bifurcation* takes place when the real part of the growth rate of the most dangerous mode crosses zero from below as r is varied. The *critical* condition is thus given by $\sigma_1(r) = 0$ which can be solved for r to define the *instability threshold* r_c.

When the base state is unstable, the linear analysis is valid only as long as the system stays sufficiently close to its base state, which does not hold long since some perturbations diverge exponentially. Terms neglected up until now in the perturbation expansion must be taken into account, expressing the interaction of the unstable modes between themselves and with all other modes.

In this domain, the essential result is the possibility to eliminate the stable modes $\hat{\mathbf{X}}_s$ ("s" for "stable"), with amplitudes collectively called $\mathbf{A}_s$, and only keep the *driving* modes $\hat{\mathbf{X}}_c$ with amplitudes $\mathbf{A}_c$ ("c" for "critical" or "central", i.e. either stable or unstable but near-marginal. The reason is that the former are not independent of the latter but rather *enslaved* to them. This elimination, sketched in Figure 4.3, p. 130, leads one to define an *effective dynamics* only coupling the central modes between themselves:

$$\tfrac{d}{dt}\mathbf{A}_c = \mathcal{L}_c \mathbf{A}_c + \mathcal{N}_{c,\text{eff}}(\mathbf{A}_c)\,.$$

Apart from during a brief transient (exponential relaxation of stable modes), the system then evolves on a manifold defined by some relation

$$\mathbf{A}_s = \mathcal{G}_{\text{eff}}(\mathbf{A}_c)\,,$$

that can usually be determined asymptotically in the long time limit and for small departure from the critical conditions by a well-defined technique called the *centre manifold reduction*.

Technically, a gap in the spectrum has to be assumed, such that $\sigma_s < 0$ and $\min|\sigma_s| \gg |\sigma_c|$ with $\sigma_c \sim 0$. But, in practice, the domain of interest of this reduction is larger than it seems because the *normal form* that is obtained has most of its structure imposed by the symmetries of the problem and the nature of the critical modes. This *universality* opens the door to a *phenomenological modelling* that relies only on these ingredients. This is particularly interesting because the number of possible forms and the manner to perturb them away from criticality are not many. Let us mention the most frequent ones.

The first one relates to the bifurcation of a single real amplitude A displaying a symmetry with respect to the change $A \mapsto -A$. At criticality it is governed by:

$$\tfrac{\mathrm{d}}{\mathrm{d}t} A = -A^3,$$

and, upon adding the most general perturbation, by:

$$\tfrac{\mathrm{d}}{\mathrm{d}t} A = rA - A^3 + h, \tag{9.4}$$

where r measures the distance to the bifurcation point in the control parameter space and h is a small perturbation that breaks the initial symmetry. (In this equation r and h are two real parameters.) This bifurcation, illustrated in Figure 4.4, p. 136, is the dynamical translation of one of Thom's seven *elementary catastrophes* listed on p. 137 and was introduced earlier in the theory of thermodynamic phase transitions by Landau.

The second example is the *Hopf bifurcation* of a pair of complex conjugate modes that accounts for the birth of self-sustained oscillations, e.g. in the van der Pol system. Close to the threshold, the normal form reads

$$\tfrac{\mathrm{d}}{\mathrm{d}t} Z = (r - i\omega_\mathrm{c})Z - g|Z|^2 Z, \tag{9.5}$$

where $Z = A_\mathrm{r} + iA_\mathrm{i}$ is the complex amplitude of one of the two interacting modes, ω_c is the angular frequency at threshold, r the distance to threshold and g a complex coefficient. Its real part, g_r, describes the saturation of the amplitude beyond threshold while its imaginary part, g_i, the nonlinear dispersion coefficient, accounts for the dependence of the frequency on the amplitude, a typical feature of nonlinear oscillators, see Figure 4.5, p. 138.

The story then continues with the emergence of more complex behaviour, still crucially involving stability considerations. The state resulting from this first bifurcation can indeed be considered as a new base state, that may become unstable with respect to secondary modes, and then tertiary modes, etc.

This point of view refers to a situation where an ever larger number of modes initially in the stable group ($\sigma_n < 0$) become critical and next unstable ($\sigma_n > 0$), therefore increasing the dimension of the effective dynamics. A stationary mode (*fixed point* in state space) described using a single real amplitude, e.g. (9.4), is then usually followed by periodic behaviours (*limit cycle* attractors) governed by (9.5), which in turn may become unstable with respect to other oscillatory modes yielding quasi-periodic behaviour (*torus* attractor), etc.

Ruelle and Takens have shown that, contrary to the original intuition of Landau who argued in terms of *superposition* of oscillatory behaviours, complexity arises from the *interaction* of modes, with *chaos* as a nontrivial consequence. These two competing scenarios were sketched in Figure 1.7, p. 17.

In order to understand the emergence of chaos from a regular behaviour, it is most appropriate to take the *Poincaré section* of trajectories, a procedure that generalises the stroboscopic analysis of forced periodic behaviours and, at any rate, reduces the initial continuous time system to an iteration (*discrete time* system).

The essential property of chaotic systems is the *instability of trajectories* on the corresponding attractor, then said to be *strange*, which expresses the loss of long term predictability resulting from the *sensitivity to initial conditions* and small perturbations. This property is understood most easily when considering the dyadic iteration, the simplest model of chaotic system:

$$X_{k+1} = 2X_k \pmod{1},$$

that perfectly illustrates the (exponential/geometrical) growth of the distance between neighbouring trajectories: $\delta X_k = 2^k \delta X_0 = \exp(n \ln 2) \delta X_0$. In the general case, this divergence can be measured by means of *Lyapunov exponents* that extend the notion of growth rate initially introduced to deal with the stability of regular trajectories. For one-dimensional iterations in the form

$$X_{k+1} = \mathcal{F}(X_k),$$

one then defines:

$$\lambda = \lim_{K \to \infty} \frac{1}{K} \sum_{k=0}^{K-1} \ln\left(\left|\frac{\mathrm{d}\mathcal{F}}{\mathrm{d}X}(X_k)\right|\right),$$

where $\{X_k, k = 0, 1, \ldots\}$ is a reference trajectory, so that the distance between neighbouring trajectories varies as $(\exp \lambda)^k$ for k large enough. In addition, strange attractors generally display a *fractal* structure contrasting with the smooth aspect of regular attractors. Finally, it has been seen that the transition to chaos develops through one out of a small number of universal *scenarios*, the most celebrated being the *subharmonic cascade* that leads to an aperiodic behaviour at the end of a series of period doublings (period 2, 4, ..., 2^m,...) with successive thresholds geometrically converging towards a finite value, as illustrated in Figure 4.17, p. 161.

9.2 Continuous Media, Instabilities and Turbulence

For a discrete system, (9.1) is a finite-dimensional differential system. For a continuous medium, it is a system of partial differential equations. The linearised operator defined by (9.2) also contains partial derivatives with respect to (physical) space coordinates.

In order to simplify the analysis, one often considers the limit of translationally invariant systems that allows one to solve (9.3) by means of a spatial Fourier transform, by making the change

$$\hat{\mathbf{X}} \mapsto \tilde{\mathbf{X}} \exp(ikx).$$

Branches of normal modes are then parameterized by the wavevector k. The relation $s = s_n(k; r) = \sigma_n(k; r) + i\omega_n(k; r)$ is called the *dispersion relation* relative to mode n. The condition $\sigma_n(k; r) = 0$, once solved for r, gives the corresponding *marginal stability condition*:

$$r = r_n^{(m)}(k).$$

As before, one is interested in the most dangerous mode, the mode with the largest σ. Let us assume that mode '1' is the most dangerous for all values of k (no branch exchange) and further drop the subscript. Usually, the wavevector of the perturbation cannot be controlled and the growth rate displays a quadratic maximum around some specific k_c called the *critical wavevector*. Generically the real part can thus be written as:

$$\tau_0 \sigma = r - r_c - \xi_0^2 (k - k_c)^2. \tag{9.6}$$

In this equation, τ_0 is the natural evolution time of the fluctuations and ξ_0, homogeneous to a length, the *coherence length*. This relation asserts the optimal character of couplings implied in the instability mechanism for the specific value $k = k_c$ (the growth rate of neighbouring modes is smaller). The marginal stability conditions close to the *instability threshold* (r_c, k_c) then reads

$$r = r_c + \xi_0^2 (k - k_c)^2. \tag{9.7}$$

The values of r_c, k_c, τ_0, ξ_0 have to be determined on a case-by-case basis and can be grossly evaluated by physical and dimensional arguments, but the universal contents of (9.7) remains.

The instability is stationary or oscillatory depending on whether the critical angular frequency ω_c is identically zero or not. A stationary instability with $k_c \neq 0$ is called *cellular*, while when $\omega_c \neq 0$ it is a *wave*. In this context, the study of thermal (Rayleigh–Bénard) convection in Chapter 3 has shown that it is a simple example of cellular instability, with a particularly intuitive mechanism involving buoyancy and dissipation. As to shear flows, the theory shows rather easily that they are unstable against spanwise-uniform streamwise-propagating waves when the base flow displays an inflection point.

In the case of open flows, a stationary instability with respect to a framework at rest with the mean flow is trivially seen as a wave in the laboratory frame. This leads to an important distinction according to whether the mean advection is able or not to evacuate perturbations downstream or not, whether the instability is *convective* or *absolute*, respectively (Figure 7.12, p. 283). Except for a few immediate elementary facts, we got past the corresponding theory, owing to its delicate analytical character.

When the continuous medium is not translationally invariant, usually due to end effects, the role of confinement has to be discussed. To this aim, one defines *aspect ratios*:

$$\Gamma = \ell/\lambda_c,$$

where ℓ is the typical size of the system in the direction of the instability wavevector, and $\lambda_c = 2\pi/k_c$ is the critical wavelength of the instability.

When $\Gamma \sim 1$ (*confined systems*), the instability mechanism produces a small number of elementary cells that play the role of effective degrees of freedom since confinement effects are strong enough to maintain space coherence. *Adiabatic elimination* of stable modes evoked earlier helps one to derive an effective low-dimensional dissipative dynamical system, which makes the theory of deterministic chaos fully relevant, especially in regard of its universal aspects, see §3.2.3 where the case of natural convection was presented in detail.

The opposite case of *extended systems*, $\Gamma \gg 1$ corresponds to a situation where the bifurcated solution remains coherent at a local scale only, typically a few critical wavelengths. Another strategy has to be employed to account for them, in terms of *envelopes* describing slow spatiotemporal modulations brought to an ideally uniform nonlinear solution. Technically these envelopes are obtained by means of *multiple-scale analysis*, which however can be short-circuited through a phenomenological approach making an immoderate use of the *universal properties* of the system close to its instability threshold. This universality is linked to its physical symmetries and the nature of the primary bifurcation. The system is then modelled by unfolding the normal form that governs the amplitude of the critical mode, thus becoming an envelope. This is achieved by adding a dynamics, often simply diffusive, driving spatial modulations to the amplitude, which yields equations of complex-Ginzburg–Landau type, see §6.4.

While being restricted to the weakly nonlinear range, these theoretical developments, sketched in Chapter 6, satisfactorily account for topological defects and universal secondary instabilities. Defects (singular points in the envelopes) and *phase* instabilities linked to the local translation and rotation invariance in space, both contribute to the large-time long-distance disorganisation of the *patterns* at the heart of *spatiotemporal chaos* as a route to turbulence in extended systems.

This route to turbulence is relatively progressive and can thus be called "globally supercritical" in the sense that the bifurca*ting* state (which develops beyond the bifurcation point) stays close to the bifurca*ted* state (which is unstable) and replaces it. Some physical systems, and especially convection even at large aspect ratios, become turbulent in that way.

The case of shear flows is relatively more delicate. As a matter of fact, one has to distinguish flows that are *mechanically unstable*, i.e. unstable at the inviscid limit, from those that remain stable in this limit. The prototype of the first class is the shear layer experiencing a Kelvin–Helmholtz instability related to the presence of an inflection point in the base profile. The second class of flow is formed with those that can possibly be unstable but through a subtle process in which viscous dissipation is crucially involved, producing *Tollmien–Schlichting waves*.

The consequences of this classification in the strongly nonlinear regime are important. In particular, whereas the flows belonging to the first class follow a globally *supercritical* route to turbulence that develops at moderate Reynolds numbers, in the second case, the linear instability threshold, if it exists at all, is high, so that there is room for other typically nonlinear solutions to the Navier–Stokes equation at intermediate Reynolds numbers. This explains that the transition to turbulence, globally *subcritical* according to this scenario also called the *bypass transition*, appears to be "wilder" than in the first case. Actually, *turbulent spots* concentrate, locally in space amid laminar flow, that special kind of solution that does not derive continuously from the base state, see Figure 7.19, p. 296.

Practical situations are often complicated: convective/absolute instability, super/sub-critical bifurcation, regular/chaotic motion, modulations and spatiotemporal chaos. A good understanding of the processes at stake can help in mastering them, that is to say, controlling the transition to turbulence. A toolbox has been presented, and its use demonstrated in some typical cases. It contains most of the implements necessary to explore the richness that surges as soon as nonlinearity cannot be neglected, whatever the field of knowledge considered.

We have said little about the regime of *turbulence developed* that sets in well beyond the transitional region. In this circumstance, the deterministic approach is no longer of help since many degrees of freedom interact in a cascade over a large range of space scales, from the large ones where the turbulent energy is extracted from the mean flow by the instability mechanisms to the small ones where it is dissipated by viscosity effects, see Figure 8.3, p. 318. Elementary dimensional arguments allowed us to estimate the rate of transfer of the energy (*per* unit mass) through the *inertial range* where viscous dissipation can be neglected and energy distribution is ruled by Kolmogorov's energy spectrum $E(k) \sim k^{-5/3}$, down to the *dissipation scale* where viscous effects cease to be negligible.

The irregular (chaotic) character of turbulent flow led us to introduce statistical methods. Two fundamental components of the flow, the *average flow* and the *fluctuations* were identified following Reynolds, but the *closure problem* arose immediately through the definition of *Reynolds stresses*. A very preliminary solution to this problem was presented, in terms of an *eddy viscosity*, itself evaluated within a *mixing length* hypothesis. The approach was then used to determine the mean profile of a fully developed turbulent boundary layer, yielding Kármán's logarithmic law, see Figure 8.6, p. 330.

The two main results obtained by dimensional arguments, Kolmogorov's spectrum and Kármán's logarithmic law are reasonably well verified by experiments but remains zero-order approximations that underestimate the role of the dynamics of eddies at different scales. Difficulties appear as soon as one tries to go beyond this elementary level. In concrete configurations of interest to applications, a numerical approach is often necessary, with a crucial modelling step to deal with the closure problem (k–ϵ model, Large Eddy Simulations, ...).

Up to now, complexity was mostly due to the *nonlinear* character of the dynamics of rather *simple* systems. More *complicated* processes can also take place, e.g. involving rotation and magnetic fields in conducting fluids (dynamo effect), combustion, etc. possibly demanding separate modelling efforts. In the concluding section we rather introduce a case where complexity arises from the interaction of a large variety of sub-systems, exchanging many physical quantities over a large range of space and time scales, the Earth's climate system, in line with the remarks made in §1.4.2, p. 27.

9.3 Approach to a Complex System: the Earth's Climate

In the traditional geographical sense, the study of climate was limited to a description of meteorological phenomena such as temperature, wind, amount of rain, etc., that characterise the mean state of the atmosphere throughout a typical year. This meaning of course remains but, in the perspective of a rapid change possibly induced by human activities, we have to enlarge the viewpoint and also study how this average situation arises from the dynamical interactions of many actors, the atmosphere, the ocean, the cryosphere (ice caps), the biosphere (mainly the vegetation), and at, the end of the chain, our society.[1]

The main physical media involved are fluids in motion, transporting energy in different forms. Energy exchanges play an important role but are constrained by rigourous physical conservation laws. A large variety of nonlinear feedback and space-time scales are involved in climate phenomena, which explains why one could naively try to tackle them using our toolbox. We therefore first extend the framework of previous approaches to evoke a few tangible applications of the concepts introduced earlier, before going to more speculative grounds.

Climate can first be understood through its meteorological dimension, variable and to a certain extent random. In French, there is a sailor saying that: *Mentira bien souvent/ Qui prédira le temps,/ Mais beaucoup moins pourtant/ S'il est bon observant*, which essentially means that you are certainly a lier when you predict weather but a much less one if you make good observations. Current wisdom of old sailors as expressed in this saying is thus an excellent anticipation of the concept of deterministic chaos, originating from *long term unpredictability* (the lie) in spite of predictable short term evolution (reasonable extrapolation of accurate observations) as it was formally introduced by Ruelle in the Seventies, not long after the informal quotation by Lorenz – a great meteorologist, by the way – in the 1960s. The key-concept of chaos theory is the indefinite amplification of tiny differences between initial conditions, that forbids long term prediction. In the case of weather, this prediction rests on the direct integration of equations governing the motions of the atmosphere. These equations are implemented in an essentially satisfactory way in the best routines of modern Meteorological Offices but at the regional scale that we consider presently, the reliability of the results depends on an accurate account of initial conditions in the bulk

---

[1] A nice introduction is [Philander (2000)]; at a more technical level one can cite [Peixoto and Oort (1992)] and [Henderson-Sellers and McGuffie (1987)] for modelling issues.

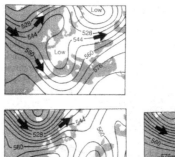

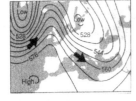 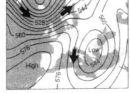 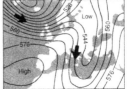

Fig. 9.1 States at day+7 (bottom line) of three simulations starting from initial conditions all compatible with the same ensemble of physical observations (top). After Palmer, in [Hall (1992)].

of the atmosphere (pressure, winds, temperature, humidity) and boundary conditions at the surface (sea surface temperature, relief,...). This data is known with a limited precision at the nodes of a sparse network of observation points, whereas the problem is set in terms of partial differential equations, which leaves aside all local events of small amplitude. Figure 9.1 illustrates the fact that radically different predictions are obtained beyond some predictability range, today still much less than fifteen days (Lorenz's *butterfly effect*). As a first step, one indeed has to proceed to *data assimilation*, i.e. put data in an appropriate form to start the simulation, after some pre-treatment optimising the short-time evolution and making the dynamical extrapolation reliable over a longer period of time, in accordance with chaos theory that tells us that the prediction (in state space) is more accurate if initial conditions are better known, but that small initial uncertainties always reveal themselves.

Recent progress may lead us to think that the detailed mid-term (few days) prediction of extreme events such as the one illustrated in Figure 9.2, their conditions of occurrence and their development, is within reach and can only improve in parallel with computing power which allows one to better follow individual trajectories, and to consider ensembles of trajectories in order to make statistically weighted predictions, while knowing that accurate long-term predictability of a single specific trajectory will remains unreachable forever.

358   *Instabilities, Chaos and Turbulence*

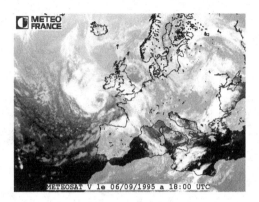

Fig. 9.2 METEOSAT image of a storm centred at the entrance of the English Channel, September 6, 1995; © Météo–France, www.meteo.fr, with permission.

Keeping the dynamical systems perspective, let us mention the use of nonlinear empirical reconstruction techniques introduced in §5.2. Takens's delayed coordinate method was in particular applied to the quantitative analysis of time series measuring the intensity of the El Niño phenomenon, as an attempt to predict its evolution one or two years in advance. El Niño is a sequence of climatic anomalies that affect firstly the inter-tropical Pacific Ocean.[2] It is characterised by a cyclic modification of heat exchanges and zonal motions of water masses together with a swing of the pressure field between the East-Pacific and the West-Pacific called the Southern Oscillation (hence the acronym ENSO meaning El Niño–Southern Oscillation). The cycle is locked on the yearly seasonal periodicity but has a variable duration between two and seven years, with warm episodes (El Niño) and cold ones (La Niña). The warm phase is characterised by warm waters off the coasts of Peru before Christmas, due to a weakening of trade winds that should push these waters further to the West. In the old days the El Niño phenomenon was considered as a gift (at the regional scale) since abundant rains made Peruvian deserts somehow fertile. Nowadays, we have a better understanding of at-a-distance effects, and intense ENSO episodes (e.g. 1982–83, 1997–98) are considered as calamities since large regions in the World suffer from them: augmented rains and floods in western South America, drought and forest fires in the West-Pacific and Australia, not mentioning suspected consequences on the global climate. The intensity of

---

[2]For an introduction, consult Chapter 9 of [Philander (2000)].

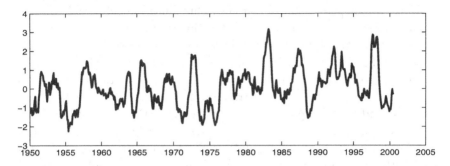

Fig. 9.3 Variations of the MEI measuring the intensity of the ENSO phenomenon.

the phenomenon is measured by several indices (Figure 9.3).[3] The aim of the game is, starting from the time series of a given index, to predict the occurrence of an intense episode six months or one year in advance. Adaptative filtering techniques (*singular spectrum analysis*, Note 12, p. 188) and nonlinear techniques of dynamics reconstruction now seem to do better than the best linear statistical extrapolation techniques.[4] In paleoclimatology, time series of the temperature in the ocean has been studied in the same way, pointing out oscillatory behaviours[5] that we will consider later.

The previous approach rests on the analysis of the time series of a single variable. A more ambitious strategy, also more computer demanding, is to account for meteorological fields in their space-time dimension. This is done by extending the classical *principal components analysis* (see [Peixoto and Oort (1992)] for the atmospheric context) by taking into account the time ordering of pictures serving to compute the principal components,

---

[3]The first one, and the oldest is the Southern Oscillation Index, SOI, that directly measures the pressure difference between Darwin, in the north of Australia and Papeete in Tahiti, see e.g. the web site of the Australian Commonwealth Bureau of Meteorology, (http://www.bom.gov.au/climate/current/) that displays monthly values of this index from 1876 on). Another one, produced by the Japan Meteorological Agency (ftp://www.coaps.fsu.edu/pub/JMA_SST_Index/), is a sliding average over five months of sea surface temperature anomalies in a portion (4°S–4°N, 150°W–90°W) of the Pacific Ocean. Data was gathered since 1949 but could be reconstructed in the past from 1868 on. A third one is elaborated by K. Wolter, at NOOA's Climate Diagnostic Center, as a compound of six different measures called MEI for Multivariate ENSO Index. It covers the contemporary period since 1950, and can be found at the web address http://www.cdc.noaa.gov/~kew/MEI/.

[4]M. Ghil, M. Kimoto, J.D. Neelin, "Nonlinear dynamics and predictability in the atmospheric sciences," Review of Geophysics Supplement (1991) 46–55.

[5]R. Vautard and M. Ghil, "Singular spectrum analysis in nonlinear dynamics, with applications to paleoclimatic time series," Physica D **35** (1989) 395–424.

in order to include dynamical aspects. This multichannel combination of singular spectrum analysis and reconstruction leads to the objective identification of *weather regimes*, centers of action, regular oscillations (30-days in the North-Atlantic), blockings (understood as trajectories being close to a saddle fixed point in state space), at-a-distance connections between regional phenomena, etc.,[6] each contributing to the understanding of climate processes and to the improvement of mid-term weather forecasts.

Up to now we have considered the meteorological facet of climate. The approach just described, local and deterministic in essence, cannot be pursued without adaptation when studying the system at a global scale and in the long term (several years to several hundred of years, not to say even more if ice ages are of interest). The problem is again of going from a *microscopic* perspective to *macroscopic* one, which is analogous to (but less well defined than) the determination of the mean flow in developed turbulence. The understanding of the Earth's climate and its evolution is however not a pure intellectual exercise, since the impact of human activities may have to be taken into account with some consequences on the political choices to be made.

As we have already mentioned, many different actors interact on a wide range of space and time scales, at the origin of the complexity of the dynamics of our natural environment. The description of this complexity relies on a hierarchy of climatic variables defined as statistical averages over geographical regions of various sizes (local, regional, global) and over different lengths of time (day, month, season, year). These variables are generally defined as sliding averages over some duration (usually thirty years) in order to suppress the short term variability. Other quantities of interest are the statistics of the departures from the means, called *anomalies*, and of extreme events. By construction, fastest meteorological fluctuations are filtered out but slower trends remain, which makes it difficult (and subject to controversy) to separate *natural* climatic variability from *anthropogenic* change.

Let us briefly examine the ingredients of this variability. The global climate system is sketched in Figure 9.4. It is composed of different tightly linked elements with various badly matched individual time constants. External forcing by the sun is essential and, while it is easy to attribute the

---

[6]G. Plaut and R. Vautard, "Spells of low-frequency oscillations and weather regimes in the northern hemisphere," J. Atmos. Sc. **51** (1994) 210–236. P.A. Michelangeli, R. Vautard and B. Legras, "Weather regimes: recurrence and quasi-stationarity," J. Atmos. Sc. **52** (1995) 1237–1256.

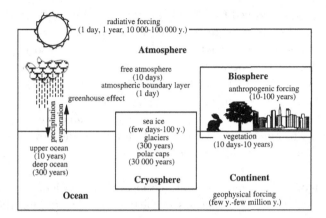

Fig. 9.4 Different components of the climate system, with an indication of their characteristic time constants.

basic driving mechanism to the pole-ward gradient of the energy exchanged with outer space – mostly received in the equatorial region while little is received and much radiated out near the poles – this does not help much understanding the details of the redistribution (somehow reminiscent of what occurs in natural convection) which mainly involves two fluids, the atmosphere and the ocean, with different physical characteristics, in a rotating framework.

Time scales introduced by the solar forcing extend over a very wide range. The highest and most obvious frequencies are of course given by the daylight and season cycles that are most important in the short term. But the sun-spot cycle (11 years) may play a role and the variation of astronomic parameters of the Earth's orbit, and in particular, the precession of the equinox, with typical periods ranging from 10^4 to 4×10^5 years, are thought to drive the ice age cycles on a geological time-scale.

The atmosphere is naturally sensitive to the daily forcing but most of its evolution takes place on a weekly basis, which corresponds to the average motion/instability of air masses at mid latitudes. On longer time-scales, it is affected by the sea surface temperatures (SST) that follow the seasonal cycle. In addition to the tidal rhythm, the ocean as a whole evolves on quite longer times, from a few years for the upper ocean to hundreds of years much below. The surface of the sea ice, important since it reflects sun light in a region that receives little, varies essentially on a seasonal basis. Polar ice caps, that store and return fresh water, change on time scales of the order of a century up to periods fixed by the longest astronomical scales. In addition to punctual effects linked to volcanic activity that introduce dust particles and gases in the atmosphere (e.g. Mount Pinatubo's eruption in 1991), geophysical phenomena linked to continental drift are also important by changing the shape of ocean basins or erecting mountain chains over tens of millions of years, but the effect of closing or opening straits may have more immediate consequences by changing oceanic currents.

The biosphere is of course sensitive to the seasonal cycle. On periods of order of tens to hundreds of years, the flora follows the slow displacement of climatic zones and its role should not be neglected in the analysis of feedback rooting the climate variability. As to the human activity, its effects can be felt on short time scales at the local scale (agricultural practices, deforestation), or on longer scales in particular by modifying the chemical composition of the atmosphere, introducing greenhouse gases in non-natural proportions, especially since the middle of the Nineteenth Century. The greenhouse gases (GG) are all gases with molecules containing at least three atoms (mainly water vapour, carbon dioxide, methane) even at the trace level (CFCs). Their climatic effect is linked to infra-red absorption properties due to rotation-vibration resonance. The molecules are transparent to visible and ultra-violet radiation in the incoming solar flux but opaque to the infra-red thermal reemission from the ground. The resulting accumulation of heat in the lower atmosphere is called the *greenhouse effect*, first put forward by Arrhenius (1896).

Paleoclimate studies focus on the very long term variability of climate. Our knowledge of past climates rests on the record of data derived from the study of cores drilled in ocean or lake sediments and in polar ice caps. Quantitative information is obtained on climatic conditions in given regions by analysing the isotopic composition of samples[7] or the correlation of the plankton/pollen contents of sediments to the climate. The relation between the GG concentrations and the temperature has also been studied from the air bubbles captured in the ice. Though the number/accuracy of these records is limited, their time series yield a precise identification of the most important events, as seen in Figure 9.5.

The alternation of glacial (cold) and interglacial (warm) episodes at mid and high latitudes, with an approximate period of order 10^5 years, shows that the variability may have a large amplitude. Though it is generally believed that this alternation is related to insulation fluctuations induced

---

[7] For example, the oxygen isotopic fraction O^{18}/O^{16} depends on the physical conditions at the time of formation of sediments/ice and gives information on the temperature or the global ice volume.

by the already mentioned small variations of the Earth's orbital parameters (obliquity, eccentricity, position of the perihelion) – Milankovitch's theory – the reason why the system entered this alternating regime, and the mechanisms controlling the periodicity are not fully understood and still subject to controversy. Direct forcing at 10^5 years is however extremely weak. The most important astronomic contributions lie at 41, 19, and 23 kilo-years, and are effectively found in the Fourier spectrum of the reconstructed temperature signal. But this idea of a superposition of simple responses to some external forcing is linear in essence, which makes it suspect in the present context. In fact, as we know, a forced nonlinear system far from equilibrium is likely to experience partial locking or chaos. In addition, the small scales of the system may play the role of a noise that can contribute to erase or amplify the response to the forcing depending on its amplitude.[8] The astronomical theory is also powerless to account for rapid climate changes recently pointed out, and to which we now turn.

The detailed study of the last ice cycle, that begins 115000 BP years ago and culminates 21000 BP, has indeed pointed out to shorter periods superposed to the large amplitude saw-tooth modulation. Figure 9.5 suggests a period of the order of 6000 years. This picture is given to prove that variability may also have fast variability yet to explain. It may however be misleading in that the oscillation pointed out seems smooth, an effect of the filtering procedure, whereas the variations are rapid, a feature which took time to be recognised.[9] These fast variations have been identified in the isotopic analysis of ice cores, as well as by carefully examining the sediment granularity. At roughly regular intervals, the North-Atlantic Ocean witnesses massive iceberg surges identified through layers of materials scratched on the North-American continent and transported as far as the point of melting, where they are incorporated in the sediments (Heinrich events). These episodes are seemingly well correlated to temperature fluctuations over Greenland, called *Dansgaard–Oeschger oscillations*.

---

[8] Discussing stochastic systems is much beyond our purpose. The phenomenon alluded to here is called "stochastic resonance." For a review, see L. Gammaitoni *et al.*: "Stochastic resonance," Rev. Mod. Phys. **70** (1998) 223–287. This phenomenon – the enhanced response of a periodically forced bistable system (here glacial/interglacial) – was suggested as an explanation of the occurrence of the 10^5-years period by R. Benzi *et al.*: "Stochastic resonance in climatic change," Tellus **34** (1982) 10–16, but without decisive argument in its favour.

[9] For a historical presentation with references, consult S. Weart: "The discovery of rapid climate change," Physics Today **56** (2003) 30–36, also:
http://www.aip.org/history/climate/.

364   *Instabilities, Chaos and Turbulence*

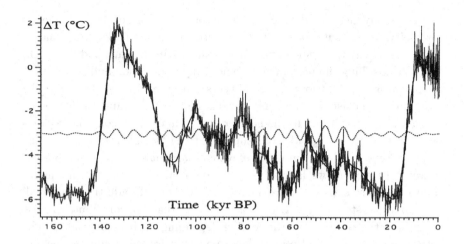

Fig. 9.5 Variation of the temperature in Antarctica during the last ice cycle as derived from isotopic measurements from a core drilled at the Vostok site. Analysis of the data by Takens delayed coordinate method (§5.2.2) combined with singular spectrum analysis (Note 12, p. 188), allows one to isolate the main components of the signal. The continuous thin line through the fluctuating data is the signal reconstructed from a projection onto the five first components. The projection onto the two next components displays a clear oscillation with modulated amplitude (dotted line). Courtesy P. Yiou, LSCE, Saclay.

Several indicators testifying for these rapid changes are displayed in Figure 9.6. Shaded vertical bands through the data signal sediment granularity anomalies. The upper curve (isotopic fraction anomaly for oxygen in Greenland ice) is interpreted in terms of the temperature at which the snow is formed over the land. The two other curves are related to the waters in the North-Atlantic ocean, respectively indicative surface temperature (middle) and salinity (bottom), and both derived from the Oxygen isotopic composition of specific plankton micro-organisms (iceberg melting brings fresh waters and changes the salinity; plankton composition is finely tuned to temperature and salinity). Most recent evidences of these oscillations give a period of 1500 years with cycles skipped (hence 6000 years would appear as a filtered subharmonic). The origin of these oscillations is searched in abrupt changes of the oceanic circulation, and convincing sophisticated models have been proposed.[10] Later we shall examine a much simpler *conceptual* model. Closer to the present time, important climate fluctuations have been observed, just to mention the *climatic optimum* around

---

[10] A. Ganopolski and S. Rahmstorf, "Abrupt glacial climate changes due to stochastic resonance," Phys. Rev. Lett. **88** (2002) 038501. See also Note 8.

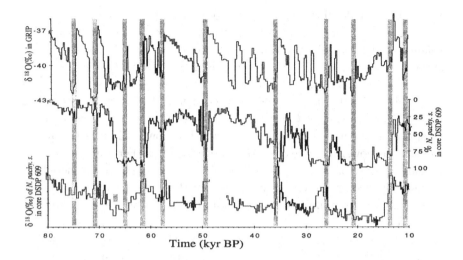

Fig. 9.6 Dansgaard–Oeschger oscillations and Heinrich events. Courtesy D. Paillard, LSCE, Saclay.

1000 AD and the *little ice age* from the Sixteenth to Eighteenth Century. At the human lifetime scale, climate variability finds its trace, most of the time catastrophic, in the already mentioned ENSO phenomenon, persistent droughts implying some extension of the deserts (Sahel), frequency change of cyclones, exceptional floods, and so on.

Understanding climate variability and learning to cope with it, in the perspective of global change, with converging pieces of evidence for warming, is an important objective of climate research. Let us examine different types of approaches in which nonlinear science might be involved *a priori*. First it should be noticed that basic science (mechanics, physics, chemistry, biology) surely have their word to add in accounting for exchanges from the most microscopic stage to the macroscopic level. However the most detailed description is presumably impossible and further useless, so that some sort of reduction is necessary. This reduction cannot attain the level of rigour reached in dynamical systems theory that allows the asymptotic separation of slow active variables from fast enslaved variables, but it is appealing to try to understand climate as some effective dynamics with time scales much slower than those of individual contributing processes. Classification of variables is however not easy. For example atmospheric modes are only partially entrained by oceanic modes *via* sea surface temperatures, and the ocean responds in a filtered way to wind stresses generated by unstable fast atmospheric modes, hence a complicated feedback.

Climate modelling can be developed at various levels, from simple *energy balance* models (EBM) governing few variables averaged over the whole surface of the Earth, to *general circulation* models (GCM) that try to mimic the dynamics of one of the main elements at closest (partial differential equations), the atmosphere or the world ocean, as forced by the other and *vice versa*, and to *coupled models* that involve also the cryosphere and the biosphere in the most general context, which is now made possible by the tremendous power increase of present-day computers. For a first approach, consult [Henderson-Sellers and McGuffie (1987)]. The level of modelling is tightly linked to the degree of realism of the *description* to which one aims, and perhaps in inverse proportion to the degree of *understanding* one would reach.

EBMs are more specifically adapted to the study of large scale equilibria, hypothesised fixed points of a global climate system (see Exercise 9.4). If they can help make up one's mind about the sensitivity of climate to different variations, e.g. the fraction of what is reflected toward space to what is received called *albedo*, by nature they can hardly account for any time dependence. One dimensional extensions, i.e. zonal models in one space-dimension, the latitude, with averaging on the longitude and the altitude) improve the situation slightly but cannot account for features of geographical origin.[11] These effects can be partially incorporated in so-called *box models*, whose variables characterise large geographic units and interact through selected processes supposed to be relevant and important to the phenomenon under study. Such models are often said to be *conceptual*, in order to insist on their heuristic role. By construction they present themselves as systems of ordinary differential equations and therefore perfectly fit the framework of low-dimensional dissipative dynamical systems theory, bifurcation and chaos theory.

As an example, let us consider the conceptual model of ocean circulation and ice dynamics sketched in Figure 9.7 and introduced by Paillard and Labeyrie[12] as an element of explanation to Heinrich events mentioned earlier. The idea rests on relaxation oscillations linked to a bistable dynamics of the North-Canada ice cap during a glacial period: a slow accumulation

---

[11] For a detailed discussion, see: G.R. North, "Lessons from energy balance models," in *Physically based modelling and simulation of climate and climatic change*, M.E. Schlesinger, Ed. (Kluwer, 1988), vol. II, pp. 627–651.

[12] D. Paillard and L. Labeyrie, "Role of the thermohaline circulation in the abrupt warming after Heinrich events," Nature **372** (1994) 162–164.

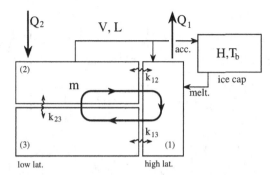

Fig. 9.7 Conceptual model of interaction between the North-Atlantic ocean and the Canadian ice cap during a glacial episode (after Paillard, 1994).

regime and a fast ice surge stage, coupled to the thermohaline circulation in the North-Atlantic ocean. The oceanic circulation is an important vector of heat transfer between low and high latitudes. It is induced by density differences due to temperature and salinity variations between large water bodies (hence the name "thermohaline", see also Exercise 3.3.3 in Chapter 3 where this circumstance was considered as a possible source of instability).

At the level of fractions of ocean basins, water bodies can be identified by their physical properties as subsystems called "boxes", exchanging mass, momentum, heat, salt, with their neighbours.[13] Here the model has three boxes. Box (2) and (3) correspond to low latitudes, box (2) to the upper ocean that receives an amount Q_2 of solar energy, and box (3) to the deep ocean. Box (1) account for high latitude waters, extends over the whole ocean depth, and carries out the sinking of surface waters coming from the low latitudes owing to gravitational instability (warm waters become denser while cooling). An energy Q_1 is returned to space from this box. The variable m measures the intensity of the circulation through these three boxes (advection). Effective transfer coefficients describe the turbulent exchanges at their common borders.

Another element of this model is a fictitious atmosphere transporting heat and humidity towards the North. Water vapour and latent heat fluxes delivered to the ice cap are measured by variables V and L.

Finally, fresh water can be stored in the last element of the model, an

---

[13] Such models were introduced by H.M. Stommel, "Thermohaline convection with two stable regimes of flow," Tellus **13** (1961) 224–230. For a more sophisticated example, see R.-X. Huang and H.M. Stommel, "Convective flow patterns in an eight-box cube driven by combined wind stress, thermal and saline forcing," J. Geophys. Res. **97** (1992) 2347–2364.

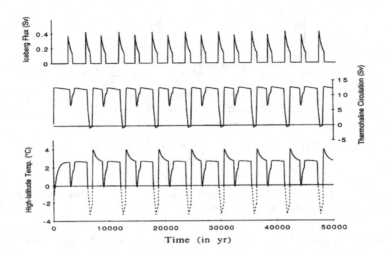

Fig. 9.8 The permanent regime of the model sketched in Figure 9.7 is rapidly reached. For the considered values of its parameters, an alternation of events with different amplitudes is observed. The graph on the top line accounts for ice surges more or less visible in Figure 9.4. Notice the temporary heating when the thermohaline circulation starts again. (Data provided by D. Paillard, LSCE, CEA-Saclay)

ice cap characterised by its thickness H and whose evolution displays the two regimes of slow accumulation and fast surge. The change from one regime to the other is controlled by an auxiliary variable, the temperature T_b of the ice on the bedrock, that decides whether it is rough or slippery.

Once the model is constructed the rest of the study is a standard problem of dynamical systems theory, ending in a bifurcation diagram depending on the phenomenological parameters introduced.[14] When the latter have realistic order of magnitudes, after elimination of the transients (that appear to be very short), one obtains a periodic behaviour like that illustrated in Figure 9.8. The fresh water cycle is seen to be enslaved to the dynamics of accumulation/surge periodicity, which accounts for the sudden heating of the inland temperature of Greenland just after an event. At first sight, this seems paradoxical since the ocean has just been cooled by the melting ice but this is boosted by the thermohaline circulation that grows rapidly and brings warmer waters. Different more or less simple behaviours can be obtained by tuning the models parameters (existence of oscillations with alternatively large and small amplitudes). This kind of approach may thus

---

[14] D. Paillard, "The hierarchical structure of glacial climatic oscillations: interaction between ice-sheets and climate," Climate Dynamics **11** (1995) 162–177 (and PhD dissertation).

help us *understand* the stability of the global oceanic circulation and some observed counter-intuitive phenomena, provided that we accept tuning parameters appropriately.

Next, an intermediate level of modelling is possible, more detailed than box models but with still a regional geographical scope, in terms of (simplified) partial differential equations. For the ENSO phenomenon, see the review article by Neelin et al.[15] But, in order to study the general dynamics of climate and not only a specific phenomenon, EBMs and box models are not refined enough, regional models are not global enough, so that coupled GCMs have to be considered. They explicitly solve as many processes as possible over the full range of scales available. These models are close to the primitive equations of hydrodynamics but, like in turbulence, and in an acute way, the problem of subgridscale parameterization has to be faced, i.e. how to account for events at scales that cannot (and will never be) represented explicitly. This irreducible part of modelling is particularly difficult because of the variety of processes acting inside each grid element.

For example, whereas a simplified description of radiative exchange within the atmosphere may be sufficient for accurate forecasts of its motion over a few days, it is not the same in the long term and a slightly defective parameterization will lead to an unacceptable systematic drift that will show up on, say, monthly averages. On the time scales at which climate evolves, these energetic exchanges are particularly sensitive to the amount of greenhouse gases (GG) and to the locally prevalent kind of clouds. Obtaining a robust parameterization of these two factors is an important challenge in view of a correct extrapolation of the present state, the more so as their effects are intimately linked.

Apart from CFCs that are now banned (owing to their catalytic effects on the destruction of the stratospheric ozone), most GGs have a natural origin. However, their abundance may be largely modified by human activity. The most important GG is water vapour; carbon dioxide and methane are also important but their effects are indirect, through a positive feedback: more of them increases the temperature of air that in turn can contain more water vapour, which increases the greenhouse effect. The problem is a difficult one, because the cycle of water is complicated. To stay with the atmosphere, the greenhouse effect of clouds depends on their nature. In equatorial regions, strongly convective thick clouds made of water droplets are thought to reflect the solar radiation toward space (cooling effect). On

---

[15] J.D. Neelin, M. Latif and F.-F. Jin, "Dynamics of coupled ocean-atmosphere models. The tropical problem," Ann. Rev. Fluid Mech. **26** (1994) 617–659.

the other hand, high altitude thin ice clouds (cirrus) should take part in the greenhouse effect by letting the solar radiation enter and stopping the infrared emission from the ground.

Other elements are equally difficult to take into account, for example dust in the high atmosphere, in general and especially after a violent volcanic eruption (cooling observed after Mount Pinatubo's eruption already mentioned), the mechanical effects of geographical relief and vegetation cover, the level of biomass production through photosynthesis, etc.

Presently, coupling of the atmosphere and the ocean often remains a problem owing to the difference in their respective time constants. Whereas, on a given period, each CGM (atmospheric or oceanic) works reasonably well when forced in the appropriate way, the coupled models may drift when they are not called back toward some observed climate state. But then, what about coping with an unknown future? In view of parameterization reliability tests, the models can be run in conditions corresponding to different reference periods, e.g. the last glacial maximum (21000 BP) or the climatic optimum (6000 BP) for which geological records are available to compare with. Another type of experiment is the so-called 'doubling of the carbon dioxide concentration', for which different models with different implementations of the subgrid processes may give outputs differing by a factor of two on a global scale and even more locally, which is considerable. In the theory of dynamical systems, the behaviour of the solution may change qualitatively close to a bifurcation point where the system considered looses robustness. The question therefore arises as to whether the dispersion of the results depends only quantitatively on the parameterization (that should be improved or better tuned) or it is the trace of a deeper qualitative problem relative to the operating point of the climatic system, besides its already known and much studied glacial/interglacial bistability.

The plausibility of climate evolution scenarios for the Twenty-first Century and beyond,[16] taking into account the *anthropogenic* perturbations brought to the *natural* operation of the climate system, is a crucial stake of the development of models. Even if general circulation models, more focused on *describing* than *understanding* the machinery, seem the better placed to bring useful answers, one might think that dynamical systems theory should get better involved. Of course it applies strictly speaking to low-dimensional systems, the conceptual models alluded to above. But its

---

[16]See the successive 'Assessment Reports' of the *Intergovernmental Panel on Climate Change* (IPCC, http://www.ipcc.ch), and the numerous Web sites dedicated to 'global change'.

demand of robustness, the distinction between enslaved modes and driving modes it makes, and concept of effective dynamics that it proposes, should not be discarded *a priori*. It is indeed tempting to understand the climate as the effective slow dynamics of a coupled system atmosphere-ocean-cryosphere-biosphere whose fast variables (atmosphere, biosphere) would be eliminated to the benefit of fewer slow variables (ocean, ice). Computer technology permits us to consider ever more sophisticated models but the risk of a forward escape towards ever more complexity really exists. Dynamical theory provides an alternate framework apt to structure our reasoning and criticise it from the inside.

Difficulties that arise in the study of climate are but one example of those that appear due to nonlinearity at work in complex systems. Our limited intuition of their effects, potentially unattended and/or exacerbated, should incite us to devote a sustained attention with a tinge of humility and caution to them.

9.4 Exercise: Ice ages as catastrophes

We consider an ultra-simplified model of climate attempting to describe it in terms of a single quantity: the average temperature T at the surface of the Earth.

The average temperature is determined by an *energy balance* that reads:

$$C\dot{T} = R_{\text{uv}} - R_{\text{ir}}, \tag{9.8}$$

expressing the fact that the power stored in the atmosphere (l.h.s.; dimensionally, C is some effective specific heat) is the difference between the power absorbed in the short wavelengths R_{uv} and the power emitted in the long wavelengths R_{ir}.

The absorbed power R_{uv} is a fraction of the power received from the Sun, the so-called solar constant, here noted Q_0:

$$R_{\text{uv}} = Q_0(1-\alpha) \tag{9.9}$$

On general grounds, the fraction α directly reflected toward outer space, called the *albedo*, is a function of the nature of the ground and the degree of cloud cover. Here it is taken to be a function of T only. The reflection capacity of the ground is indeed mostly a function of the presence of ice or snow, itself function of the temperature. Here we assume that α is continuous and linear by parts (*Sellers' hypothesis*[17]):

$$T < T_i : \qquad \alpha = \alpha_i ,$$
$$T_i < T < T_s : \qquad \alpha = \alpha_i + \beta(T - T_i) \qquad \beta = \frac{\alpha_s - \alpha_i}{T_s - T_i} ,$$
$$T_s < T : \qquad \alpha = \alpha_s .$$

Taking into account previous remarks, one expects $\alpha_s < \alpha_i$. Constant T_i corresponds to a temperature below which snowy ground is surely found (α_i high) and T_s a temperature above which the ground is free of ice or snow (α_s low).

One could think that the Earth behaves as a black body that radiates according to the Stephan law $R_{ir} \propto \sigma T^4$ but the actual situation is more complex since the atmosphere is not transparent to IR radiation (greenhouse gases absorb part of this radiation). The Stephan law is therefore replaced by the simplified relation

$$R_{ir} = a + b(T - T_i), \qquad (9.10)$$

where a and b are two empirical constants (*Budyko's hypothesis*[18]). We assume $b > 0$ so that R_{ir} remains a growing function of T.

1) The parameters in (9.9) being given, draw on the same figure R_{uv} and R_{ir} as functions of T and discuss the number of fixed points of system (9.8–9.10) graphically, depending on the value of a at fixed b, $b < \beta Q_0$ (the variation of a could be due to some extrinsic, e.g. anthropogenic, modification of the greenhouse effect).

2) Discuss this again while assuming that the parameters of R_{ir} are fixed and that R_{uv} varies through Q_0 (e.g. astronomic forcing).

3) Examine the stability of the different solutions to (9.8) when it has three roots (the difference $R_{uv} - R_{ir}$ will be drawn as a function of T). Discuss the physical origin of the mechanisms that guarantee the stability of extreme solutions (i.e. interpret the negative character of the feedback).

4) Show that when the orbital parameters (term Q_0) or the greenhouse effect (coefficient a) vary with a sufficient amplitude but quasi-statically (i.e. slowly enough so that T is the steady-state solution to Problem (9.8) at the corresponding instantaneous value of its parameters) the climate of the Earth can pass from a cold solution (ice-age) to a warm solution (interglacial state) and *vice versa*.

---

[17] W.D. Sellers, "A global climate model based on the energy balance of the earth-atmosphere system," J. Appl. Meteor. **8** (1969) 396–400.
[18] M.I. Budyko, "The effect of solar radiation variations on the climate of the earth," Tellus **21** (1969) 611–619.

The Budyko–Sellers model (9.8–9.10), completed by a noise term, was used in the specific work mentioned in Note 8 to exemplify stochastic resonance.

Appendix A
Linear Algebra

We recall here a few elements of linear algebra frequently used in applications. Two perspectives are taken, the first one focuses on the decomposition of vectors on eigen-bases, at the heart of linear stability theory (§A.2), the other tackles essentially the same questions but from an "energetic" point of view *via* the definition of appropriate scalar products (§A.3). Beforehand, Section A.1 is devoted to the enumeration of a few elementary properties linked to the matrix representation of operators and changes of bases. We conclude this appendix by a brief introduction to linear control in state space (§A.4). Among the many reference books on the subject, let us just mention [Hirsch and Smale (1974)] which fits well our purpose.

A.1 Vector Spaces, Bases, and Linear Operators

We suppose that the definition and immediate properties of vector spaces are known. So, let us consider such a space, noted $\mathbb{X}$ with elements (vectors) **X**. Scalars x entering linear combinations can be real or complex. To every vector space on $\mathbb{R}$, one can associate a complex extension formed with the same vectors but in which the scalars (and thus also the components of the vectors) are in $\mathbb{C}$.

A vector $\mathbf{X} \in \mathbb{X}$ can be specified by its components in a given basis $\{\mathbf{E}_i, i = 1, \ldots, d\}$:

$$\mathbf{X} = \sum_{i=1}^{d} x_i \mathbf{E}_i, \tag{A.1}$$

where d is the dimension of $\mathbb{X}$. A *linear form* is a linear function that associates a scalar to every vector in the space. The set of forms also has a vector space structure called the *dual space*. Forms extracting every component of the vectors on a given basis makes the *dual basis*.

A linear operator $\mathcal{L}$ can be defined by its action on the basis vectors:
$$\mathcal{L}\mathbf{E}_j = \sum_i l_{ij}\mathbf{E}_i\,,$$
which defines a $(d \times d)$, two-dimensional array called a *matrix*, here denoted $[\mathrm{L}]$; by convention, the first and second subscripts labels the lines and columns, respectively. The column vectors of $[\mathrm{L}]$ are the images of the basis vectors. The result of the action of $\mathcal{L}$ on an arbitrary vector then reads:
$$y_j = (\mathcal{L}\mathbf{X})_j = \sum_i l_{ji}x_i\,. \tag{A.2}$$
The representation of a linear operator changes with the basis. Let us define the new basis vectors $\{\mathbf{F}_j, j = 1,\ldots,d\}$ by their components in the old basis through:
$$\mathbf{F}_j = \sum_i t_{ij}\mathbf{E}_i\,. \tag{A.3}$$
This builds a matrix $[\mathrm{T}]$ with elements t_{ij} representing the operator $\mathcal{T}$ expressing the change of basis. $\mathcal{T}$ is a linear *invertible* operator. Its column vectors are the components of the new basis vectors in the old basis. Components $\{x_i\}$ and $\{x'_i\}$ of some given vector $\mathbf{X}$ in the two bases $\{\mathbf{E}_i\}$ and $\{\mathbf{F}_i\}$ are then related by:
$$x_i = \sum_j t_{ij}x'_j\,, \tag{A.4}$$
while the new components are given as a function of the old ones by:
$$x'_j = \sum_k u_{jk}x_k\,, \tag{A.5}$$
where the u_{ij} are the elements of the matrix $[\mathrm{U}]$ inverse of $[\mathrm{T}]$ and denoted $[\mathrm{T}]^{-1}$. For this reason, the components x_i are said to be *contra-variant*.

Concretely, the inverse of a given square matrix $[\mathrm{A}]$ can be obtained by hand as the transposed of the matrix of cofactors, divided by the determinant $\det([\mathrm{A}])$ of the matrix. The cofactor of element a_{ij} is $\alpha_{ij} = (-1)^{i+j}\det([\tilde{\mathrm{A}}_{ij}])$, where $[\tilde{\mathrm{A}}_{ij}]$ is the $(d-1)\times(d-1)$ matrix obtained by suppressing line i and column j of matrix $[\mathrm{A}]$. Matrix $[\mathrm{U}]$ can be obtained from $[\mathrm{T}]$ in this way.

From (A.1–A.4), one readily gets:
$$y'_j = \sum_k u_{jk}\sum_l l_{kl}\sum_i t_{li}x'_i = \sum_i l'_{ji}x'_i\,,$$
hence
$$l'_{ji} = \sum_k\sum_l u_{jk}l_{kl}t_{li}\,, \tag{A.6}$$

so that the elements of the matrix representing $\mathcal{L}$ in the new basis are given by $l'_{li} = \sum_k \sum_j u_{lk} m_{kj} t_{ji}$, i.e.

$$[\mathrm{L}'] = [\mathrm{T}]^{-1}[\mathrm{L}][\mathrm{T}].$$

Matrices $[\mathrm{L}']$ et $[\mathrm{L}]$ are said to be *similar*. It is easy to see that

$$[\mathrm{L}']^k = [\mathrm{T}]^{-1}[\mathrm{L}]^k[\mathrm{T}].$$

Remark. While developing Quantum Mechanics, Dirac introduced notations useful to linear algebra. In his setting, vectors are denoted as $|\mathbf{X}\rangle$ and linear forms as $\langle \mathbf{Y}|$. Let $\{|\mathbf{E}_i\rangle, i = 1, \ldots, d\}$ be a basis and $\{\langle \mathbf{E}_i|, i = 1, \ldots, d\}$ the dual basis. The projector on basis vector $|\mathbf{E}_i\rangle$ reads $|\mathbf{E}_i\rangle\langle \mathbf{E}_i|$, and the fact that the basis is complete leads to $\sum_i |\mathbf{E}_i\rangle\langle \mathbf{E}_i| = \mathcal{I}$, where $\mathcal{I}$ is the identity operator in $\mathbb{X}$. Coming back to (A.1), one gets $|\mathbf{X}\rangle = \left[\sum_i |\mathbf{E}_i\rangle\langle \mathbf{E}_i|\right] |\mathbf{X}\rangle$, so that, by identification, $\langle \mathbf{E}_i|\mathbf{X}\rangle$ is the ith contra-variant component of $|\mathbf{X}\rangle$.

The action of the linear operator $\mathcal{L}$ on a vector $|\mathbf{X}\rangle$ gives the vector

$$|\mathbf{Y}\rangle = \mathcal{L}|\mathbf{X}\rangle = \left[\sum_j |\mathbf{E}_j\rangle\langle \mathbf{E}_j|\right] \mathcal{L} \left[\sum_i |\mathbf{E}_i\rangle\langle \mathbf{E}_i|\right] |\mathbf{X}\rangle,$$

which allows the identification $l_{ji} = \langle \mathbf{E}_j|\mathcal{L}|\mathbf{E}_i\rangle$, cf. (A.2).

In a change of basis, vectors of the new basis are defined from the old one by $|\mathbf{F}_j\rangle = \sum_i |\mathbf{E}_i\rangle\langle \mathbf{E}_i|\mathbf{F}_j\rangle$, which defines operator $\mathcal{T}$. Its inverse $\mathcal{U} = \mathcal{T}^{-1}$ is represented by the matrix with elements $u_{ji} = \langle \mathbf{F}_j|\mathbf{E}_i\rangle$. The transformation (A.4) of the components of vector $|\mathbf{X}\rangle$ is then immediately obtained. In the same way the elements of the matrix representing $\mathcal{L}$ in basis $\{|\mathbf{F}_j\rangle\}$ are given by: $\langle \mathbf{F}_j|\mathcal{L}|\mathbf{F}_i\rangle = \sum_k \sum_l \langle \mathbf{F}_j|\mathbf{E}_k\rangle\langle \mathbf{E}_k|\mathcal{L}|\mathbf{E}_l\rangle\langle \mathbf{E}_l|\mathbf{F}_i\rangle$, which is easily identified to (A.6). At this stage Dirac's notations are essentially of mnemotechnical interest.

MATLAB's notations can also be useful. In this framework, all kinds of arrays are placed between square brackets. The comma (or a blank space) is used to separate elements in a line, the semi-colon to indicate the line change (equivalent to a "carriage return"). A vector, traditionally represented in column form is thus a $(d \times 1)$ array $\mathbf{V} = [V_1; V_2; \ldots]$. The line representation, transposed of the column representation then reads: $[V_1, V_2, \ldots] \equiv [V_1; V_2; \ldots]^t$. With these conventions, the canonical scalar product can be written $\mathbf{V} \cdot \mathbf{W} = [V_1, V_2, \ldots][W_1; W_2; \ldots] = \sum V_k W_k$. In contrast, the tensorial product of two vectors $\mathbf{V} \otimes \mathbf{W}$ is a $d \times d$ array with elements $V_i W_j$, which one can write $[V_1; V_2; \ldots][W_1, W_2, \ldots]$.

A.2 Structure of a Linear Operator

As we have seen in Chapter 2, and more specifically in Section 2.2, the solution to the initial value problem for a linear differential system:

$$\tfrac{\mathrm{d}}{\mathrm{d}t}\mathbf{X} = \mathcal{L}\mathbf{X}, \quad \mathbf{X}(t=0) = \mathbf{X}_0, \qquad (A.7)$$

leads to the evaluation of

$$\mathbf{X}(t) = \exp(t\mathcal{L})\mathbf{X}_0 \qquad (A.8)$$

with, by definition:

$$\exp \mathcal{Z} \equiv \mathcal{I} + \sum_{k=1}^{\infty} \frac{1}{k!} \mathcal{Z}^k \qquad (A.9)$$

for an arbitrary linear operator $\mathcal{Z}$. The computation of this exponential is made easier by resolving the structure of the operator $\mathcal{Z}$, that is to say by decomposing the full space $\mathbb{X}$ into a direct sum of nontrivial *invariant subspaces*.

One says that a vector subspace $\mathbb{X}'$ of $\mathbb{X}$ is invariant for $\mathcal{L}$ if the image of every $\mathbf{X} \in \mathbb{X}'$ is in $\mathbb{X}'$. Trivial invariant subspaces are $\mathbb{X}$ itself, the null space $\mathbb{O}$, the image of $\mathbb{X}$ by $\mathcal{L}$ (set of vectors $\mathbf{Y} = \mathcal{L}\mathbf{X}$ for all $\mathbf{X} \in \mathbb{X}$) and the kernel of $\mathcal{L}$ (set of vectors $\mathbf{Z} \in \mathbb{X}$ such that $\mathcal{L}\mathbf{Z} = \mathbf{0}$).

The kernel of $\mathcal{L}$ is reduced to $\mathbb{O}$ when $\mathcal{L}$ is invertible, which is the case if its determinant is nonzero. One then says that the matrix has maximal rank. (The rank of a general matrix, i.e. not necessarily square, is the dimension of the largest square sub-matrix with nonzero determinant that can be extracted from it by suppressing lines and columns.)

The simplification brought by the so-obtained decomposition of space $\mathbb{X}$ is apparent through the change of basis that makes it explicit: Let $\mathbb{X} = \mathbb{X}_1 \oplus \mathbb{X}_2$ and $\mathcal{L}$ the operator for which $\mathbb{X}_1$ is invariant, then for $\mathbf{X}_1 \in \mathbb{X}_1$, $\mathcal{L}\mathbf{X}_1 \in \mathbb{X}_1$ and, in a basis of $\mathbb{X}$ formed with a basis of $\mathbb{X}_1$ appropriately completed by a basis in the supplementary space $\mathbb{X}_2$, $\mathcal{L}$ takes on a block structure:

$$\mathcal{L} = \begin{bmatrix} \mathcal{L}_{11} & \mathcal{L}_{12} \\ \mathcal{O} & \mathcal{L}_{22} \end{bmatrix},$$

where $\mathcal{O}$ is the null operator ($\mathcal{O}\mathbf{X} = \mathbf{0}$ whatever $\mathbf{X}$). If in addition $\mathbb{X}_2$ is also invariant, $\mathcal{L}$ acquires a block-diagonal structure:

$$\mathcal{L} = \begin{bmatrix} \mathcal{L}_{11} & \mathcal{O} \\ \mathcal{O} & \mathcal{L}_{22} \end{bmatrix}.$$

The eigen-direction attached to an eigenvalue s of $\mathcal{L}$, such that:

$$\mathcal{L}\mathbf{X} = s\mathbf{X}$$

is the prototype of the sought invariant subspaces (hence the kernel of operator $\mathcal{M} = \mathcal{L} - s\mathcal{I}$). The eigenvalues are the roots of the *characteristic polynomial* obtained by expanding the determinant:

$$\det(\mathcal{L} - s\mathcal{I}) = 0\,.$$

The *fundamental theorem of algebra* asserts that, in dimension d, this degree-d polynomial has d roots in $\mathbb{C}$, possibly degenerate. One can thus write:

$$\det(\mathcal{L} - s\mathcal{I}) = a_0 + a_1 s + \ldots + a_{d-1} s^{d-1} + s^d = 0 = \prod_j (s - s_j)^{d_j}\,,$$

where d_j is the multiplicity of eigenvalue s_j, with $\sum_j d_j = d$.

In order to treat complex eigenvalues ($s \in \mathbb{C}$) of a linear operator acting in a vector space on $\mathbb{R}$, one must work with its *complex extension*. For example a rotation in a two-dimensional subspace is associated to a pair of simple conjugate purely imaginary roots, solutions to $s^2 + \omega^2 = 0$, and, though there are no real eigenvectors, there are two eigen-directions in the complex extension.

In the general case, at least one eigenvector $\widehat{\mathbf{X}}_j$ can be found for a given (real or complex) eigenvalue s_j in the complex extension. If s_j is non degenerate, the associated eigen-subspace is one-dimensional. It is generated by the eigenvector $\widehat{\mathbf{X}}_j$ solution to $(\mathcal{L} - s_j \mathcal{I})\widehat{\mathbf{X}}_j = \mathbf{0}$ (see Exercise A.5).

If s_j is degenerate, which corresponds in dynamics to a linear resonance condition, the problem is more complicated and one must search for a special basis in which the matrix representing the restriction of the operator to this subspace is in its *normal* form.

A.2.1 Jordan normal form

The main tool to determine the Jordan normal form of an operator is the theory of operator polynomials, i.e. polynomials in the form $\mathcal{Q} = \sum_{n=0}^{n_{\max}} a_n \mathcal{L}^n$. The decomposition of the vector space $\mathbb{X}$ into a direct sum of invariant subspaces rests on the determination of the kernels of operators $(\mathcal{L} - s_j \mathcal{I})^{d'}$, where d' is a trial dimension, as a generalisation of the eigen-direction defined as the kernel of $\mathcal{L} - s_j \mathcal{I}$. These kernels are called the *generalised eigen-subspaces* associated to eigenvalue s_j.

Without giving the derivation, we now give some practical results about them. First they form a series of embedded subspaces of increasing dimensions, the kernel of $(\mathcal{L} - s\mathcal{I})^2$ containing that of $\mathcal{L} - s\mathcal{I}$, etc. The largest of them, called the *principal subspace*, has dimension d_j equal to the multiplicity of the eigenvalue, it is the kernel of $(\mathcal{L} - s_j\mathcal{I})^{d_j}$. Next, the Cayley–Hamilton theorem stipulates that $\mathcal{L}$ fulfils its own characteristic polynomial, that is:

$$\prod_j (\mathcal{L} - s_j \mathcal{I})^{d_j} \equiv 0,$$

which can be understood as the most direct translation of this decomposition into a direct sum of principal subspaces.

The numerical approach to multiples eigenvalues as it is implemented in usual computer routines can lead to difficulties in concrete cases. It may therefore be interesting to present the main analytical steps of the computation. Let $\mathbb{X}_j$ be the principal subspace associated to s_j, with dimension d_j, and $\mathcal{L}_j$ the restriction of $\mathcal{L}$ to $\mathbb{X}_j$. Define also $\mathcal{M}_j = \mathcal{L}_j - s_j \mathcal{I}_{d_j}$, where $\mathcal{I}_{d_j}$ is the identity operator in $\mathbb{X}_j$. Then:

- The *index* of a principal vector $\mathbf{X}$ ($\in \mathbb{X}_j$) is defined as the smallest nonnegative integer m such that $\mathcal{M}_j^m \mathbf{X} = 0$. This index m is such that $1 \leq m \leq d_j$; the lower bound $m = 1$ corresponds to the case of the eigenvector ($\mathcal{M} \mathbf{X} = 0$); the upper bound d_j is because $\mathbf{X}$ belongs to $\mathbb{X}_j$ by assumption ($\mathcal{M}_j^{d_j} \mathbf{X} = 0$). It can easily be seen that when $\mathbf{X}$ is a vector with index m then $\mathcal{M}_j^n \mathbf{X}$ as index $m - n$ for all $n < m$.

- Exploring $\mathbb{X}_j$, one begins with finding the maximum index $m_{\max}$ of vectors in this subspace by computing the kernel of $\mathcal{M}^{d'}$ for increasing trial dimensions d'. When $m_{\max} = d_j$, one gets a vector with maximal index $\mathbf{Y}_{d_j}$ by explicitly solving the linear problem

$$(\mathcal{L}_j - s_j \mathcal{I}_{d_j})^{d_j} \mathbf{X} = 0.$$

- One then considers the series of d_j vectors $\mathbf{Y}_{d_j}$, $\mathbf{Y}_{d_j - 1} = \mathcal{M}_j \mathbf{Y}_{d_j}$, $\mathbf{Y}_{d_j - 2} = \mathcal{M}_j \mathbf{Y}_{d_j - 1} = \mathcal{M}_j^2 \mathbf{Y}_{d_j}$, ..., $\mathbf{Y}_1 = \mathcal{M}_j \mathbf{Y}_2 = \mathcal{M}_j^{d_j - 1} \mathbf{Y}_{d_j}$, that one can write in the more condensed form:

$$\left\{ \mathbf{Y}_{d_j - n} = \mathcal{M}_j^n \mathbf{Y}_{d_j}; \quad n = 0, 1, \ldots, d_j - 1 \right\},$$

noticing that $\mathbf{Y}_0 = \mathcal{M}_j \mathbf{Y}_1 = \mathcal{M}_j^{d_j} \mathbf{Y}_{d_j} \equiv 0$ since $\mathbf{Y}_{d_j}$ has maximal index by assumption. The series of vectors $\{\mathbf{Y}_1, \ldots, \mathbf{Y}_{d_j}\}$ is an acceptable basis for the principal subspace $\mathbb{X}_j$ attached to s_j since one has d_j linearly independent vectors. This property can easily be checked by showing that any linear combination $\mathbf{Z} = \sum_{n=1}^{d_j} \mu_n \mathbf{Y}_n$ cancels if and only if all its coefficients cancel. Computing first $\mathcal{M}_j^{d_j-1}\mathbf{Z}$, one indeed gets a single non-identically zero term $\mu_{d_j}\mathbf{Y}_{d_j}$, which implies $\mu_{d_j} = 0$. Next, the computation of $\mathcal{M}^{d_j-2}\mathbf{Z}$ gives $\mu_{d_j-1} = 0$, and so on, down to μ_1.

- The Jordan normal form directly derives from the matrix representation of $\mathcal{L}_j$ in this basis: Vector $\mathbf{Y}_1$ (index 1) is along eigenvector $\widehat{\mathbf{X}}_j$ and, by construction of the $\mathbf{Y}_n$, one gets:

$$\mathbf{Y}_{n-1} = (\mathcal{L}_j - s_j \mathcal{I}_{d_j})\mathbf{Y}_n,$$

that is:

$$\mathcal{L}_j \mathbf{Y}_n = s_j \mathbf{Y}_n + \mathbf{Y}_{n-1}.$$

The expression of the elementary Jordan block directly derives from this expression:

$$\begin{array}{c} \\ n-1 \\ n \\ n+1 \end{array} \begin{array}{ccc} n-1 & n & n+1 \\ \end{array} \left(\begin{array}{cccccc} & 0 & \vdots & & & \\ & 1 & 0 & \vdots & & \\ & s_j & 1 & 0 & & \\ & 0 & s_j & 1 & & \\ & \vdots & 0 & s_j & & \\ & & \vdots & 0 & & \end{array} \right)$$

Introducing the operator $\mathcal{N}_{d_j}$ represented in this basis by a matrix with a series of ones just above the diagonal and zeroes everywhere else, one gets

$$\mathcal{L}_j = s_j \mathcal{I}_{d_j} + \mathcal{N}_{d_j}.$$

- When the maximal index is such that $m_{\max} < d_j$, one first constructs the subspace generated by the $\mathbf{Y}_n$ just found. This subspace has dimension $m_{\max}$, and one has to redo the same work in the supplementary subspace with dimension $d_j - n_{\max}$, i.e. find the maximal index of vectors in this new subspace, find one vector with maximal index, build the associated basis, etc. up to the point where the full subspace $\mathbb{X}_j$ is decomposed. Using the so-obtained partial bases to build the basis of $\mathbb{X}_j$ one can represent $\mathcal{L}_j$ as a direct sum of Jordan blocks aligned along the diagonal. This result is illustrated in Figure A.1(a). In applications, one rarely deals analytically

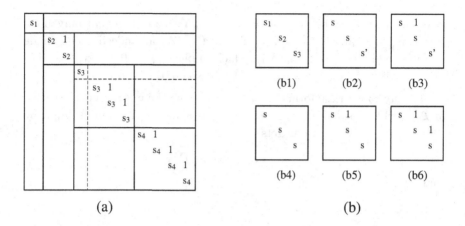

Fig. A.1 (a) Jordan block structure in the basis of generalised eigenvectors, example: s_1 is non-degenerate; s_2 is double but without diagonal form; s_3 is with multiplicity four, decomposed into two subspaces with dimensions one and three; s_4 is also with multiplicity four (zero entries are left blank). (b) Possible cases in dimension $d = 3$: Three distinct eigenvalues, possibly only in the complex extension if two eigenvalues out of the three are complex conjugate (b1). Two real distinct eigenvalues, in diagonal form (b2) or not (b3). Eigenvalue with multiplicity three and vectors with maximum index $n_{\max} = 1, 2,$ or 3, giving three Jordan blocks of order 1 (b4), one of order 1 and one of order 2 (b5), or one of order 3 (b6).

with problems in a dimension higher than three. The work involved in this systematic approach is therefore never as fastidious as it could seem. Figure A.1(b) illustrates all the possible cases for $d = 3$. Notice also that in case (b1), when the three distinct roots are s_1 plus a pair of complex eigenvalues $s_\pm = \sigma \pm i\omega$, the diagonal form holds in the complex extension only, otherwise one has a square block, e.g.

$$\begin{bmatrix} s_1 & 0 & 0 \\ 0 & \sigma & -\omega \\ 0 & +\omega & \sigma \end{bmatrix}.$$

The two-dimensional case was illustrated in Chapter 2 without mentioning the concept of index. This gap is filled in Exercise A.5.2.

A.2.2 Exponential of a matrix

As far as the linear dynamics evolution governed by (A.7) is concerned, the aim of all that precedes is to turn operator $\mathcal{L}$ into the best adapted form in view of the computation of its exponential (A.8) defined as a power series

(A.9). In the basis where $\mathcal{L}$ is in its Jordan form one gets:
$$\exp(t\mathcal{L}_j) = \exp\left[t\left(s_j \mathcal{I}_{d_j} + \mathcal{N}_{d_j}\right)\right] = \exp\left(ts_j \mathcal{I}_{d_j}\right)\exp\left(t\mathcal{N}_{d_j}\right).$$
Here the exponential of the sum is simply the product of the exponentials because $\mathcal{I}_{d_j}$ commutes with any operator and thus also with $\mathcal{N}_{d_j}^n$. One obviously gets $\exp\left(ts_j \mathcal{I}_{d_j}\right) = \exp(ts_j)\mathcal{I}_{d_j}$, and one is left with the computation of the second exponential, which is an easy matter since operator $\mathcal{N}_{d_j}$ is *nilpotent*, i.e. $\left(\mathcal{N}_{d_j}\right)^{d_j} = \mathcal{O}$, so that we are left with a finite number of terms in the power series. Considering the case $d_j = 3$ as an example and dropping all useless indices, we get:
$$\mathcal{N} = \begin{bmatrix} 0 & 1 & 0 \\ 0 & 0 & 1 \\ 0 & 0 & 0 \end{bmatrix}, \quad \mathcal{N}^2 = \begin{bmatrix} 0 & 0 & 1 \\ 0 & 0 & 0 \\ 0 & 0 & 0 \end{bmatrix}, \quad \mathcal{N}^3 = \begin{bmatrix} 0 & 0 & 0 \\ 0 & 0 & 0 \\ 0 & 0 & 0 \end{bmatrix},$$
so that
$$\exp(t\mathcal{N}) = \mathcal{I} + t\mathcal{N} + \tfrac{1}{2}t^2 \mathcal{N}^2 = \begin{bmatrix} 1 & t & \tfrac{1}{2}t^2 \\ 0 & 1 & t \\ 0 & 0 & 1 \end{bmatrix}$$
and thus
$$\exp(t\mathcal{L}) = \begin{bmatrix} \exp(st) & t\exp(st) & \tfrac{1}{2}t^2 \exp(st) \\ 0 & \exp(st) & t\exp(st) \\ 0 & 0 & \exp(st) \end{bmatrix}.$$
The terms $(t^j/j!)\exp(st)$ are said to be *secular*. They account for a slow algebraic drift with respect to a dominant exponential behaviour.

A.2.3 Perturbation of a linear problem

Degeneracy of eigenvalues is a singularity that usually demands to be raised by introducing perturbations in the most general way. It turns out that, the most general perturbation $\delta\mathcal{L}$ brought to some operator $\mathcal{L}$, i.e. $\mathcal{L}' = \mathcal{L} + \delta\mathcal{L}$ may be decomposed in two parts, one that leaves its structure unchanged and another one ("true" perturbations) that does something nontrivial, either by changing its eigenvalues or the fact that it can be turned into diagonal form or not.

Examples of operations that do nothing to a linear system are changes of bases, and more generally invertible transformations $\mathcal{T}$, such that $\mathcal{L}' = \mathcal{T}^{-1}\mathcal{L}\mathcal{T}$, in which case $\mathcal{L}$ and $\mathcal{L}'$ are said to be *similar*. In such operations, the characteristic polynomial is indeed left invariant since:
$$\det\left[\left(\mathcal{T}^{-1}\mathcal{L}\mathcal{T}\right) - s\mathcal{I}\right] = \det\left[\mathcal{T}^{-1}(\mathcal{L} - s\mathcal{I})\mathcal{T}\right]$$
$$= \det\left[\mathcal{T}\mathcal{T}^{-1}(\mathcal{L} - s\mathcal{I})\right] = \det(\mathcal{L} - s\mathcal{I}).$$

Only the part of perturbation $\delta\mathcal{L}$ that makes $\mathcal{L}'$ depart from staying similar to $\mathcal{L}$ is of interest. The complementary part, which leaves them similar, is sought for as a transformation $\mathcal{T}$ close to identity:

$$\mathcal{T} = \mathcal{I} + \Gamma,$$

i.e. with γ_{ij} small so that $\mathcal{T}$ remains invertible. The transformed operator is then given by

$$\mathcal{L}' = \mathcal{T}^{-1}\mathcal{L}\mathcal{T} = (\mathcal{I} + \Gamma)^{-1}\mathcal{L}(\mathcal{I} + \Gamma).$$

We need the inverse of $\mathcal{I} + \Gamma$ but, by identification, it can be checked that

$$(\mathcal{I} + \Gamma)^{-1} = \mathcal{I} - \Gamma + \Gamma^2 - \cdots + (-1)^k \Gamma^k + \ldots$$

so that neglecting all terms beyond first order in Γ, by substitution and in full generality we get:

$$\mathcal{L}' = \mathcal{L} + (\Gamma\mathcal{L} - \mathcal{L}\Gamma).$$

The operator $\Gamma\mathcal{L} - \mathcal{L}\Gamma = [\Gamma, \mathcal{L}]$ is called the *commutator* of $\mathcal{L}$ and Γ. The "true" perturbation is therefore $\delta\mathcal{L} - [\Gamma, \mathcal{L}]$ and one must next ask for the number of independent parameters on which it depends.

Nontrivial answers are already found in dimension two. Let

$$\mathcal{L}' = \mathcal{L} + \delta\mathcal{L},$$

where $\delta\mathcal{L}$ is represented by the most general matrix:

$$\delta\mathcal{L} = \begin{bmatrix} \delta l_{11} & \delta l_{12} \\ \delta l_{21} & \delta l_{22} \end{bmatrix} = \delta l_{11}[\mathrm{U}_1] + \delta l_{12}[\mathrm{U}_2] + \delta l_{21}[\mathrm{U}_3] + \delta l_{22}[\mathrm{U}_4],$$

where

$$\left\{ [\mathrm{U}_1] = \begin{bmatrix} 1 & 0 \\ 0 & 0 \end{bmatrix}; \quad [\mathrm{U}_2] = \begin{bmatrix} 0 & 1 \\ 0 & 0 \end{bmatrix}; \quad [\mathrm{U}_3] = \begin{bmatrix} 0 & 0 \\ 1 & 0 \end{bmatrix}; \quad [\mathrm{U}_4] = \begin{bmatrix} 0 & 0 \\ 0 & 1 \end{bmatrix} \right\}$$

form a canonical basis of the space of (2×2) matrices. Operator Γ is further defined as

$$\Gamma = \begin{bmatrix} \gamma_{11} & \gamma_{12} \\ \gamma_{21} & \gamma_{22} \end{bmatrix}, \tag{A.10}$$

whose coefficients are adjustable parameters (in contrast with the δl_{ij} which are given).

Let us first consider the case when $\mathcal{L}$ can be turned into diagonal form:

$$\mathcal{L} = \begin{bmatrix} s_1 & 0 \\ 0 & s_2 \end{bmatrix}.$$

Computation of the commutator leads to:

$$[\,\Gamma,\mathcal{L}\,] = \begin{bmatrix} 0 & \gamma_{12}(s_2 - s_1) \\ \gamma_{21}(s_1 - s_2) & 0 \end{bmatrix},$$

and therefore

$$\delta\mathcal{L} - [\,\Gamma,\mathcal{L}\,] = \begin{bmatrix} \delta l_{12} & \delta l_{12} - \gamma_{12}(s_2 - s_1) \\ \delta l_{21} - \gamma_{21}(s_1 - s_2) & \delta l_{22} \end{bmatrix}.$$

When $s_1 = s_2$, $\mathcal{L}$ is a multiple of $\mathcal{I}$ and commutes with any Γ. The part of $\delta\mathcal{L}$ that derives from a similarity is thus identically zero so that any perturbation $\delta l_{jj'} \neq 0$ modifies the dynamics, generically by raising the degeneracy (as an exercise one can look for cases where only the diagonalisable character is broken).

In contrast, when $s_1 \neq s_2$, there exists a continuous two-parameter family of perturbations that gives an operator similar to the original one since one can choose $\gamma_{jj'} = \delta l_{jj'}/(s_{j'} - s_j)$ for $(jj') = (12)$ or (21). True perturbations form a complementary two-parameters family $(\delta l_{11}, \delta l_{22})$ that modifies each eigenvalue separately.

Things are a little less trivial when $\mathcal{L}$ cannot be turned to diagonal form, but only into the Jordan form:

$$\mathcal{L} = \begin{bmatrix} \bar{s} & 1 \\ 0 & \bar{s} \end{bmatrix}$$

The commutator then reads:

$$[\,\Gamma,\mathcal{L}\,] = \begin{bmatrix} -\gamma_{21} & \gamma_{11} - \gamma_{22} \\ 0 & \gamma_{21} \end{bmatrix}$$

and we see that γ_{12} has disappeared, so that there remains only two parameters γ_{21} and $\gamma_{11} - \gamma_{22}$ to cancel as many terms as possible in:

$$\delta\mathcal{L} - [\,\Gamma,\mathcal{L}\,] = \begin{bmatrix} \delta l_{11} + \gamma_{21} & \delta l_{12} - (\gamma_{11} - \gamma_{22}) \\ \delta l_{21} & \delta l_{22} - \gamma_{21} \end{bmatrix}.$$

The term δl_{12} can always be suppressed by choosing $\gamma_{11} - \gamma_{22}$ appropriately, and there is enough freedom to cancel one of the two other terms, either δl_{11} or δl_{22}, whereas term δl_{21} can never be suppressed. Choosing $\gamma_{21} = -\delta l_{11}$, $\gamma_{22} = 0$, $\gamma_{11} = \delta l_{12}$, setting $\delta l_{21} = -\eta$ and $\delta l_{22} + \delta l_{11} = -\eta'$, one gets the most general perturbation operator *unfolding the singularity* in the form $\delta\mathcal{L} - [\,\Gamma,\mathcal{L}\,] = -\eta\mathcal{U}_3 - \eta'\mathcal{U}_4$, yielding:

$$\mathcal{L}' = \mathcal{L} - \eta\mathcal{U}_3 - \eta'\mathcal{U}_4 \equiv \begin{bmatrix} \bar{s} & 1 \\ -\eta & \bar{s} - \eta' \end{bmatrix}. \quad (A.11)$$

The characteristic polynomial then reads:

$$(s - \bar{s})(s - \bar{s} + \eta') + \eta = s^2 - (2\bar{s} - \eta')s + \bar{s}(\bar{s} - \eta') + \eta = 0\,.$$

The discriminant of this quadratic polynomial is $\Delta = {\eta'}^2 - 4\eta$. How the degeneracy is raised is illustrated in Figure A.2, and one can see that, depending on the path in the parameter space, one can get either real or complex eigenvalues, which account for the dynamics in the vicinity of an improper node (Figure 2.7(b), p. 54): trajectories resemble those close to a focus as well as close to a node (Figure 2.5(a) and Fig. 2.6(a)).

The generalisation in dimension d:

$$\begin{bmatrix} \bar{s} & 1 & 0 & \cdots & 0 & 0 \\ 0 & \bar{s} & 1 & \cdots & 0 & 0 \\ \vdots & \vdots & \vdots & \ddots & \vdots & \vdots \\ 0 & 0 & 0 & \cdots & \bar{s} & 1 \\ -\eta_1 & -\eta_2 & -\eta_3 & \cdots & -\eta_{d-1} & \bar{s} - \eta_d \end{bmatrix}$$

is called the *Jordan–Arnold normal form*.

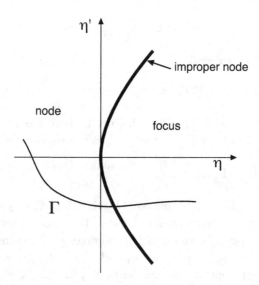

Fig. A.2 Raising the degeneracy at a double non-diagonalisable eigenvalue. The most general perturbation depends on parameters η and η' ($\ll |s|$). Γ is a path in parameter space.

A.3 Metric Properties of Linear Operators

A.3.1 *Scalar products, adjoints, non-normal operators*

The definition of a *scalar product* adds an Euclidean structure to vector spaces on $\mathbb{R}$. A scalar product is a bilinear form with general expression:

$$\mathcal{E}(\mathbf{X}, \mathbf{Y}) = \sum_i \sum_j g_{ij} x_i y_j,$$

symmetric ($g_{ij} = g_{ji}$) and such that $\mathcal{E}(\mathbf{X}, \mathbf{X})$ is definite positive, i.e. $\mathcal{E}(\mathbf{X}, \mathbf{X}) \geq 0$ and $\mathcal{E}(\mathbf{X}, \mathbf{X}) = 0$ if and only if $\mathbf{X} = \mathbf{0}$.

The canonical scalar product relative to a given basis simply reads:

$$\mathcal{E}(\mathbf{X}, \mathbf{Y}) = \sum_i x_i y_j.$$

By construction, the basis is orthonormal with respect to the so-defined scalar product since one has: $\mathcal{E}(\mathbf{E}_i, \mathbf{E}_j) = \delta_{ij}$, where δ_{ij} is the Kronecker symbol ($\delta_{ii} = 1$ and $\delta_{ij} = 0$ for $i \neq j$).

The elements of the matrix $[\mathrm{T}]$ representing the change from an orthonormal basis $\{\mathbf{E}_i\}$ to another $\{\mathbf{F}_j\}$ are given by $t_{ij} = \mathcal{E}(\mathbf{E}_i, \mathbf{F}_j)$, those of the inverse matrix $[\mathrm{U}]$ by

$$u_{ij} = \mathcal{E}(\mathbf{F}_i, \mathbf{E}_j) = \mathcal{E}(\mathbf{E}_j, \mathbf{F}_i) = t_{ji}.$$

Such matrices are said to be *orthogonal*.

In the case of vector spaces on $\mathbb{C}$, the scalar product is defined through a Hermitian form

$$\mathcal{H}(\mathbf{X}, \mathbf{Y}) = \sum_i \sum_j g_{ij} x_i^* y_j,$$

whose coefficients g_{ij} now fulfil $g_{ij} = g_{ji}^*$, so that

$$\mathcal{H}(\mathbf{X}, \mathbf{Y}) = \mathcal{H}(\mathbf{Y}, \mathbf{X})^*.$$

The quadratic form $\mathcal{H}(\mathbf{X}, \mathbf{X})$ must also be definite positive (Hermitian norm). The canonical scalar product relative to a given basis reads:

$$\mathcal{H}(\mathbf{X}, \mathbf{Y}) = \sum_i x_i^* y_j.$$

The change from an orthonormal basis to another one yields

$$u_{ij} = \mathcal{H}(\mathbf{F}_i, \mathbf{E}_j) = \mathcal{H}(\mathbf{E}_j, \mathbf{F}_i)^* = t_{ji}^*.$$

Such transformations are said to be *unitarian*.

Let us consider the general case of a Hermitian scalar product. An operator $\mathcal{L}$ being given, its *adjoint* $\mathcal{L}^\dagger$ is defined from the relations:

$$\mathcal{H}(\mathbf{Y}, \mathcal{L}^\dagger \mathbf{X}) = \mathcal{H}(\mathcal{L}\mathbf{Y}, \mathbf{X}) = \mathcal{H}(\mathbf{X}, \mathcal{L}\mathbf{Y})^*. \tag{A.12}$$

Adjoint operators are thus represented by matrices such that: $(l^\dagger)_{ij} = l^*_{ji}$.

An operator $\mathcal{L}$ is said to be *normal* if it commutes with its adjoint, i.e. $\mathcal{L}^\dagger \mathcal{L} \equiv \mathcal{L}\mathcal{L}^\dagger$. It is said to be *self-adjoint* or *Hermitian* if it is identical to its adjoint: $\mathcal{L}^\dagger \equiv \mathcal{L}$.

It can be shown that a normal operator $\mathcal{L}$ possesses an orthogonal eigen-basis in which it is represented by a diagonal matrix. If in addition it is Hermitian, all its eigenvalues are real.

When the operator is *non-normal* (i.e. does not commute with its adjoint), these properties are lost. Not only the operator can have complex eigenvalues, but there is no orthogonal basis in which it is represented by a diagonal matrix (Jordan normal form, p. 381). On the other hand, it is possible to find a double series of basis vectors $\{\mathbf{E}_j, \mathbf{F}_j, j = 1,\ldots,d\}$ such that $\{\mathbf{E}_i\}$ is an eigen-basis of $\mathcal{L}$ and $\{\mathbf{F}_j\}$ an eigen-basis of $\mathcal{L}^\dagger$ with:

$$(s_j^* - s_i)\mathcal{H}(\mathbf{F}_j, \mathbf{E}_i) = 0,$$

i.e. vectors $\mathbf{E}_i$ and $\mathbf{F}_j$ are orthogonal except when s_j and s_i are complex conjugate. The series is said to be *bi-orthogonal*. Unfortunately, this says nothing about scalar products $\mathcal{H}(\mathbf{E}_j, \mathbf{E}_i)$ and $\mathcal{H}(\mathbf{F}_j, \mathbf{F}_i)$, see Exercise A.5.

Remark: Transient growth of the energy. Let a linear differential dynamical system be governed by

$$\tfrac{\mathrm{d}}{\mathrm{d}t}\mathbf{X} = \mathcal{L}\mathbf{X},$$

one can interpret the canonical Hermitian norm

$$\mathcal{H}(\mathbf{X}, \mathbf{X}) = 2K$$

where K is an energy. This interpretation is particularly appropriate when considering systems of mechanical origin, whose dependent variables are velocities, e.g. fluids governed by the Navier–Stokes equations linearised around some base state for which the kinetic energy

$$K = \frac{1}{2}\int_\mathcal{V} (u'^2 + v'^2 + w'^2)\,\mathrm{d}x\,\mathrm{d}y\,\mathrm{d}z$$

contained in a perturbation has all the properties requested for a norm.

On general grounds, asymptotic stability properties (long time limit) of the fixed point at the origin are obtained from the spectrum of eigenvalues s of $\mathcal{L}$ (i.e. stability, neutrality, or instability for $\mathcal{R}e(s) < 0, = 0$ or > 0, respectively). When the considered system is unstable, the energy contained in an arbitrary (but small) perturbation may initially decrease when the trajectories are called back toward the fixed point along strongly attracting eigen-directions but always ultimately increases exponentially since the long time behaviour is controlled by eigenvalues with positive real parts. The case of asymptotically stable systems is less transparent because, while they always approach the fixed point at late stage, trajectories may initially depart from it. This going away of trajectories, in the sense of the norm, can be interpreted as a transient growth of the energy for some initial conditions. One can indeed have $\frac{d}{dt}K > 0$ in some sectors of the tangent phase space, which may be surprising for a stable system. This is because the quadratic form:

$$\frac{d}{dt}K = \tfrac{1}{2}[\mathcal{H}(\tfrac{d}{dt}\mathbf{X}, \mathbf{X}) + \mathcal{H}(\mathbf{X}, \tfrac{d}{dt}\mathbf{X})]$$
$$= \tfrac{1}{2}[\mathcal{H}(\mathcal{L}\mathbf{X}, \mathbf{X}) + \mathcal{H}(\mathbf{X}, \mathcal{L}\mathbf{X})] = \mathcal{R}e(\mathcal{H}(\mathbf{X}, \mathcal{L}\mathbf{X}))$$

is not necessarily definite negative when $\mathcal{L}$ is non-normal, even when all eigenvalues are negative or have negative real parts. A simple example is treated in Exercise A.5.4.

The interest of this property in hydrodynamics derives from the fact that nonlinearity conserves the perturbation kinetic energy, so that the initial growth of the energy in the perturbations around the base state is controlled by the linear part of the tangent operator whatever their amplitude.[1] In practice, transient effects are all the more marked that diagonal terms are small when compared to non-diagonal terms. If one notices that diagonal terms mainly account for viscous dissipation, one understands that the situation is highly degenerate at large Reynolds numbers, so that this growth has essentially the same origin as the sub-dominant algebraic evolution called 'secular' in the study of the Jordan normal form. See also the introduction of Chapter 7 and §7.3.4.

A.3.2 Fredholm Alternative

Let us now consider the role played by the adjoints in the resolution of inhomogeneous linear problems (as they appear for example in perturbation theory, Chapter 2, p. 67ff):

$$\mathcal{L}\mathbf{X} = \mathbf{F}. \qquad (A.13)$$

The solution is unique and given by $\mathbf{X} = \mathcal{L}^{-1}\mathbf{F}$ for all $\mathbf{F}$, as long as $\mathcal{L}$ is invertible, that is to say as long as its kernel is trivial ($\mathcal{L}$ of maximal rank).

---
[1]D.S. Henningson and S.C. Reddy, "On the role of linear mechanisms in transition to turbulence," Phys. Fluids **6** (1994) 1396–1398.

The homogeneous problem

$$\mathcal{L}\mathbf{X} = \mathbf{0} \qquad (A.14)$$

then admits only the null vector $\mathbf{X} = \mathbf{0}$ as a solution. When this is not the case, i.e. $\mathcal{L}$ is not of maximal rank and has a nontrivial kernel, the adjoint homogeneous problem

$$\mathcal{L}^\dagger \tilde{\mathbf{X}} = \mathbf{0} \qquad (A.15)$$

also has nontrivial solutions $\tilde{\mathbf{X}} \neq \mathbf{0}$ (*Fredholm alternative*). The inhomogeneous problem (A.13) then has solutions only if the right hand side $\mathbf{F}$ is orthogonal to the kernel of the adjoint (*Fredholm theorem*). Vector $\tilde{\mathbf{X}}$ being in the kernel of $\mathcal{L}^\dagger$, this reads

$$\mathcal{H}(\tilde{\mathbf{X}}, \mathbf{F}) = 0, \qquad (A.16)$$

see Exercise A.5.3. When this condition is fulfilled the solution exists but is not unique. This indeterminacy can be raised by imposing, e.g. the orthogonality of the solution to the kernel of $\mathcal{L}^\dagger$, which, in perturbations problems presents the solution to the inhomogeneous problem as a true correction (in the sense of the scalar product) brought to the solution to the homogeneous problem.

A.3.3 *Boundary value problems and adjoint operators*

Up to now we have considered only finite dimensional spaces. Let us have a look at spaces whose elements are real or complex functions defined on some interval $[a, b]$. The natural extension of the sum appearing in the definition of the scalar product is now an integral:

$$\mathcal{H}(g, f) = \int_a^b g(x)^* f(x) \, \mathrm{d}x.$$

The case of interest to us is when these functions fulfil a differential boundary value problem of order n, with the differential operator $\mathcal{L}$ written as:

$$\mathcal{L}f = \sum_{m=0}^{n} a_m(x) f^{(m)},$$

$f^{(m)}$ denoting the m-th derivative of f with respect to x and $f^{(0)} \equiv f$. In order to formulate a well posed problem, we must add n boundary conditions:

$$\sum_{m=0}^{n-1} \alpha_{jm} f^{(m)}(a) + \beta_{jm} f^{(m)}(b) = u_j, \qquad j = 1, \ldots, n.$$

In many practical cases, these boundary conditions apply separately at each end of the interval, that is $\beta_{jm} = 0$ when $\alpha_{jm} \neq 0$ and conversely. When $u_j = 0$ for all j, boundary conditions are said to be *homogeneous*.

Returning to the definition of the adjoint operator (A.12), we see that the expression of $\mathcal{L}^\dagger$ has to be determined from:

$$\int_a^b g^*(x)\mathcal{L}^\dagger f(x)\, \mathrm{d}x = \int_a^b (\mathcal{L}g(x))^* f(x)\, \mathrm{d}x,$$

by removing g^* from the action of the differential operator on the right hand side of this identity. This is done by means of a series of integrations by parts:

$$\int_a^b a_m^* g^{*(m)} f\, \mathrm{d}x = \left[g^{*(m-1)} a_m^* f \right]_a^b - \int_a^b g^{*(m-1)} (a_m^* f)'\, \mathrm{d}x.$$

This operation progressively lowers the order of derivatives acting on g^* and must be pursued down to zero. The formal expression of $\mathcal{L}^\dagger$ is then easily obtained as:

$$\mathcal{L}^\dagger f = \sum_{m=0}^n (-1)^m (a_m^* f)^{(m)}.$$

However, the integrations by parts leaves us with a complicated boundary form made of integrated terms evaluated at the boundaries and involving derivatives of f, the a_m^*, and g^*. Inserting the boundary conditions of the direct problem in this linear form and imposing the cancellation of the residue gives the set of boundary conditions to be applied to the functions of the adjoint problem.

A.4 Application to linear control in state space

Results recalled above can be used to perform *linear control in state space*, basically how to obtain from a given system $\{\mathbf{X} \in \mathbb{X}, \mathcal{F}\}$ a specific *output* $\mathbf{Y}$ by forcing it with a well chosen *input* $\mathbf{U}$.[2] In a continuous-time framework this reads:

$$\frac{\mathrm{d}}{\mathrm{d}t}\mathbf{X} = \mathcal{F}(\mathbf{X}, \mathbf{U}), \tag{A.17}$$

$$\mathbf{Y} = \mathcal{H}(\mathbf{X}, \mathbf{U}), \tag{A.18}$$

expressions in which it is tacitly assumed that $\mathbf{X}$ and $\mathbf{U}$ are evaluated at t (no delay). Let d, d' and d'' be the number of components of the state $\mathbf{X}$, the control signal $\mathbf{U}$ and the output $\mathbf{Y}$. A SISO system (= Single-Input/Single-Output) thus correspond to $d' = d'' = 1$, and $d' > 1$, $d'' > 1$

---

[2] A thorough presentation in the spirit of this introduction is given in [Sontag (1998)].

to a MIMO system (= Multiple-Input/Multiple-Output). Here we only consider linear systems, so that $\mathbf{X} \in \mathbb{R}^d$, $\mathbf{U} \in \mathbb{R}^{d'}$, $\mathbf{Y} \in \mathbb{R}^{d''}$, and functions $\mathcal{F}$ and $\mathcal{H}$ are linear operators:

$$\tfrac{d}{dt}\mathbf{X} = \mathcal{A}\mathbf{X} + \mathcal{B}\mathbf{U}, \tag{A.19}$$

$$\mathbf{Y} = \mathcal{C}\mathbf{X} + \mathcal{D}\mathbf{U}, \tag{A.20}$$

represented by constant matrices $\mathcal{A}$, $\mathcal{B}$, $\mathcal{C}$, and $\mathcal{D}$, with respective dimensions $d \times d$, $d \times d'$, $d'' \times d$, and $d'' \times d'$

System (A.19, A.20) operates in *open loop* conditions. In *closed loop* conditions, the input is a function of the state itself:

$$\mathbf{U} = \mathcal{F}\mathbf{X} + \mathbf{U}', \tag{A.21}$$

which yields

$$\tfrac{d}{dt}\mathbf{X} = (\mathcal{A} + \mathcal{B}\mathcal{F})\mathbf{X} + \mathcal{B}\mathbf{U}', \tag{A.22}$$

$$\mathbf{Y} = (\mathcal{C} + \mathcal{D}\mathcal{F})\mathbf{X} + \mathcal{D}\mathbf{U}'. \tag{A.23}$$

The main problem is to design the $\mathcal{F}$ introduced in (A.21) which endows system (A.22, A.23) with the best possible performances.[3]

• **Evolution, response and transfer functions.** Open-loop integration of System (A.19) needs the specification of input $\mathbf{U}(t)$ and initial condition $\mathbf{X}_0$ at $t_0 = 0$. Introducing the change of variable $\mathbf{X} = \exp(t\mathcal{A})\mathbf{Z}$, one obtains:

$$\tfrac{d}{dt}\mathbf{Z} = \exp(-t\mathcal{A})\mathcal{B}\mathbf{U} \tag{A.24}$$

which yields

$$\mathbf{Z}(t) = \mathbf{Z}(0) + \int_0^t \exp(-t'\mathcal{A})\mathcal{B}\mathbf{U}(t')dt'.$$

with $\mathbf{Z}(0) = \mathbf{X}_0$. Back to $\mathbf{X}$, one gets

$$\mathbf{X}(t) = \exp(\mathcal{A}t)\mathbf{X}_0 + \int_0^t \exp[(t-t')\mathcal{A}]\mathcal{B}\mathbf{U}(t')dt'$$

which allows one to compute $\mathbf{Y}(t) = \mathcal{C}\mathbf{X}(t) + \mathcal{D}\mathbf{U}(t)$.

The *impulse* response is defined as

$$t < 0 : \; \mathcal{G}(t) = 0,$$
$$t \geq 0 : \; \mathcal{G}(t) = \mathcal{C}\exp(t\mathcal{A})\mathcal{B} + \mathcal{D}\delta(t),$$

---

[3] The case of a discrete-time system is similar, but with geometrical series replacing exponential functions.

so that
$$\mathbf{Y}(t) = \int_0^t dt' \mathcal{G}(t-t')\mathcal{U}(t').$$

The *transfer function* is then obtained from $\mathcal{G}(t)$ using the Laplace transform:
$$\widehat{\mathcal{G}}(s) = \int_0^\infty \mathcal{G}(t)\exp(-st)dt.$$

The initial value problem with $\mathbf{X}(0) = 0$ is then solved as
$$s\widehat{\mathbf{X}}(s) = \mathcal{A}\widehat{\mathbf{X}}(s) + \mathcal{B}\widehat{\mathbf{U}}(s),$$
$$\widehat{\mathbf{Y}}(s) = \mathcal{C}\widehat{\mathbf{X}}(s) + \mathcal{D}\widehat{\mathbf{U}}(s),$$

hence
$$(s\mathcal{I} - \mathcal{A})\widehat{\mathbf{X}}(s) = \mathcal{B}\widehat{\mathbf{U}}(s),$$

so that:
$$\widehat{\mathbf{Y}}(s) = \left[\mathcal{C}\frac{1}{s\mathcal{I}-\mathcal{A}}\mathcal{B} + \mathcal{D}\right]\widehat{\mathbf{U}}(s) \equiv \widehat{\mathcal{G}}(s)\widehat{\mathbf{U}}(s).$$

The response to periodic forcing at frequency ω is obtained from $\widehat{\mathcal{G}}(s)$ by setting $s = i\omega$.

- **Stability.** The linear unforced system $\frac{d}{dt}\mathbf{X} = \mathcal{A}\mathbf{X}$ is *stable* when all the eigenvalues of $\mathcal{A}$ have negative real parts (*Hurwitz property*). The forced system
$$\tfrac{d}{dt}\mathbf{X} = \mathcal{A}\mathbf{X} + \mathcal{B}\mathbf{U},$$
can be *stabilised* if one is able to find a feedback $\mathbf{U} = \mathcal{F}\mathbf{X}$ such that
$$\tfrac{d}{dt}\mathbf{X} = (\mathcal{A} + \mathcal{B}\mathcal{F})\mathbf{X}$$
is stable.

Using the distance to the origin as a measure of stability, and considering a *positive definite* quadratic form $\mathcal{M}$
$$\mathcal{M}(\mathbf{X}) = \mathbf{X}^t \mathcal{Q} \mathbf{X} \quad \text{with} \quad \mathcal{Q}^t = \mathcal{Q},$$
one readily observes that the system is stable if, as a function of $\mathbf{X}$, $\frac{d}{dt}\mathcal{M}$ is itself a *negative definite* quadratic form:
$$\tfrac{d}{dt}\mathcal{M} = \left(\tfrac{d}{dt}\mathbf{X}^t \mathcal{Q}\mathbf{X}\right) + \left(\mathbf{X}^t \mathcal{Q}\tfrac{d}{dt}\mathbf{X}\right) = \mathbf{X}^t(\mathcal{A}^t\mathcal{Q} + \mathcal{Q}\mathcal{A})\mathbf{X} = \mathbf{X}^t \mathcal{Q}'\mathbf{X}.$$

Requiring this to be valid for all $\mathbf{X}$ one obtains the *Lyapunov equation*:
$$\mathcal{A}^t \mathcal{Q} + \mathcal{Q}\mathcal{A} = \mathcal{Q}'.$$

• **Controllability and observability.** A dynamical system is said to be *controllable at time t* if, for every initial state $\mathbf{X}(0) = \mathbf{X}_0$, and any final state $\mathbf{X}_f$, one is able to find an input signal $\mathbf{U}(t')$, $0 \leq t' \leq t$ such that $\mathbf{X}(t) = \mathbf{X}_f$. It is *controllable* if it is controllable for at least one time t. State $\mathbf{X}_f$ is then *accessible* from $\mathbf{X}_0$ using input signal $\mathbf{U}(t)$.

Next, the system is *observable* if, from the time series of input $\mathbf{U}(t')$ and output $\mathbf{Y}(t')$ over $[0, t]$, one can determine the initial state $\mathbf{X}_0$. Two initial states $\mathbf{X}_0^{(1)}$ and $\mathbf{X}_0^{(2)}$ cannot be distinguished if, for all $t \geq 0$, output signals $\mathbf{Y}^{(1)}$ et $\mathbf{Y}^{(2)}$ are identical for any admissible input signal $\mathbf{U}(t)$. The system is thus observable if there exists a time t and an admissible input that make initial states distinguishable from the output at t, or equivalently, if the output responses to the same input are different when the initial states are different.

For linear systems, a criterion for the controllability of system $\frac{d}{dt}X = \mathcal{A}\mathbf{X} + \mathcal{B}\mathbf{U}$ is given by the rank of Kalman's *controllability matrix*

$$[\mathcal{B}, \mathcal{A}\mathcal{B}, \mathcal{A}^2\mathcal{B}, ..., \mathcal{A}^{d-1}\mathcal{B}], \tag{A.25}$$

of size $d \times (dd')$, where MATLAB's conventions, p. 377, have been used. The system is then controllable if the rank is maximum, i.e. $= d$.

An important property of controllable systems is that one can fix eigenvalues of the system $\mathcal{A} - \mathcal{B}\mathcal{F}$ arbitrarily (possibly by complex conjugate pairs) by choosing an appropriate feedback $\mathcal{F}$. Let $\mathbf{X}_r$ be a reference trajectory of system $(\mathcal{A}, \mathcal{B})$ under some control $\mathbf{U}_r$, and $\mathbf{X}_a$ be the response to control $\mathbf{U}_a$. One thus gets

$$\tfrac{d}{dt}\mathbf{X}_r = \mathcal{A}\mathbf{X}_r + \mathcal{B}\mathbf{U}_r$$
$$\tfrac{d}{dt}\mathbf{X}_a = \mathcal{A}\mathbf{X}_a + \mathcal{B}\mathbf{U}_a$$

The error ($\delta\mathbf{X} = \mathbf{X}_a - \mathbf{X}_r, \delta\mathbf{U} = \mathbf{U}_a - \mathbf{U}_r$) is thus governed by

$$\tfrac{d}{dt}\delta\mathbf{X} = \mathcal{A}\delta\mathbf{X} + \mathcal{B}\delta\mathbf{U}$$

so that, letting $\delta\mathbf{U} = -\mathcal{F}\delta\mathbf{X}$, one gets

$$\tfrac{d}{dt}\delta\mathbf{X} = (\mathcal{A} - \mathcal{B}\mathcal{F})\delta\mathbf{X}.$$

If system $(\mathcal{A}, \mathcal{B})$ is controllable, a feedback $\mathcal{F}$ can be found, so that all the eigenvalues of $\mathcal{A} - \mathcal{B}\mathcal{F}$ have negative real parts which means that the error decays exponentially to zero.

In contrast to controllability, linked to the characteristics of matrices $\mathcal{A}$ and $\mathcal{B}$ of system (A.20), observability is associated to the properties of matrices $\mathcal{A}$ and $\mathcal{C}$. Controllability and observability are dual properties and the latter results from a criterion which is the dual of (A.25). With $d'' = \dim(\mathbf{Y})$ and $d = \dim(\mathbf{X})$, using MATLAB's conventions again, one defines the *observability matrix* as

$$[\mathcal{C}; \mathcal{CA}; \mathcal{CA}^2; \ldots; \mathcal{CA}^{d-1}].$$

Matrices $\mathcal{C}$ and $\mathcal{A}$ being $d'' \times d$ and $d \times d$, respectively, this matrix is thus a column of d blocks of size $d'' \times d$, hence $dd'' \times d$. Observability is fulfilled if and only if this matrix has maximal rank $(= d)$.

On general grounds, when the ranks of the controllability and observability matrices are not maximal, the state space $\mathbb{X}$ can be decomposed into orthogonal subspaces, controllable–observable, controllable–non-observable, non-controllable–observable, and non-controllable–non-observable. A change of basis can then put the system into the form

$$\begin{bmatrix} \frac{d}{dt}\mathbf{X}_{co} \\ \frac{d}{dt}\mathbf{X}_{\bar{c}o} \\ \frac{d}{dt}\mathbf{X}_{\bar{c}\bar{o}} \\ \frac{d}{dt}\mathbf{X}_{\bar{c}\bar{o}} \end{bmatrix} = \begin{bmatrix} \mathcal{A}_{co} & 0 & \mathcal{A}_{13} & 0 \\ \mathcal{A}_{21} & \mathcal{A}_{\bar{c}o} & \mathcal{A}_{23} & \mathcal{A}_{24} \\ 0 & 0 & \mathcal{A}_{\bar{c}o} & 0 \\ 0 & 0 & \mathcal{A}_{43} & \mathcal{A}_{\bar{c}\bar{o}} \end{bmatrix} \begin{bmatrix} \mathbf{X}_{co} \\ \mathbf{X}_{\bar{c}o} \\ \mathbf{X}_{\bar{c}o} \\ \mathbf{X}_{\bar{c}\bar{o}} \end{bmatrix} + \begin{bmatrix} \mathcal{B}_{co} \\ \mathcal{B}_{\bar{c}o} \\ 0 \\ 0 \end{bmatrix} \mathbf{U},$$

$$\mathbf{Y} = [\mathcal{C}_{co}; 0; \mathcal{C}_{\bar{c}o}; 0] \begin{bmatrix} \mathbf{X}_{co} \\ \mathbf{X}_{\bar{c}o} \\ \mathbf{X}_{\bar{c}o} \\ \mathbf{X}_{\bar{c}\bar{o}} \end{bmatrix},$$

where subscripts "c", "$\bar{c}$", "o" and "$\bar{o}$" respectively mean "controllable", "non-controllable", "observable" and "non-observable".

A.5 Exercises

Among the results recalled in this Appendix, several find a nontrivial illustration already in dimension two. The following elementary exercises may help one better understand the generalisations stated without derivation in the text. Vectors $\mathbf{X}$ are represented in the canonical basis by column arrays:

$$\mathbf{X} = \begin{bmatrix} x_1 \\ x_2 \end{bmatrix}$$

and the linear operator $\mathcal{L}$ by a matrix $[\mathrm{L}]$:

$$\mathcal{L} = \begin{bmatrix} l_{11} & l_{12} \\ l_{21} & l_{22} \end{bmatrix}, \tag{A.26}$$

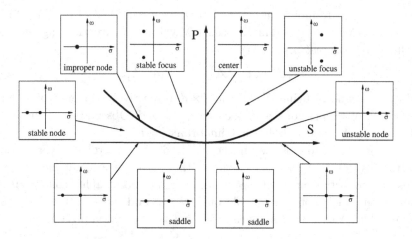

Fig. A.3 Nature of the roots of the characteristic equation of the two-dimensional problem as a function of S and P.

The complex extension of the vector space on $\mathbb{R}$ (i.e. $x_1 \in \mathbb{R}$, $x_2 \in \mathbb{R}$) is the set of vectors with complex coefficients (i.e. $x_1 \in \mathbb{C}$, $x_2 \in \mathbb{C}$). When necessary, the canonical scalar product is defined as:

$$\mathcal{H}(\mathbf{Y}, \mathbf{X}) = y_1^* x_1 + y_2^* x_2.$$

A.5.1 *Eigenvalues and eigenvectors, non-degenerate case*

1) Write down the eigenvalue problem $\mathcal{L}\mathbf{X} = s\mathbf{X}$ and the corresponding characteristic equation. Rewrite the latter using the sum S and the product P of the roots ($S = \text{trc}(\mathcal{L}) = l_{11} + l_{22}$, $P = \det(\mathcal{L}) = l_{11}l_{22} - l_{21}l_{12}$. Discuss the number and the nature of the roots in the plane of parameters (S, P) and identify the possible bifurcations upon variation of a control parameter r, $S = S(r)$ and $P = P(r)$. [See Fig. A.3.]

2) Identify the domain of parameters corresponding to two real distinct eigenvalues. Compute the eigenvalues, determine the corresponding eigenvectors and the matrix expressing the change from the canonical basis to the eigen-basis. Compute the inverse of this matrix.

3) Determine trajectories of $\frac{d}{dt}\mathbf{X} = \mathcal{L}\mathbf{X}$ with $\mathbf{X}(0) = \mathbf{X}_0$ in the canonical basis by changing back and forth to the eigen-basis. (As a concrete example, take $\frac{d}{dt}X_1 = 2X_1 + X_2$, $\frac{d}{dt}X_2 = 4X_1 - X_2$.)

4) Find the adjoint $\mathcal{L}^\dagger$ to $\mathcal{L}$ for the canonical scalar product (general case,

[L] with coefficients in $\mathbb{C}$). Show that when [L] has real coefficients, $\mathcal{L}^\dagger$ is normal ($\mathcal{L}\mathcal{L}^\dagger \equiv \mathcal{L}^\dagger\mathcal{L}$) only when it is symmetric ($l_{12} = l_{21}$) or such that $l_{11} = l_{22}$ and $l_{12} = -l_{21}$. Find the eigenvectors ($\mathbf{E}_1, \mathbf{E}_2$) of $\mathcal{L}$ and ($\mathbf{F}_1, \mathbf{F}_2$) of $\mathcal{L}^\dagger$. Check that, in all cases, $\mathcal{H}(\mathbf{F}_1, \mathbf{E}_2) = 0 = \mathcal{H}(\mathbf{F}_2, \mathbf{E}_1)$ but that one has $\mathcal{H}(\mathbf{E}_1, \mathbf{E}_2) = 0 = \mathcal{H}(\mathbf{F}_1, \mathbf{F}_2)$ only when $\mathcal{L}$ is normal.

5) When there are two complex conjugate eigenvalues, $s = \sigma \pm i\omega$, find a basis in which the system is represented by (2.31), Chapter 2, p. 52.

A.5.2 Degenerate case and Jordan normal form

1) Find the condition fulfilled by the l_{ij} when the characteristic polynomial has a double root. Find eigenvectors and determine the condition for any vector of the plane to be an eigenvector.

2) In the general case, this condition is not satisfied and there is only one eigen-direction, generated by a single vector $\mathbf{E}$, the components of which will be determined. Compute the image of an arbitrary vector $\mathbf{X}$ and check that $\mathbf{Y} = (\mathcal{L} - \bar{s}\mathcal{I})\mathbf{X}$ is parallel to $\mathbf{E}$. Derive from this an illustration of the *Cayley–Hamilton theorem*.

3) Show that in the basis $\{\mathbf{Y} = (\mathcal{L} - \bar{s}\mathcal{I})\mathbf{X}; \mathbf{X}\}$, the matrix representing $\mathcal{L}$ is in canonical Jordan form, which justifies writing a two-dimensional dynamical system with degenerate eigenvalues as (2.35, 2.36), p. 53.

4) Unfold the singularity by setting $s_{(\pm)} = \bar{s} \pm \delta s$, where δs is a small perturbation, and consider the dynamical system:

$$\tfrac{d^2}{dt^2}X - 2\bar{s}\,X + \left(\bar{s}^2 - \delta s^2\right)X = \left(\tfrac{d}{dt} - s_{(+)}\right)\left(\tfrac{d}{dt} - s_{(-)}\right)X = 0,$$

integrate it explicitly by solving successively $\left(\tfrac{d}{dt} - s_{(+)}\right)Y = 0$ and next $\left(\tfrac{d}{dt} - s_{(-)}\right)X = Y$, then show how secular terms associated to linear resonance appear by taking the limit $\delta s \to 0$.

A.5.3 Fredholm Alternative

1) Compute the kernel of operator $\mathcal{L}$ in (A.26).

2) Solve the inhomogeneous problem $\mathcal{L}\mathbf{X} = \mathbf{F}$ taken in the form

$$l_{11}x_1 + l_{22}x_2 = f_1,$$
$$l_{21}x_1 + l_{22}x_2 = f_2,$$

when the kernel of $\mathcal{L}$ is not trivially reduced to the null vector. Observe that the problem has solutions only when the two equations are proportional and express this condition.

3) Determine the adjoint $\mathcal{L}^\dagger$ of $\mathcal{L}$ and its kernel. Then check that the existence condition obtained can be written in the form (A.16).

A.5.4 Transient growth of the energy

Consider $\frac{d}{dt}\mathbf{X} = \mathcal{L}\mathbf{X}$ defined by:

$$\frac{d}{dt}X_1 = -aX_1 + X_2,$$
$$\frac{d}{dt}X_2 = -bX_2,$$

where constants a and b are positive and $b > a$ (already considered in Exercise 4.4.3, p. 164).

1) This system is non-normal, and thus has no orthogonal eigen-basis. Find the eigenvalues and the eigenvectors of $\mathcal{L}$ and its adjoint $\mathcal{L}^\dagger$ for the canonical scalar product.

2) Consider the energy $K = \frac{1}{2}(X_1^2 + X_2^2)$ and determine the condition on a and b rendering $\frac{d}{dt}K$ definite negative (monotonic stability). When this condition is not satisfied, identify the region of the (X_1, X_2) plane where $\frac{d}{dt}K > 0$ and the direction $\rho_{\max} = X_2/X_1$ that maximise the initial growth of the energy. Examine in particular the case $a \ll 1$, $b \ll 1$, so that the behaviour of the system is dominated by the out-of-diagonal term.

3) Compute the solution (X_1, X_2) issued from the initial condition $X_{1,0}, X_{2,0} = \rho X_{1,0}$, the evolution of the energy associated to this trajectory and the value of ρ that maximises $K(t)/K(0)$.

Appendix B

Numerical Approach

Present-day computers make numerical simulations of complex physical systems feasible. Their flexibility and accuracy allow one to gather quantitative information on situations of practical interest but also to obtain more qualitative hints about the nature of interactions at work. In this appendix we focus on the simplest, low-cost, approaches most particularly appropriate to the second aspect, leaving the first one to specialists. We present some classical numerical schemes used in simulations of *initial value problems* defined in terms of *ordinary* (ODEs) or *partial* (PDEs) *differential equations*. The level is kept very elementary and theoretical discussions are avoided, though numerical stability properties derive directly from the application of techniques developed in other parts of the present work, as can be seen by working out the exercises proposed.

In practice, the evolution of the system under consideration is determined at a series of regularly spaced times $t_n = n\,\Delta t$. The corresponding integration methods, appropriate to ODEs as well as PDEs, are introduced in §B.1. The physical-space dependence inherent in PDEs implies a specific coupling between degrees of freedom. This can be treated in the most naive way by estimating the values of the different fields at the nodes of a discrete lattice, $x \to x_j = j\,\Delta x$. Replacing partial derivatives by their *finite differences* yields the numerical schemes presented in §B.2.1. Another description can be achieved in terms of projection of the dynamics on a functional basis spanning the whole physical space. These so-called *spectral* methods are briefly introduced, using Fourier modes in §B.2.2.

Finite element methods and their variants specifically developed for fluid mechanics are left completely aside despite their accuracy and their versatility regarding complex configurations encountered in technical applications for which commercial codes are fully adapted. Instead of relying on such black-boxes, we prefer present simple algorithms that can be straightforwardly translated into programming languages such as MATLAB. To our belief, the required investment is light but rewarding in terms of the insight gained in the complexity arising from nonlinearity.

B.1 Treatment of the Time Dependence

We consider the following general initial value problem:
$$\tfrac{\mathrm{d}}{\mathrm{d}t} v = f(v,t) \quad \text{with} \quad v(t_0) = v_0 \,. \tag{B.1}$$
here written as a non-autonomous ODE but the case of PDEs is the same as far as time dependence is concerned.

Expanding the solution as a Taylor series in the neighbourhood of t_n, we get:
$$v_{n+1} = v_n + \Delta t \tfrac{\mathrm{d}}{\mathrm{d}t} v \big|_n + \tfrac{1}{2} \Delta t^2 \tfrac{\mathrm{d}^2}{\mathrm{d}t^2} v \big|_n + \tfrac{1}{6} \Delta t^3 \tfrac{\mathrm{d}^3}{\mathrm{d}t^3} v \big|_n + \cdots \tag{B.2}$$
where the successive derivatives of v, total derivative with respect to t, can be evaluated at t_n recursively, using (B.1):
$$\tfrac{\mathrm{d}}{\mathrm{d}t} v \big|_n = f_n \,,$$
$$\tfrac{\mathrm{d}^2}{\mathrm{d}t^2} v \big|_n = \tfrac{\mathrm{d}}{\mathrm{d}t} \tfrac{\mathrm{d}}{\mathrm{d}t} v \big|_n = \tfrac{\mathrm{d}}{\mathrm{d}t} f(v,t) \big|_n$$
$$= \left(\partial_v f \tfrac{\mathrm{d}}{\mathrm{d}t} v + \partial_t f \right) \big|_n = (\partial_v f_n) f_n + \partial_t f_n$$
$$\tfrac{\mathrm{d}^3}{\mathrm{d}t^3} v \big|_n = \cdots ,$$
provided that we can evaluate the successive partial derivatives of $f(v,t)$ analytically, which is hardly conceivable beyond the lowest order without the help of some formal algebra software. The idea is thus to look for numerical approximations to these derivatives.[1]

Truncating expression (B.2) beyond first order, we get the most elementary formula (Euler, cf. Fig. B.1):
$$v_{n+1} = v_n + \Delta t\, f(v_n, t_n) \,. \tag{B.3}$$
This integration scheme is said to be *first order* because it is exact up to correction of order Δt^2, and *explicit* because the knowledge of v and $f(v,t)$ at previous steps is sufficient to compute the solution at the next step.

Another first order scheme is obtained by evaluating v_n using quantities computed at t_{n+1}:
$$v_n = v_{n+1} - \Delta t\, \tfrac{\mathrm{d}}{\mathrm{d}t} v \big|_{n+1} + \tfrac{1}{2} \Delta t^2 \tfrac{\mathrm{d}^2}{\mathrm{d}t^2} v \big|_{n+1} - \tfrac{1}{6} \Delta t^3 \tfrac{\mathrm{d}^3}{\mathrm{d}t^3} v \big|_{n+1} + \cdots , \tag{B.4}$$

---
[1] For elementary formulas consult [Abramowitz and Stegun (1972)]. A detailed presentation can be found in [Lapidus and Seinfeld (1971)].

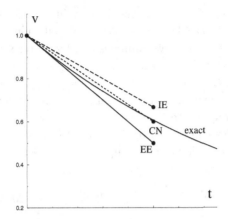

Fig. B.1 Solution of $\frac{d}{dt}v = -v$ at $t = \Delta t = 0.5$ with $v(0) = 1$: Exact result compared to estimates from the explicit Euler (EE), implicit Euler (IE) and Crank–Nicholson (CN) formulas, Eqs. (B.3), (B.5), and (B.6), respectively.

which leads to:
$$v_{n+1} = v_n + \Delta t\, f(v_{n+1}, t_{n+1}).$$
This scheme, still exact up to corrections $\mathcal{O}(\Delta t^2)$, is termed *implicit* since v_{n+1} is obtained from an implicit equation:
$$v_{n+1} - \Delta t\, f(v_{n+1}, t_{n+1}) = v_n. \tag{B.5}$$
Subtracting (B.4) from (B.2) and noticing that
$$\left.\tfrac{d^2}{dt^2}v\right|_{n+1} = \left.\tfrac{d^2}{dt^2}v\right|_n + \Delta t \left.\tfrac{d^3}{dt^3}v\right|_n + \mathcal{O}(\Delta t^2),$$
one can see that the term $\mathcal{O}(\Delta t^2)$ disappears, so that:
$$v_{n+1} = v_n + \frac{\Delta t}{2}\bigl(f(v_n, t_n) + f(v_{n+1}, t_{n+1})\bigr) + \mathcal{O}(\Delta t^3). \tag{B.6}$$
This *implicit second order* formula is called the *Crank–Nicholson* scheme (cf. Fig. B.1).

Let us compare the error terms using $f(v) = -v$ with $v_0 = 1$ as a simple example. We get $v = \exp(-t)$ and after one time step

Exact $(\exp(-\Delta t))$: $\quad 1 - \Delta t + \tfrac{1}{2}\Delta t^2 - \tfrac{1}{6}\Delta t^3 + \ldots$

Explicit Euler (B.3): $\quad 1 - \Delta t$

Implicit Euler (B.5): $\quad \dfrac{1}{1+\Delta t} = 1 - \Delta t + \Delta t^2 - \Delta t^3 + \ldots,$

Crank–Nicholson (B.6): $\quad \dfrac{1-\tfrac{1}{2}\Delta t}{1+\tfrac{1}{2}\Delta t} = 1 - \Delta t + \tfrac{1}{2}\Delta t^2 - \tfrac{1}{4}\Delta t^3 + \ldots,$

which clearly shows the improvement brought by the Crank–Nicholson scheme (cf. Exercise B.3.1). The obvious computational drawback of implicit iteration schemes, largely compensated by better stability properties as shown later, is that the solution at t_{n+1} is not expressed simply in terms of the solution at t_n. The inversion is trivial only if v is a single scalar variable and f is linear as in the example chosen. When this is not the case, specific techniques are required, see §B.2.1.

Up to now we have used only *one-step* formulas, requiring the knowledge of the solution at a single time. A simple improvement of (B.3) is obtained by inserting:

$$\left.\frac{d^2}{dt^2}v\right|_n = \frac{1}{\Delta t}(\tfrac{d}{dt}v|_n - \tfrac{d}{dt}v|_{n-1}) + \mathcal{O}(\Delta t)$$

in (B.2), which leads to the second-order *Adams–Bashford* formula:

$$v_{n+1} = v_n + \Delta t \left(\tfrac{3}{2}f(v_n, t_n) - \tfrac{1}{2}f(v_{n-1}, t_{n-1})\right), \tag{B.7}$$

exact up to $\mathcal{O}(\Delta t^3)$. This formula is one of the simplest examples of *multi-step* schemes, computing v_{n+1} from the knowledge of v at several previous times, here t_n and t_{n-1}.

An equally simple possibility is to write (B.4) at times t_{n-1} and t_n instead of t_n and t_{n+1}, which yields

$$v_{n-1} = v_n - \Delta t \, \tfrac{d}{dt}v|_n + \tfrac{1}{2}\Delta t^2 \tfrac{d^2}{dt^2}v|_n - \tfrac{1}{6}\Delta t^3 \tfrac{d^3}{dt^3}v|_n + \ldots,$$

so that after subtraction of (B.2) one gets another second order formula, also exact up to $\mathcal{O}(\Delta t^3)$:

$$v_{n+1} = v_{n-1} + 2\,\Delta t\, f(v_n, t_n), \tag{B.8}$$

called the *leap-frog* scheme.

Applying the Adams–Bashford scheme to $f(v) = -v$, after one time step one gets:

$$\text{Adams–Bashford (B.7)}: \quad 1 - \Delta t + \frac{1}{2}\Delta t^2 + \frac{1}{4}\Delta t^3 + \ldots$$

with an error $5\Delta t^3/12$ instead of $-\Delta t^3/12$ for the Crank–Nicholson scheme. For the leap-frog scheme, assuming that v_{-1} is known exactly, i.e. $1 + \Delta t + \tfrac{1}{2}\Delta t^2 + \tfrac{1}{6}\Delta t^3 + \ldots$, one obtains:

$$\text{leap-frog (B.8)}: \quad 1 - \Delta t + \frac{1}{2}\Delta t^2 + \frac{1}{6}\Delta t^3 + \ldots$$

and a quite large error, $\tfrac{1}{3}\Delta t^3$, which may even be much amplified if v_{-1} is less well known, owing to the bad stability properties of that scheme.

In addition to accuracy requirements, the choice of an integration scheme is indeed also subjected to stability requirements. As a matter of fact, the numerical scheme must not amplify round-off errors so that the solution obtained is the physical one and not some parasitic solution. The problem is important since when, for example, one uses a two-step formula to integrate a first-order differential equation, one obtains an iteration that starts with two initial conditions instead of one for the time-continuous problem. The added mode must be damped in order that the simulation keeps its meaning. Let us compare schemes (B.7) and (B.8) used to integrate $\frac{d}{dt} v = -v$:

- Iteration (B.7) reads

$$v_{n+1} = v_n - \Delta t \left(\tfrac{3}{2} v_n - \tfrac{1}{2} v_{n-1} \right)$$

or, better,

$$u_{n+1} = v_n,$$
$$v_{n+1} = \tfrac{1}{2} \Delta t \, u_n + \left(1 - \tfrac{3}{2} \Delta t\right) v_n,$$

which, from standard linear stability analysis (Exercise B.3.2), leads to the characteristic equation:

$$s^2 - s\left(1 - \tfrac{3}{2} \Delta t\right) - \tfrac{1}{2} \Delta t = 0.$$

At lowest order in Δt, the two roots are $s_{(+)} = 1 - \Delta t$ and $s_{(-)} = -\tfrac{1}{2}\Delta t$. The first one clearly corresponds to the physical solution and the second one to the numerical mode. Recalling that the differential equation is here replaced by a discrete-time system like those introduced in Chapter 3 (see in particular Figure 4.11, p. 150) we obtain that the latter mode is damped provided that $|s_{(-)}| < 1$, which can be achieved by taking Δt sufficiently small, here $\Delta t < 2$. This does not mean that the numerical solution obtained with Δt just below the limit is a good approximation, but just that it is not polluted by the numerical mode. The Adams–Bashford scheme is said to be *conditionally stable*.

- Iteration (B.8) reads:

$$v_{n+1} = v_{n-1} - 2 \Delta t \, v_n,$$

that is:

$$u_{n+1} = v_n,$$
$$v_{n+1} = u_n - 2 \Delta t \, v_n,$$

which leads to
$$s^2 + 2\Delta t\, s - 1 = 0,$$
so that $s_{(+)} \simeq 1 - \Delta t$ as before. But now one has $s_{(-)} \simeq -1 - \Delta t$ and since $|s_{(-)}| > 1$, the numerical mode is always unstable. The leap-frog scheme is thus always *unconditionally unstable*. Since $s_{(-)}$ is negative, successive iterates alternate from positive to negative values. This subharmonic instability is due to the fact that the scheme leaves even time steps uncoupled to odd ones. Such a behaviour is frequently the signature of a *numerical instability* (see later Figure B.16). Scheme (B.8) is however little demanding and its use is not forbidden provided the growth of errors is controlled. In general, the numerical mode is tamed by averaging over two iterations every p computation, with $p \gg 1$ but still small enough, so that its amplitude remains small when compared to the truncation errors.

High order accuracy can be reached with sophisticated multi-step formulas. Their drawback stems from the need to generate numerical initial conditions from the physical ones, especially when the time step has to be changed for accuracy reasons. Explicit formulas of *Runge–Kutta* type avoid this problem. The idea is to use the basic first-order Euler formula but with a better estimate of the slope. Second order accuracy is obtained by averaging the slopes evaluated at t_n and t_{n+1} but the resulting (Crank–Nicholson) scheme is implicit. To avoid such a difficulty one can first estimate v_{n+1} with the first-order explicit formula:
$$v' = v_n + \Delta t\, f(v_n, t_n), \tag{B.9}$$
then compute the slope at this point $f(v', t_{n+1})$, and take the average with the slope at (v_n, t_n) as an improved guess. This yields:
$$v_{n+1} = v_n + \tfrac{1}{2}\Delta t \big(f(v_n, t_n) + f(v', t_{n+1})\big). \tag{B.10}$$
The correction is sufficient to achieve second order accuracy while keeping an explicit scheme (Exercise B.3.1). This case is illustrated in Figure B.2, left.

Another way to achieve the same goal is to introduce an intermediate time $t_{n+1/2}$, to compute v' with the first-order formula
$$v' = v_n + \tfrac{1}{2}\Delta t f(v_n, t_n), \tag{B.11}$$
and to use the slope at that point to extrapolate the solution at time t_{n+1}, as shown in Figure B.2, right:
$$v_{n+1} = v_n + \Delta t\, f(v', t_{n+1/2}). \tag{B.12}$$

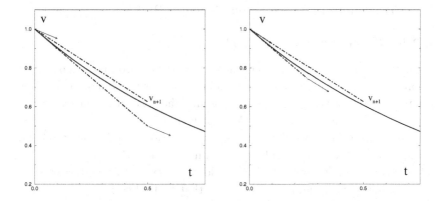

Fig. B.2 Achieving second order accuracy with Runge–Kutta formulas. Left: (B.9, B.10). Right: (B.11, B.12).

High-order formulas combine these two basic techniques to get the best estimate of the slope. The classical fourth-order scheme

$$v_{n+1} = v_n + \tfrac{1}{6}\Delta t \big(f(v_n, t_n) + 2f(v', t_{n+1/2}) + 2f(v'', t_{n+1/2}) + f(v''', t_{n+1})\big) \tag{B.13}$$

involves three intermediate estimates:

$$v' = v_n + \tfrac{1}{2}\Delta t f(v_n, t_n),$$

$$v'' = v_n + \tfrac{1}{2}\Delta t f(v', t_{n+1/2}),$$

$$v''' = v_n + \Delta t f(v'', t_{n+1/2}).$$

One full time step thus requires four evaluations of f, which may be numerically demanding, but the scheme stays explicit and starts with a single initial condition so that the time step can easily be changed.

Predictor-corrector methods are multi-step schemes that are often preferred to Runge–Kutta methods because they require a smaller number of evaluations of f to reach the same accuracy though it is more difficult to change the time step. For more information, consult [Lapidus and Seinfeld (1971)] or [Press et al. (1986)].

Explicit methods can easily be extended to treat differential systems coupling several variables, see the examples considered in §B.4.

B.2 Treatment of Space Dependence in PDEs

B.2.1 *Finite difference methods*

The derivation of an integration scheme is made of two main steps. First, a consistent approximation of space differential operators has to be found. Next a time-integration scheme is chosen. The resulting space-time iteration is then studied for stability. A supplementary step would be to demonstrate convergence in the limit of infinitesimal space and time steps. For *linear* partial differential problems, it can be shown that *consistence* and *stability* implies *convergence*. Here we do not worry about convergence and just feel happy with numerical solutions that look physical.

We consider only systems that extend in one space direction. The solution is sought for at regularly spaced nodes $\{x_j\}$ of a periodic lattice with spacing $\Delta x = \ell/N$ where ℓ is the length of the domain and $N+1$ the number of points, including end points. The numerical solution $v_{n,j}$ is thus noted with two subscripts but the first one, corresponding to time, will be dropped where it is not necessary.

B.2.1.1 *Space discretization and consistence*

Finite difference approximations to space derivatives are obtained in the same way as those relative to time.[2] For example, combination of two first order approximations of the first order derivatives:

$$\partial_x v|_j = (v_{j+1} - v_j)/\Delta x \quad \text{and} \quad \partial_x v|_j = (v_j - v_{j-1})/\Delta x \quad \text{(B.14)}$$

yields the second order centred formula:

$$\partial_x v|_j = (v_{j+1} - v_{j-1})/2\Delta x. \quad \text{(B.15)}$$

In the same way, the expression of the second derivative at order Δx^2 reads:

$$\partial_{xx} v|_j = (v_{j+1} - 2v_j + v_{j-1})/\Delta x^2, \quad \text{(B.16)}$$

Higher order accuracy requires more points, e.g. five at fourth order:

$$\partial_{xx} v|_j = (-v_{j+2} + 16v_{j+1} - 30v_j + 16v_{j-1} - v_{j-2})/12\Delta x^2. \quad \text{(B.17)}$$

The second-order approximation of the fourth derivative reads:

$$\partial_{xxxx} v|_j = (v_{j+2} - 4v_{j+1} + 6v_j - 4v_{j-1} + v_{j-2})/\Delta x^4. \quad \text{(B.18)}$$

The order of *consistency* of some finite difference approximation is the value of the exponent that controls the decay of the global truncation error

---

[2] Useful formulas are found in [Abramowitz and Stegun (1972)]. A complete presentation is given by [Richtmyer and Morton (1967)]. Consult also [Acton (1970)].

when Δx decreases. It thus serves little to approximate different derivatives appearing in a model by formulas at different orders. In practice, the representation of differential operators by finite differences is not very good since truncation errors decay only algebraically, as powers of Δx. From this point of view spectral methods briefly introduced in Section B.2.2 behave in a much better way.

B.2.1.2 *Boundary conditions*

Consistency considerations also affect the treatment of boundary conditions when they involve derivatives (*Neumann conditions*). Since it is in general easy to achieve consistency at order Δx^2 for interior points —for example by preferring (B.15) to (B.14) for ∂_x— it would be a pity to spoil the quality of an approximation by a bad account of boundary conditions, even if their effects do not propagate beyond some narrow numerical boundary layer.

Consider for example the condition $\partial_x v = 0$ at one end point of the x interval. Using (B.14) we get an expression accurate at order Δx, which simply implies $v_j = v_{j+1}$, with $j = 0$ and 1 for the boundary point and the first interior point, respectively. To achieve second order consistency, we can use:

$$\partial_x v|_j = (-3v_j + 4v_{j+1} - v_{j+2})/2\Delta x \qquad (B.19)$$

so that cancelling $\partial_x v$ at the boundary gives $3v_0 - 4v_1 + v_2 = 0$. Another possibility is to add a fictitious point outside, $j = -1$ and associate a variable v_{-1} whose value is forced to that of the first interior point at all times, $v_{-1} = v_1$, as dictated by formula (B.15).

B.2.1.3 *Time discretization and stability*

For more specificity we consider the one-dimensional *diffusion equation*:

$$\partial_t v = \partial_{xx} v, \qquad (B.20)$$

with boundary conditions that need not be specified at this stage. The temporal scheme is obtained according to one of the recipes of section B.1 (here we restore subscript n). For space we assume second order consistency, using (B.16) to discretize $\partial_{xx} v$. The most important question is that of the numerical stability of the scheme, as determined from the growth rate of the noise generated by truncation and round-off errors. Here, we only compare the case of the two simplest schemes deriving from (B.3) and (B.5). In the *explicit* case, the space derivative $\partial_{xx} v$ is evaluated at t_n,

that is:
$$v_{n+1,j} = v_{n,j} + \frac{\Delta t}{\Delta x^2}(v_{n,j+1} - 2v_{n,j} + v_{n,j-1}), \quad (B.21)$$
while it is evaluated at t_{n+1} in the *implicit* case, which yields:
$$-\frac{\Delta t}{\Delta x^2}v_{n+1,j+1} + \left(1 + \frac{2\Delta t}{\Delta x^2}\right)v_{n+1,j} - \frac{\Delta t}{\Delta x^2}v_{n+1,j-1} = v_{n,i}. \quad (B.22)$$

The explicit method is straightforward and requires less computational work than the implicit method that requires the inversion of a linear system with a band structure and only three near-diagonal terms. As we will see, this amount of work is rewarding.

The stability analysis of finite-difference systems is performed in the same way as that of physical systems (§3.1.3 and §4.2). The evolution of a perturbation $\delta v_{n,j}$ to a given numerical solution $v_{j,n}$ is studied by means of a discrete Fourier transform:
$$\delta v_{n,j} = \delta\tilde{v}_{n,k}\exp(ikj).$$
Inserting this expression into equations (B.21) and (B.22), we obtain:

Explicit: $\quad \delta\tilde{v}_{n+1,k} = \left(1 - \dfrac{2\Delta t}{\Delta x^2}(1 - \cos(k))\right)\delta\tilde{v}_{n,k}, \quad$ (B.23)

Implicit: $\quad \left(1 + \dfrac{2\Delta t}{\Delta x^2}(1 - \cos(k))\right)\delta\tilde{v}_{n+1,k} = \delta\tilde{v}_{n,k}, \quad$ (B.24)

both of the general form:
$$\delta\tilde{v}_{n+1,k} = \xi(k)\,\delta\tilde{v}_{n,k},$$
and, as we know, the perturbation grows when $|\xi(k)| > 1$ and decays when $|\xi(k)| < 1$.

For the explicit scheme, this gives:
$$1 - \frac{4\Delta t}{\Delta x^2} \leq \xi_{\exp}(k) = 1 - \frac{2\Delta t}{\Delta x^2}(1 - \cos(k)) \leq 1,$$
The upper bound condition brings nothing, whereas the lower bound implies $-1 < \xi_{\exp}(\pi)$ for stability. This leads to:
$$-1 < 1 - \frac{4\Delta t}{\Delta x^2} \Rightarrow \frac{4\Delta t}{\Delta x^2} < 2 \Rightarrow \Delta t < \frac{\Delta x^2}{2},$$
which means that the time step must be small enough at given spacing: the explicit scheme is thus *conditionally stable*.

By comparison, the growth rate of perturbations with the implicit scheme:
$$\xi_{\text{imp}}(k) = \left(1 + \frac{2\Delta t}{\Delta x^2}(1 - \cos(k))\right)^{-1}$$

is such that $0 < \xi_{\text{imp}}(k) \leq 1$ for all k since $(1 - \cos(k)) \geq 0$. This scheme is therefore *unconditionally stable*, so that Δt can be chosen for temporal accuracy requirement only, whatever the value of Δx.

For the explicit scheme the most unstable perturbations are those with $k = \pi$. Going back to physical space, one finds $\delta v_{n,j} = \delta \tilde{v}_n (-1)^j$: the behaviour of the solution at even nodes is the opposite of that at odd nodes, which is typical of numerical instabilities. At this stage, the instability can be tamed by decreasing the time step until the stability condition is fulfilled but it is better to turn to an implicit scheme, especially when higher derivatives are involved (Exercise B.3.3). Higher order time discretization would be studied in the same way.

B.2.1.4 Efficient treatment of implicit schemes

The supplementary work needed to find the solution at time step t_{n+1} by the implicit scheme (B.22) is easily achieved by a simple numerical *double sweep* method since the matrix to be inverted has only a few terms close to the diagonal. This is in fact a special case of the classical LU decomposition used to solve linear systems.

Let the initial system be $\mathcal{M}\mathbf{V} = \mathbf{F}$, the operator $\mathcal{M}$ is then written as the product $\mathcal{L}\mathcal{U}$ of two operators, $\mathcal{L}$ represented by a lower triangular matrix $[\mathrm{L}]$ with elements $l_{ij} = 0$ when $i < j$, and an operator $\mathcal{U}$ represented by an upper triangular matrix $[\mathrm{U}]$ such that $u_{ij} = 0$ when $i > j$. The decomposition is unique upon requiring that the diagonal elements of $[\mathrm{U}]$ are scaled to unity, $u_{ii} = 1$. The starting matrix $[\mathrm{M}]$ is three-diagonal, i.e. $m_{ij} \neq 0$ for $j = i$ and $j = i \pm 1$, $[\mathrm{L}]$ and $[\mathrm{U}]$ have a band structure with two non-zero diagonals only: $l_{ij} \neq 0$ for $j = i$ and $j = i - 1$, $u_{jj} = 1$, and $u_{ij} \neq 0$ for $j = i + 1$.

The problem which reads:

$$(\mathcal{L}\mathcal{U})\mathbf{V} = \mathcal{L}(\mathcal{U}\mathbf{V}) = \mathcal{L}\mathbf{W} = \mathbf{F}$$

is then replaced by two equations

$$\mathcal{L}\mathbf{W} = \mathbf{F} \quad \text{and} \quad \mathcal{U}\mathbf{V} = \mathbf{W},$$

that are solved by means of two first-order recursion relations, first a *forward* iteration to obtain $\mathbf{W}$:

$$l_{11} w_1 = f_1,$$
$$l_{i\,i-1} w_{i-1} + l_{ii} w_i = f_i, \qquad i = 2, \ldots, N,$$

and second a *backward* iteration to get **V** from **W**:

$$v_N = w_N,$$
$$v_i + u_{i\,i+1} v_{i+1} = w_i, \qquad i = N-1,\ldots,1.$$

where the coefficients l_{ij} and u_{ij} are computed once for all using two simple first-order recursion relations. The general term of the product reads $\sum_k l_{ik} u_{kj} = m_{ij}$, with two special cases for $i=1$ and $j=N$. One readily obtains

$$l_{11} = m_{11},$$
$$l_{11} u_{12} = m_{12},$$

then for $i = 2$ to $N-1$

$$l_{ii-1} = m_{ii-1},$$
$$l_{ii-1} u_{i-1\,i} + l_{ii} = m_{ii},$$
$$l_{ii} u_{ii+1} = m_{ii+1},$$

and finally for $i = N$

$$l_{NN-1} = m_{NN-1},$$
$$l_{NN-1} u_{N-1\,N} + l_{NN} = m_{NN}.$$

These relations are easily implemented numerically and can be generalised to the case with any finite and small number of elements close to the diagonal. This happens for example when the model contains fourth order derivatives, hence five terms from (B.18), which leads to similar but second-order recursion relations, Exercises B.3.3 and B.4.4.2 (*Swift–Hohenberg model*).

For problems that are two-dimensional in space, implicit schemes lead to linear systems where the matrix [M] is sparse with a block-diagonal structure. Instead of generalising the previous algorithms to block matrices, which is possible but requires a lot of storage, one usually turns to iterative methods. The description of corresponding algorithms, over-relaxation, conjugate-gradients, etc., goes beyond our present purpose.[3]

B.2.1.5 Treatment of nonlinear terms

Up to now we have considered linear problems. In most cases of interest, nonlinearity is present, so let us write formally $\partial_t v = \mathcal{L}v + \mathcal{N}(v)$. The linear

---
[3] Consult for example J.H. Wilkinson: "Solution of linear algebraic equations and matrix problems by direct methods," in *Digital computer user's handbook*, M.K. Klerer and G.A. Korn, eds. (McGraw-Hill, 1967), Chapter 2.2.

part is often the most dangerous one from the point of view of numerical stability since it usually contains partial derivatives of higher order than the nonlinear part. In order to preserve second order accuracy while avoiding fully implicit treatment of nonlinearity, one can develop *quasi-linearised* schemes of the Crank–Nicholson type by replacing v_{n+1} by $v_n + (v_{n+1} - v_n)$ in the nonlinear term, expanding it to first order in $(v_{n+1} - v_n)$ assumed to be small enough:

$$\mathcal{N}(v)\big|_{n+1} \approx \mathcal{N}(v)\big|_n + \mathcal{Q}_n(v_{n+1} - v_n),$$

where the functional derivative of $\mathcal{N}$ with respect to v, $\mathcal{Q}_n = \delta\mathcal{N}/\delta v\big|_n$, is evaluated at time t_n. When nonlinearity is not numerically dangerous, another possibility is to treat the linear and nonlinear terms separately, using an implicit Crank–Nicholson scheme for the linear part and an explicit Adams–Bashford scheme for the nonlinear part

$$v_{n+1} = v_n + \tfrac{1}{2}\Delta t \left[\left(\mathcal{L}v\big|_{n+1} + \mathcal{L}v\big|_n\right) + \left(3\mathcal{N}(v)\big|_n - \mathcal{N}(v)\big|_{n-1}\right)\right].$$

B.2.2 *Spectral methods*

Spectral methods rest on a conversion of the partial differential problem into an ordinary differential problem for the amplitudes of modes obtained by projection onto a complete functional basis. For simplicity, we consider here only the case of periodic boundary conditions at the ends of an interval of length ℓ, which permits the use of Fourier modes:

$$v(x) = \sum_{m \in \mathbb{Z}} \hat{v}_m \exp(2\pi i m x/\ell), \tag{B.25}$$

with $\hat{v}_{-m} = \hat{v}_m^*$ when $v \in \mathbb{R}$. The Fourier coefficients are given by:

$$\hat{v}_m = \frac{1}{\ell} \int_0^\ell v(x) \exp(-2\pi i m x/\ell)\, dx, \quad m \in \mathbb{Z}. \tag{B.26}$$

We recall that:

$$\frac{1}{\ell} \int_0^\ell \exp\left(2\pi i (m'-m)x/\ell\right) dx = \delta_{mm'}, \tag{B.27}$$

where $\delta_{mm'}$ is the Kroneker symbol ($= 1$ when $m = m'$ and zero otherwise), and that, conversely,

$$\sum_{m=-\infty}^{+\infty} \exp\left(2\pi i m(x-x')/\ell\right) = \delta_{\mathrm{D}}(x - x'), \tag{B.28}$$

where $\delta_D(x)$ is the Dirac generalised function such that:

$$\int_{-\infty}^{+\infty} f(x')\delta_D(x-x')\,dx = f(x).$$

For the diffusion equation (B.20), the evolution of the solution is easily obtained by integrating the differential equations for the Fourier coefficients $\hat{v}_m = \hat{v}_m(t)$:

$$\tfrac{d}{dt}\hat{v}_m = -k_m^2 \hat{v}_m \quad \text{with} \quad k_m = 2\pi m/\ell,$$

which yields:

$$\hat{v}_m(t) = \hat{v}_m(0)\exp(-k_m^2 t), \quad m \in \mathbb{Z}.$$

Let us notice that, here, the differential operator is diagonal in Fourier space. This may not be the case with other boundary conditions requiring other basis functions, e.g. Chebyshev polynomials, but, in general, an exact representation of the differential operator in spectral space can still be found.

In practice, the series is truncated beyond some maximal value $N/2$, so that $\sum_{m=-\infty}^{+\infty}$ is replaced by $\sum_{-N/2+1}^{+N/2}$, see Fig. B.3(a, b). This truncation introduces some approximation forbidding us to resolve the dynamics of structures with wavelengths shorter than $\lambda_{\min} = 2\pi/k_{\max} = 2\ell/N$, which can be understood as the result of sampling the solution on a regular lattice with spacing $\Delta x = \ell/N$. One may say that the method is of *infinite order*

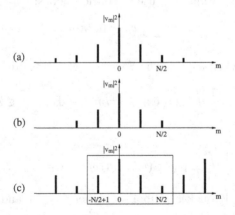

Fig. B.3 (a) Spectrum $|\hat{v}_m|^2$ of a periodic function. (b) Same spectrum, truncated beyond $N/2$. (c) Discrete Fourier transform of the function that has the same value as the solution at the collocation points.

since the error term decreases exponentially with N, as $\exp\left(-(N\pi/\ell)^2 t\right)$ at given t, faster than any power of Δx.

In fact we have not yet said how integral (B.26) is computed. In practice the solution is only represented at the nodes of the lattice mentioned above and called the *collocation points*. This allows us to pass from continuous Fourier transforms to discrete ones.[4] Let $\{x_j = j\Delta x; j = 0, \ldots, N-1\}$ be the set of these points, (B.26) is then modified to read[5]

$$\hat{v}_m^{(d)} = \frac{1}{N} \sum_{j=0}^{N-1} v_j \exp\left(-2\pi i \frac{mj}{N}\right), \quad \text{(B.29)}$$

while the solution (B.25) is written as

$$v_j = \sum_{m=-N/2+1}^{+N/2} \hat{v}_m^{(d)} \exp\left(2\pi i \frac{mj}{N}\right). \quad \text{(B.30)}$$

Relations (B.27, B.28) then read:

$$\frac{1}{N} \sum_{j=0}^{N-1} \exp\left(2\pi i \frac{(m'-m)j}{N}\right) = \delta_{mm'}$$

and

$$\sum_{m=-N/2+1}^{N/2} \exp\left(2\pi i \frac{m(j'-j)}{N}\right) = N\delta_{jj'}.$$

It is important to notice that the series of Fourier coefficients $\hat{v}_m^{(d)}$ is periodic with period N since:

$$\hat{v}_{m+N}^{(d)} = \sum_{j=0}^{N-1} v_j \exp[-2\pi i(m+N)j/N] \equiv \sum_{j=0}^{N-1} v_j \exp(-2\pi i\, mj/N) = \hat{v}_m^{(d)}, \quad \text{(B.31)}$$

whereas $\hat{v}_m$ is not periodic, as illustrated in Figure B.3. When v is a real quantity, it is also observed that not only $\hat{v}_0^{(d)}$ is real (like $\hat{v}_0$), but that $\hat{v}_{N/2}^{(d)}$ is also real since one has $\left(\hat{v}_{N/2}^{(d)}\right)^* = \hat{v}_{-N/2}^{(d)} = \hat{v}_{-N/2+N}^{(d)} = \hat{v}_{N/2}^{(d)}$. The Fourier transform thus changes the N real quantities v_j, $j = 0, \ldots, N-1$, into a

---

[4] A brief but nice introduction to discrete Fourier transforms is given by R.J. Higgins: "Fast Fourier transform: an introduction with some mini computer experiments," *Am. J. Physics* **44** (1976) 766–773. For more information, of course consult [Gottlieb and Orszag (1977)] or [Canuto et al. (1988)].

[5] Notice that, depending on the computer program, the $1/N$ factor may be placed in (B.30) or in (B.29), or one can find a factor $1/\sqrt{N}$ in each.

set of 2 real quantities, $\hat{v}_0^{(d)}$ and $\hat{v}_{-N/2}^{(d)}$, and $N/2 - 1$ complex quantities $\hat{v}_m^{(d)}$, $m = 1, \ldots, N/2 - 1$, so that the total number of degrees of freedom describing the solution is preserved.

The periodicity of the discrete spectrum induces what is known as the *aliasing phenomenon*: two Fourier modes distant by N in the spectrum are aliases of each other. This is another way of saying that, having no information on modes with $m > N/2$, one cannot resolve structures with scales smaller than the lattice spacing.

Going from v_j to $\hat{v}_m^{(d)}$ and *vice versa* make use of *fast* algorithms with operation counts of order $N \log N$ and not N^2. The *Fast Fourier Transform* (FFT) takes advantage of standard trigonometric relations to organise the information flow so as to minimise the number of arithmetic operations. Commercial codes such as MATLAB are effective in computing FFTs when N is a power of two but also when N can be decomposed as $N = 2^{n_2} 3^{n_3} 5^{n_5} \ldots$ with prime factors that are not too large. This possibility of a fast passage from physical to spectral space and back allows one to treat nonlinearity in a clever way by means of so-called *pseudo-spectral* methods, provided that aliasing is appropriately dealt with.

The difficulty with nonlinearity is that products in physical space are transformed into convolution sums in spectral space, as exemplified here using quadratic nonlinearity vw. For the exact solution we have:

$$\widehat{vw}_m = \sum_{m'=-\infty}^{+\infty} \hat{v}_{m-m'} \hat{w}_{m'},$$

which derives from the Fourier analysis of

$$v(x)w(x) = \left(\sum_{m'=-\infty}^{+\infty} \hat{v}_{m'} \exp(2\pi i \, m' x/\ell) \right) \left(\sum_{m''=-\infty}^{+\infty} \hat{w}_{m''} \exp(2\pi i \, m'' x/\ell) \right)$$

$$= \sum_{m'=-\infty}^{+\infty} \sum_{m''=-\infty}^{+\infty} \hat{v}_{m'} \hat{w}_{m''} \exp[2\pi i \, (m' + m'') x/\ell],$$

see Figure B.4(a).

In the perspective of a truncation beyond $\pm N/2$, one should only keep the combinations such that m', m'', and $m = m' + m''$ belong to the considered interval (Fig. B.4(b), grey-shaded region). The computation of the Fourier transform of a formally quadratic term thus involves N sums with N^2 products each. A clever way to get around the rapid growth of the operation count with N is to take advantage of fast transforms to compute

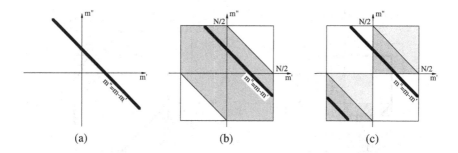

Fig. B.4 (a) The computation of the convolution involves terms such that $m = m'+m''$. (b) For the truncated system only modes with $|m'| < N/2$, $|m''| < N/2$, $|m| < N/2$ should be kept (the grey-shaded region). (c) For the discrete transformation, aliasing send each term in the triangular grey-shaded corners in the corresponding region close to the centre, which perturbs the result.

the product in physical space instead of staying in the spectral space, with a reduction factor of order $\log(N)/N$.

Things are however not simple since one must take care of aliasing. If nothing special is done, the Fourier evaluation of a product computed in physical space contains spurious terms (Fig. B.4c). Let us indeed start with:

$$v_j = \sum_{m=-N/2+1}^{N/2} \hat{v}_m^{(d)} \exp\left(2\pi i \frac{mj}{N}\right), \quad w_j = \sum_{m'=-N/2+1}^{N/2} \hat{w}_{m'}^{(d)} \exp\left(2\pi i \frac{m'j}{N}\right),$$

next define:

$$\widehat{vw}_m^{(d)} = \frac{1}{N} \sum_{j=0}^{N-1} (v_j w_j) \exp\left(-2\pi i \frac{mj}{N}\right)$$

and expand the right hand side. We get:

$$\widehat{vw}_m^{(d)} = \frac{1}{N} \sum_{j=0}^{N-1} \sum_{m'=-N/2+1}^{N/2} \sum_{m''=-N/2+1}^{N/2} \hat{v}_{m'}^{(d)} \hat{w}_{m''}^{(d)} \exp\left(2\pi i \frac{(m'+m''-m)j}{N}\right)$$

with $(m', m'') \in [-N/2+1, N/2]$ and $m = -N+2, \ldots, N$. Terms with $m = m' + m''$ outside the band $[-N/2+1, N/2]$ are sent back inside the band by the periodicity property (B.31) so that we get

$$\widehat{vw}_m^{(d)} = \sum_{m'+m''=m} \hat{v}_{m'}^{(d)} \hat{w}_{m''}^{(d)} + \sum_{m'+m''=m\pm N} \hat{v}_{m'}^{(d)} \hat{w}_{m''}^{(d)}$$

with $m = -N/2 + 1, \ldots, 0, \ldots, N/2$.

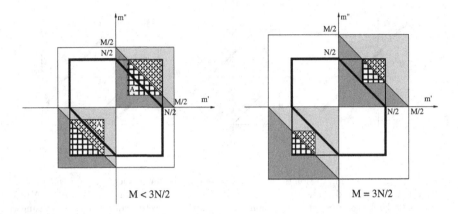

Fig. B.5 As in Figure B.4c, but now with M instead of N, the aliased terms belong to the grey-shaded triangular regions. In the regions with uniform shading, they arise from terms that are zero by construction ($|m|$ outside the $[-N/2+1, N/2]$ interval). The non-trivial aliased terms belong to hatched regions. Some still belong to the domain bounded by the thick line when $M < 3N/2$ (left), which is no longer the case for $M \geq 3N/2$ (right).

When the Fourier coefficients $\hat{v}_m^{(d)}$ and $\hat{w}_m^{(d)}$ decrease sufficiently fast as $|m|$ increases, the correction due to the spurious terms may be negligible but it is preferable to get rid of them by aliasing removal. A way to achieve this purpose is illustrated in Figure B.5. Starting with the spectrum of v and w with $-N/2+1 \leq m \leq N/2$ one turns to a spectrum with $-M/2+1 \leq m \leq M/2$ and $M > N$ by adding zeroes for $-M/2+1 \leq m \leq -N/2$ and $N/2+1 \leq m \leq M/2$ (in the figure, all products in the band outside the interior square with side N cancel). Going back to the physical space one obtains an evaluation of v and w on a lattice with M regularly spaced points, finer than the original lattice. The product vw is then computed at the nodes of this lattice and next Fourier transformed. It remains to drop the Fourier coefficients outside interval $[-N/2+1, N/2]$ to get the unaliased spectrum of the product. The most dangerous non-zero term is with $m' = m'' = N/2$. Its alias, $m' + m'' - M$, gets off the band if $m' + m'' - M < -N/2 + 1$, thus if $3N/2 - 1 < M$, so that one can take $M = 3N/2$ ("3/2 rule"). Despite manipulations implied by the size increase from N to M, the evaluation of convolutions involved in the treatment of nonlinear terms is freed from systematic errors.

B.3 Exercises

B.3.1 Truncation and round-off errors

Consider the differential equation $\frac{d}{dt}v = -v$ with initial condition $v(0) = 1$ and exact solution $v(t) = \exp(-t)$ and point out the role of the order of a numerical scheme on the precision of the numerical integration by comparing the exact result at $t = 1$, $v = 1/e$, to approximations obtained with first-order iterations (B.3) and (B.5), second-order schemes (B.6), (B.7), (B.8), (B.9, B.10) and (B.11, B.12). In particular, show explicitly that the Runge–Kutta schemes are of order Δt^2.

Write down all the corresponding programs and determine the distance $|v_N - 1/e|$ between the exact solution and the numerical solution v_N obtained performing N successive iterations of each scheme with a time-interval $\Delta t = 1/N$. Use logarithmic scales to plot this distance as a function of Δt and to identify the regime dominated by truncation errors from that dominated by the accumulation of round-off errors. Test also the fourth-order Runge–Kutta scheme (B.13).

B.3.2 Stability of multi-step schemes

Consider the integration of $\frac{d}{dt}v = f(v)$ by means of second-order schemes (B.7) and (B.8) and assume that a solution $\{v_n, n = 0, 1, \dots\}$ is known. Neighbouring solutions $\{\tilde{v}_n, n = 0, 1, \dots\}$ can be written as series of perturbations $\{u_n, n = 0, 1, \dots\}$, with $u_n = \tilde{v}_n - v_n$ governed by the linearised map $g_n = df/dv|_n$. Show that the two numerical schemes respectively lead to:

$$u_{n+1} = u_n + \tfrac{1}{2}\Delta t \left(3 g_n u_n - g_{n-1} u_{n-1}\right), \quad (B.32)$$

$$u_{n+1} = u_{n-1} + 2\Delta t \, g_n u_n. \quad (B.33)$$

Then write (B.32) and (B.33) as two-dimensional maps for the two variables u_n and $v_n = u_{n-1}$ and compute their eigenvalues. Determine which is the physical eigenvalue λ_{phys} controlling the solution to the linearised problem $\frac{d}{dt}u = g(v)u$, where v is the solution to the primitive problem $\frac{d}{dt}v = f(v)$, and which is the eigenvalue λ_{num} corresponding to the spurious numerical mode. The instability takes place when $|\lambda_{\text{num}}| > 1$. Assuming that $|g|$ is bounded by some $g_{\max}$ for values of v of interest, discuss the stability properties of the two schemes.

B.3.3 Finite difference schemes for the SH model

Derive second-order consistent finite difference approximations to the linear part of the Swift–Hohenberg (SH) model:

$$\partial_t v = rv - (\partial_{xx} + 1)^2 v \equiv (r-1)v - \partial_{xxxx}v - 2\partial_{xx}v. \qquad (\text{B.34})$$

Consider explicit–Euler, implicit–Euler and Crank–Nicholson temporal schemes and use (B.18) to discretize the fourth-order derivative. Study their stability by introducing discrete Fourier normal modes analogous to those leading to conditions (B.23) and (B.24).

Write down quasi-linearised Crank–Nicholson schemes for the original SH model, i.e. (B.34) with nonlinear term $\mathcal{N}(v) = -v^3$ added to its r.h.s. by expanding the nonlinear terms $\mathcal{N}(v)$ as:

$$\mathcal{N}(v)|_{n+1} = \mathcal{N}(v)|_n + \delta\mathcal{N}/\delta v|_n (v_{n+1} - v_n).$$

Consider also the modified SH model completed by a term $\mathcal{N}(v) = -v\partial_x v$ on the r.h.s. (i.e. an advection term $v\partial_x v$ on its l.h.s).

B.4 Case studies

The working sessions are organised around two themes: ODEs and PDEs. To deal with ODEs, we choose the simplest fourth-order Runge–Kutta scheme (B.13) and apply it to dynamical systems with a few degrees of freedom, the periodically forced pendulum (two-dimensional, non-autonomous, §B.4.1) and the Lorenz model (three-dimensional, autonomous, §B.4.2). Other examples are illustrated but not studied in detail, the Rössler system, §B.4.3.1, and the Chua circuit, §B.4.3.2. All this can serve to materialise concepts introduced in Chapters 2 and 4.

Problems arising in EDPs are illustrated using two variants of the SH model introduced in some exercises of Chapters 3 and 4, and considered in exercise B.3.3 above. They may help the understanding of pattern formation studied in Chapter 6. The approach can be extended to illustrate the generation of dissipative waves. Boundary conditions of Neumann–Dirichlet type are better adapted to finite difference methods and will mainly serve us to illustrate numerical stability problems. Periodic boundary conditions are ideal for a first approach of spectral methods used to illustrate space-time chaotic regimes. Computational Fluid Dynamics (CFD) will be left to specialists.

B.4.1 ODEs 1: Forced pendulum

We consider a pendulum submitted to a sinusoidal external force:

$$\frac{d^2}{dt^2}\theta + \eta \frac{d}{dt}\theta + \sin(\theta) = f\cos(\omega t) \qquad (B.35)$$

where θ is the angle that the pendulum makes with the vertical, η is the damping coefficient, f is the intensity of the forcing and ω its angular frequency.

Write down a second-order Runge–Kutta scheme (B.11, B.12) for (B.35) written as a system of two first-order ODEs for θ and $\phi = \dot{\theta}$, taking care of the explicit time dependence introduced by the forcing. Then turn to a fourth-order scheme (B.13).

1) Use that routine to draw the projection of the trajectories in the plane (θ, ϕ). Choose for example $\omega = \frac{2}{3}$, $\eta = 0.5$ and f variable in the range $[0.5, 3]$. Consider in particular cases $f = 1.07$ and 1.15. See Figure B.6.

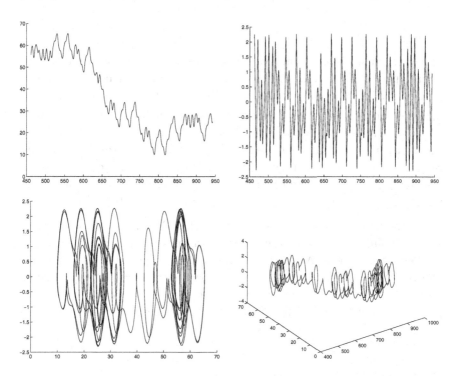

Fig. B.6 Forced pendulum for $\omega = 2/3$, $f = 1.15$. Top-left: Time series of θ. Top-right: Time series of $\dot{\theta} \equiv \phi$. Bottom-left: ϕ as a function of θ. Bottom-right: Three-dimensional representation in the space (t, θ, ϕ), ϕ-axis vertical, θ-axis to the front, t-axis to the rear.

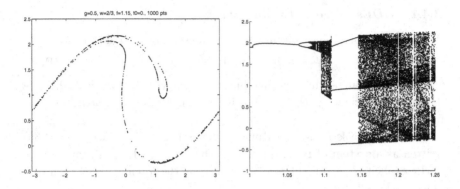

Fig. B.7 Left: Poincaré section of the forced pendulum for $f = 1.15$. Right: Bifurcation diagram for $f \in [1, 1.25]$ obtained when increasing f (the result would be slightly different when decreasing f because some bifurcations are sub-critical, different attractors may coexist in some limited ranges of the parameter f, and hysteresis takes place correspondingly).

2) Perform a stroboscopic analysis (Poincaré section) of the system at the forcing period $T = 2\pi/\omega$ and observe the attractor in particular for the previous parameter values mentioned. See Figure B.7, left.

3) When the intensity of the forcing f increases, one finds parameter ranges where the attractor is periodic and other ranges where it is chaotic. Draw the bifurcation diagram obtained by varying f by recording the values reached by variables ϕ on the surface of section (for every $\Delta t = T$), at given $\omega \in [\omega_1, \omega_2]$, for a sufficiently long trajectory and after having eliminated points corresponding to the transient. See Figure B.7, right.

In these simulations, choose the time step Δt such that $\Delta t = T/N$ so that a full oscillation of the forcing corresponds exactly to an integer number of time steps. N will be sufficiently large in order to fulfil $\Delta t \ll T$ and $\Delta t \ll T_0 = 2\pi$ (natural period of the pendulum close to its equilibrium position). A detailed study would show that the value of the bifurcation points is sensitive to the numerical resolution: compare results with the 2nd-order and 4th-order schemes.

B.4.2 ODEs 2: Lorenz model

The simplified model of nonlinear convection initially proposed by E.N. Lorenz (Note 6, p. 17) was introduced in §3.2.1. It is re-written here

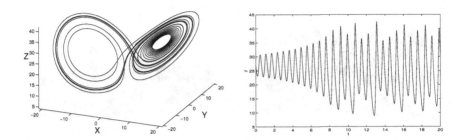

Fig. B.8 Left: The Lorenz attractor for $r = 28$ with $\Delta t = 0.01$. Initial conditions $X_0 = 6 = Y_0$, $Z_0 = 27$. Right: Time series $Z(t)$ in the same conditions.

for convenience, using standard notations:

$$\frac{d}{dt}X = \sigma(Y - X), \tag{B.36}$$

$$\frac{d}{dt}Y = -XZ + rX - Y, \tag{B.37}$$

$$\frac{d}{dt}Z = XY - bZ. \tag{B.38}$$

1) Display the solution in the three-dimensional space (X, Y, Z), draw the time series of Z and study the attractor obtained for the values initially chosen by Lorenz: $\sigma = 10$, $b = 8/3$ and $r = 28$ (Figure B.8). Then vary r on the interval $r \in [145, 170]$, a range where the chaotic attractor decays into a periodic cycle through a subharmonic cascade and returns to chaos by intermittency (Figures B.9 and B.10).

2) Determine the Lorenz map defined as $Z_{k+1} = f(Z_k)$ where Z_k is defined by the condition that $Z(t)$ goes through a maximum, i.e. $\frac{d}{dt}Z = 0$ with

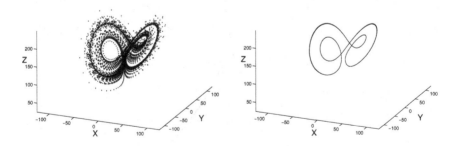

Fig. B.9 Left: "Intermittent" Lorenz attractor for $r = 166.5$ with $\Delta t = 0.01$. Initial conditions: $X_0 = 12 = Y_0$, $Z_0 = 165.5$. Lines joining successive points are not drawn to better point out the existence of a "ghost" trajectory recurrently visited by the system. Right: Limit cycle for $r = 166$ with $\Delta t = 0.01$. Initial conditions: $X_0 = 12 = Y_0$, $Z_0 = 165$. The cycle is reached after a transient that is not represented here.

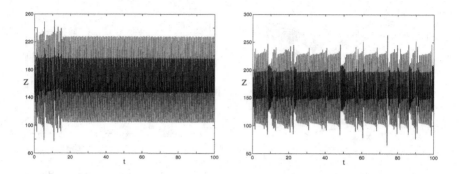

Fig. B.10 Left: $Z(t)$ for $r = 166$, periodic regime after a brief transient. Right: $Z(t)$ for $r = 166.5$, intermittent regime.

$\frac{d^2}{dt^2} Z < 0$. To do so, at each iteration check the value of $D_n = X_n Y_n - b Z_n$, identify intervals $[t_n, t_{n+1} = t_n + \Delta t]$ where D_n changes sign from positive to negative. By linear interpolation, find the time τ_n where the change of sign takes place. Restarting from the result at t_n, perform a single Runge–Kutta step with time interval $\tau_n - t_n$ to obtain the set of coordinates of the corresponding point with the same accuracy as the trajectory itself. Continue the simulation from the state at t_{n+1} previously determined. The Lorenz map for $r = 28$ is displayed in Figure B.11. The same strategy can thus be used for Poincaré sections other than that defined by the condition

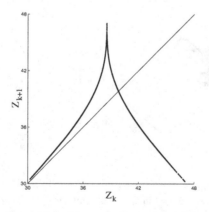

Fig. B.11 Lorenz map $Z_{k+1} = f(Z_k)$ for $r = 28$.

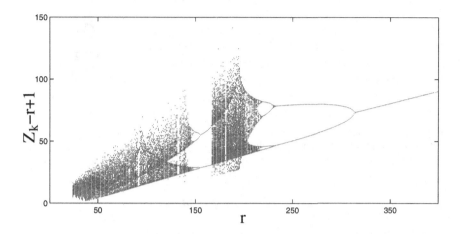

Fig. B.12 Bifurcation diagram of the Lorenz model. A periodic attractor is represented as a finite set of isolated points $Z_k - r + 1$ at given r and a chaotic attractor as a continuous unevenly distributed series of points. At high values of r a symmetric periodic attractor is obtained. After losing its symmetry at $r \approx 310$ (the two points for $r < 310$ correspond to two different maxima on the same period-1 cycle), it suffers a subharmonic cascade starting at $r \approx 230$. Apart from narrow windows of regularity, a chaotic attractor prevails down to $r \approx 163$ where it decays by intermittency into another, more complicated, period-1 cycle. A second subharmonic cascade is conspicuous around $r = 150$. The Lorenz attractor seen at $r = 28$ persists down to $r \approx 24$ where it decays into a fixed point at $Z_f = r - 1$ after a long transient.

$D = 0$, as initially proposed by Hénon.[6] A popular one in early studies of the Lorenz model was the surface defined by $Z = r - 1$ and crossed from above (for example).

3) Construct a bifurcation diagram for the Lorenz model by plotting $Z_k - r + 1$ as a function of r (Figure B.12). Sets of isolated points at given r correspond to periodic behaviour, continuous series of points to chaotic behaviour. The bifurcation around $r = 320$ is not a period doubling but a symmetry breaking bifurcation as shown by drawing trajectories in three dimensions before and after the bifurcation.

---

[6]M. Hénon, "On the numerical computation of Poincaré maps," Physica D **5** (1982) 412–414.

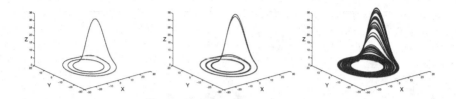

Fig. B.13 The Rössler attractor for $a = 0.1$, $b = 0.1$. From left to right: Limit cycle for $c = 12$ and after a period doubling for $c = 12.7$; chaotic attractor beyond the accumulation point of the subharmonic cascade for $c = 13.6$.

B.4.3 ODEs 3: Rössler and Chua models

B.4.3.1 Rössler model

Another simple three-dimensional autonomous dynamical system with a strange attractor is the Rössler model.[7]

$$\frac{d}{dt}X = -Y - Z,$$
$$\frac{d}{dt}Y = X + aY,$$
$$\frac{d}{dt}Z = b + (X - c)Z.$$

Notice that its definition is even simpler than that of the Lorenz model since it has a single nonlinear quadratic term and no obvious symmetry.[8] The dynamics has two ingredients: approximately periodic behaviour at some distance from an unstable spiral fixed point and fast excursions away from the (X, Y)-plane. It displays periodic and chaotic regimes. Typical trajectories are shown in Figure B.13. Use the same tools as for the Lorenz model to study its bifurcation diagram as a function of $c \in [1, 20]$.

B.4.3.2 Chua circuit

The Chua circuit is the last example proposed here. It is defined as a prototype of analogue systems displaying chaos that can be implemented as electronic circuits with good accuracy. It is also three-dimensional and autonomous. Its main properties are reviewed in a collective work edited

---

[7] O.E. Rössler: "An equation for continuous chaos," Physics Lett. A **57** (1976) 397–398; see also O.E. Rössler: "Continuous chaos - Four prototype equations," Ann. NY Acad. Sc. **316** (1979) 376–392.

[8] J.C. Sprott: "Simplest dissipative chaotic flows," Physics Lett. A **228** 271–274.

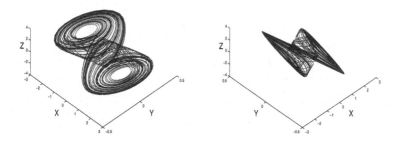

Fig. B.14 Two views of the Chua "double scroll" attractor for $\alpha = 9$, $\beta = 100/7$, $m_0 = -8/7$, $m_1 = -5/7$.

by R.N. Madan.[9] It reads

$$\frac{d}{dt}X = \alpha(Y - X - F(X)),$$
$$\frac{d}{dt}Y = X - Y + Z,$$
$$\frac{d}{dt}Z = -\beta Y.$$

where $F(X)$ is an odd function of X that is linear by part: $F(X) = m_1 X + \frac{1}{2}(m_0 - m_1)(|X+1| - |X-1|)$. The "double scroll" attractor is displayed in Figure B.14. Establish the bifurcation diagram of the Chua system from a Lorenz map of variable Z, with $\beta \in [1, 25]$ and other parameters fixed as in the caption of Figure B.14. Display other typical attractors.

B.4.4 PDEs 1: SH model, finite differences

Finite difference methods are illustrated here using the original variant of the Swift–Hohenberg model (Note 16, p. 113) with a cubic nonlinear term in one space dimension:

$$\partial_t v = rv - (\partial_{xx} + 1)^2 v - v^3 \qquad (B.39)$$

with boundary conditions

$$v(0) = 0 = v(\ell) = \partial_x v(0) = \partial_x v(\ell) \qquad (B.40)$$

The two parameters are r and ℓ. The model derives from a potential and is a good tool to study the formation of cellular structures in simple cases.

---

[9] R.N. Madan, Ed., *Chua's circuit: a paradigm for chaos* (World Scientific, 1993).

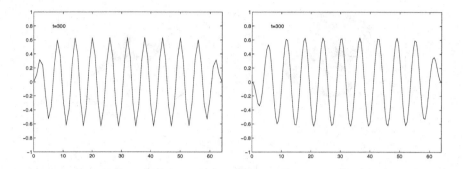

Fig. B.15 Explicit-Euler numerical simulation of the Swift–Hohenberg model with $r = 0.3$; random initial conditions with amplitude 0.01; solution at $t_f = 300$. Left: $\ell = 64$, $n = 64$, $\Delta x = 1.0$, $\Delta t = 0.1$. Right: $\ell = 64$, $n = 128$, $\Delta x = 0.5$, $\Delta t = 0.008$, numerically stable.

B.4.4.1 *Explicit scheme*

Develop the simplest possible, first order in time, second order in space, explicit numerical scheme. To maintain second order consistency at the boundaries, add fictitious exterior points enslaved to the first inner points as explained in §B.2.1.2. The effect of the space resolution is shown in Figure B.15. In order to reach the finest grid, the time step should be considerably reduced in order to avoid the numerical instability. Observe how the numerical instability develops when the condition on Δt obtained in Exercise B.3.3, namely $\Delta t < \Delta x^4/8$ for Δx small, is violated (Figure B.16).[10]

B.4.4.2 *Implicit scheme*

Develop an implicit Euler scheme, first order in time, second order in space, by solving the linear problem as explained in §B.2.1.4. Take $\ell = 64$ ($\simeq 10\lambda_{\rm c}$) in order to deal with a sufficiently extended system. The number of grid points N will be varied to check the effect of the numerical resolution (Figure B.17). Second order schemes can also be developed, e.g. quasi-linearised Crank–Nicholson (Exercise B.3.3) or Adams–Bashford.

Since most numerical problems are set aside owing to the unconditional linear stability of the scheme (§B.2.1.3, Exercise B.3.3), focus on the physics

---

[10] In certain cases depending on initial conditions, the nonlinear term may be sufficiently strong to prevent the numerical divergence but the so-obtained solution is not physical in that it has a space period very different from the expected one $\simeq \lambda_{\rm c} = 2\pi$ (in practice much shorter).

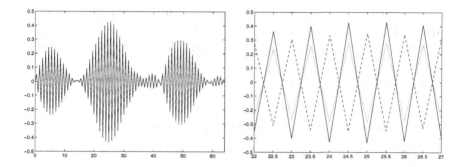

Fig. B.16 Numerical instability of the Swift–Hohenberg model with $\ell = 64$, $n = 128$, $\Delta x = 0.5$, $\Delta t = 0.01$, $r = 0.3$, random initial conditions with amplitude 0.01. Left: State at iteration #20. Right: The same numerical instability at three successive time steps, iterations #18 (dots), #19 (dashes), and #20 (solid line).

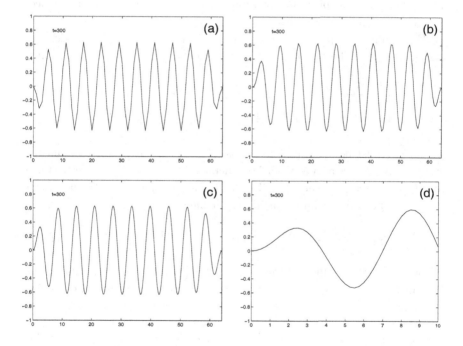

Fig. B.17 Swift–Hohenberg model with $\ell = 64$, $\Delta t = 0.1$, $r = 0.3$, random initial conditions with amplitude 0.01; solution at $t_f = 300$. (a) $n = 64$, $\Delta x = 1.0$. (b) $n = 128$, $\Delta x = 0.5$. (c) $n = 256$, $\Delta x = 0.25$. (d) zoom on the left boundary showing the accuracy of the numerical account of boundary condition $\partial_x w = 0$.

of the formation of the cellular structure. Identify the initial phase of exponential growth of perturbations selected by the linear dynamics, then the saturation phase where nonlinearity comes in the foreground, and the final stage where most effects compensate so that a slow residual motion (phase diffusion) towards the final time-independent state is observed.

B.4.5 PDEs 2: SH model, pseudo-spectral method

Consider now the modified Swift–Hohenberg model:[11]

$$\partial_t v + v \partial_x v = \left[r - (\partial_{xx} + 1)^2 \right] v, \qquad (B.41)$$

with periodic boundary conditions at the ends of an interval of length ℓ. This model, which does not derive from a potential and thus can have unsteady behaviour, is particularly appropriate to study the growth of space-time chaos in cellular structures.

Derive a Fourier pseudo-spectral algorithm implementing the model taking advantage of the fact that $v \, \partial_x v = \frac{1}{2} \partial_x (v^2)$, according to the scheme:

$$\hat{v}_m \begin{cases} \mapsto \widehat{\mathcal{L}v}_m = s_m \hat{v}_m = \left(r - (k_m^2 - 1)^2 \right) \hat{v}_m \\ \mapsto v(x) \mapsto \tfrac{1}{2} v(x)^2 \mapsto \widehat{\tfrac{1}{2} v^2}_m \mapsto \widehat{v \partial_x v}_m \equiv \tfrac{1}{2} i k_m \widehat{v^2}_m = \widehat{\mathcal{N}}_m \end{cases}$$

where $\widehat{\mathcal{N}}_m$ must be computed as discussed earlier in order to avoid aliasing errors. Next choose a time evolution scheme (§B.1). Here the linear part can be integrated exactly since it is diagonal in Fourier space. For the nonlinear part a second order Adams–Bashford scheme can be chosen, ending in

$$\hat{v}_{n+1,m} = \exp(s_m \Delta t) \hat{v}_{n,m} + \Delta t \left(\tfrac{3}{2} \widehat{\mathcal{N}}_{n,m} - \tfrac{1}{2} \widehat{\mathcal{N}}_{n-1,m} \right). \qquad (B.42)$$

When $r = 1$ the Swift–Hohenberg model happens to be a variant of the Kuramoto–Sivashinsky equation for *phase turbulence*. In a slightly different form:

$$\partial_t v + \tfrac{1}{2} (\partial_x v)^2 + \partial_{xx} v + \partial_{xxxx} v = 0, \qquad (B.43)$$

this equation appears in the treatment of turbulent interface propagation, e.g. flame fronts, as shown in Figure B.18.

---

[11] Y. Pomeau, P. Manneville: "Wavelength selection in cellular flows," Physics Letters A **75** (1980) 296–298.

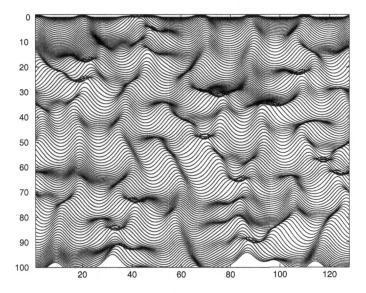

Fig. B.18 Kuramoto–Sivashinsky equation for $\ell = 64$, $n = 128$, $\Delta x = 0.5$, with a time interval $\Delta T = 1$ between two successive snapshots.

Bibliography

N.B. Some of the listed works are mentioned for further reading.

Dynamical systems, bifurcations and catastrophes, chaos

Abarbanel, H.D.I. (1996). *Analysis of Observed Chaotic Data*, (Springer).

Abraham, R.H. and Shaw, C.D. (1992). *Dynamics: the Geometry of Behavior* (Addison-Wesley). [Pictorial introduction.]

Alligood, K.T., Sauer, T.D., and Yorke, J.A. (1996). *Chaos, an Introduction to Dynamical Systems* (Springer). [Textbook.]

Baker, G.L. and Gollub, J.P. (1996). *Chaotic Dynamics, an Introduction*, 2nd Edition (Cambridge University Press).

Beck, Ch. and Schlögl, F. (1995). *Thermodynamics of Chaotic Systems* (Cambridge University Press).

Bergé, P., Pomeau, Y., and Vidal, Ch. (1986). *Order within Chaos, towards a Determinsitic Approach to Turbulence* (Wiley). [Classical reference.]

Cladis, P.E. and Palffy-Muhoray, P., Editors (1995) *Spatio-Temporal Patterns in Nonequilibrium Complex Systems* (Addison-Wesley).

Cross, M.C. and Hohenberg, P.C. (1993). Pattern formation outside equilibrium, *Rev. Mod. Phys.* **65** (1993) 851–1112.

Cvitanović, P., Editor (1989). *Universality in Chaos* (Adam Hilger) [Reprint collection.]

Devaney, R.L. (1989). *An Introduction to Chaotic Dynamical Systems*, 2nd edition (Addison-Wesley).

Drazin, P.G. (1992). *Nonlinear Systems* (Cambridge University Press).

Eckmann, J.P. and Ruelle, D. (1985). Ergodic theory of chaos and strange attractors. *Rev. Mod. Phys.* **57** (1985) 617–656.

Glansdorff, P. and Prigogine, I. (1971). *Thermodynamics of Structures, Stability and Fluctuations* (Wiley).

Guckenheimer, J. and Holmes, Ph. (1983). *Nonlinear Oscillations, Dynamical Systems, and Bifurcations of Vector Fields* (Springer).

Haken, H. (1983). *Synergetics*, 3rd edition (Springer).

Hao, B.-l., Ed. (1987-...). *Directions in Chaos* (World Scientific). [Review articles, several volumes.]

Hao B.-l., Ed. (1990). *Chaos II* (World Scientific) [Reprint collection.]
Hirsch, M.W. and Smale, S. (1974). *Differential Equations, Dynamical Systems and Linear Algebra* (Academic Press).
Kanz, H. and Schreiber, Th. (1997). *Nonlinear Time Series Analysis*, Cambridge Nonlinear Science Series 7 (Cambridge University Press).
Lefschetz, S. (1977). *Differential Equations: Geometric Theory* (Dover).
Mandelbrot, B.B. (1982). *The Fractal Geometry of Nature* (Freeman).
Moon, F.C. (1992). *Chaotic and Fractal Dynamics, an Introduction for Applied Scientists and Engineers* (Wiley). [Classical reference.]
Nayfeh, A.H. and Mook, D.T. (1979). *Nonlinear Oscillations* (Wiley).
Ott, E. (1993). *Chaos in Dynamical Systems* (Cambridge University Press).
Poston, T. and Stewart, I. (1978). *Catastrophe Theory and its Applications* (Pitman).
Rabinovich, M.I., Ezersky, A.B., and Weidman, P.D. (2000). *The Dynamics of Patterns* (World Scientific).
Schuster, H.G. (1988). *Deterministic Chaos, an Introduction* 2nd edition (VCH). [Classical reference.]
Schuster, H.G., ed. (1999). *Handbook of Chaos Control* (Wiley).
Sontag, E.D. (1998). *Mathematical Control Theory, Deterministic Finite Dimensional Systems* 2nd edition (Springer).
Steeb, W.-H. (1991). *A Handbook of Terms used in Chaos and Quantum Chaos* (BI, Mannheim). [Brief introduction.]
Strogatz, S.H. (1994). *Nonlinear Dynamics and Chaos, with Applications to Physics, Chemistry and Engineering* (Addison-Wesley). [Classical reference.]
Tabor, M. (1989). *Chaos and Integrability in Nonlinear Dynamics: An Introduction* (Wiley).
Weigend, A.S. and Gershenfeld, N.A., eds. (1993). *Time Series Prediction: Forecasting the Future and Understanding the Past* (Addison-Wesley).

Hydrodynamics, instabilities, and turbulence

Batchelor, G.K. (1967). *An Introduction to Fluid Dynamics* (Cambridge University Press). [Classical reference.]
Batchelor, G.K., Moffatt, H.K., and Worster, M.G., eds. (2000). *Perspectives in Fluid Dynamics. A Collective Introduction to Current Research* (Cambridge University Press).
Betchov, R. and Criminale, W.O. (1967). *Stability of Parallel Flows* (Academic Press).
Chandrasekhar. S. (1961). *Hydrodynamic and Hydromagnetic Stability* (Clarendon Press).
Drazin, P.G. (2002). *Introduction to Hydrodynamic stability* (Cambridge University Press).
Drazin, P.G. and Reid, W.H. (1981). *Hydrodynamic Stability* (Cambridge University Press).

Frisch, U. (2001). *Turbulence. The legacy of A.N. Kolmogorov*, 2nd edition (Cambridge University Press).
Getling, A.V. (1998). *Rayleigh–Bénard Convection, Structure and Dynamics* Advanced Series in Nonlinear Dynamics Vol. 11 (World Scientific).
Godrèche, Cl. and Manneville, P., eds. (1998). *Hydrodynamics and Nonlinear Instabilities* (Cambridge University Press).
Landau, L.D. and Lifshitz, E.M. (1987). *Fluid Mechanics - Course of Theoretical Physics, Vol. 6*, 2nd edition (Butterworth–Heinemann). [Classical reference.]
Lesieur, M. (1997). *Turbulence in Fluids* (Kluwer).
Lesieur, M., Yaglom, A., and David, F., eds. (2001). *New Trends in Turbulence.* Les Houches, session LXXIV (Springer).
Mollo-Christensen, E.L. (1972). *Flow Instabilities*, film produced by [NCFMF (1972)].
Mutabazi, I., Wesfreid, J.E., Guyon, E., eds. (2006). *Dynamics of Spatio-Temporal Cellular Structures. Henri Bénard Centenary Review.* Springer Tracts in Modern Physics 207 (Springer).
MFM (2007). *Multimedia Fluid Mechanics* DVD produced by Stanford University, Homsy, G.M. et al.. Second Edition, distributed by Cambridge University Press.
NCFMF (1972). Series of films produced by the National Commitee for Fluid Mechanics, distributed by Encyclopedia Britannica Educational Corp.; accompanying book: *Illustrated Experiments in Fluid Mechanics* (MIT Press).
Pope, S.B. (2000). *Turbulent flows* (Cambridge University Press).
Schlichting, H. (1979). *Boundary Layer Theory* (McGraw-Hill).
Schmid, P.J. and Henningson, D.S. (2001). *Stability and Transition in Shear Flows* (Springer-Verlag).
Stewart, R.W. (1972). *Turbulence*, film produced by [NCFMF (1972)].
Swinney, H.L. and Gollub, J.P., eds. (1985). *Hydrodynamic Instabilities and the Transition to Turbulence* (Springer).
Tabeling, P. and Cardoso, O., eds. (1994). *Turbulence. A Tentative Dictionary* (Plenum Press).
Tennekes, H. and Lumley, J.L. (1972). *A First Course in Turbulence* (MIT Press).
Tritton, D.J. (1990). *Physical Fluid Dynamics* (Cambridge University Press).
van Dyke, M. (1982). *An Album of Fluid Motion* (Parabolic Press).
Wilcox, D.C. (2000). *Turbulence Modeling for CFD* (DWC Industries).

Numerical modelling

Abramowitz, M. and Stegun, I.A., eds. (1972). *Handbook of Mathematical Functions* (National Bureau of Standards) [Chapter 22: Numerical interpolation, differentiation, and integration].
Acton, F.S. (1970). *Numerical Methods that Work* (Harper and Row).
Canuto, C., Hussaini, M.Y., Quarteroni, A., and Zang, T.A. (1988). *Spectral Methods in Fluid Dynamics* (Springer).

Finlayson, B.A. (1972). *The Method of Weighted Residuals and Variational Principles, with Application in Fluid Mechanics, Heat and Mass Transfer* (Academic Press).
Gottlieb, D. and Orszag, S.A. (1977). *Numerical Analysis of Spectral Methods: Theory and Applications* (SIAM Publications).
Lapidus, L. and Seinfeld, J.H. (1971). *Numerical Solution of Ordinary Differential Equations* (Academic Press).
Press, W.H., Flannery, B.P., Teukolsky, S.A., and Vetterling, W.T. (1986). *Numerical Recipes, the Art of Scientific Computing* (Cambridge University Press) [program sources available].
Richtmyer, R.D. and Morton, K.W. (1967). *Difference Methods for Initial Value Problems* (Interscience).

Miscellaneous

Anderson, P.W., Arrow, K.J., and Pines, D., eds. (1988). *The Economy as an Evolving Complex System*, SFI Studies in the Sciences of Complexity (Addison-Wesley).
Bar-Yam Y. (1997). *Dynamics of Complex Systems* (Wetsview Press). [For further reading.]
Boccara N. (2004). *Modeling Complex Systems* (Springer).
Dauxois, Th. and Peyrard M. (2006) *Physics of Solitons* (Cambridge University Press).
Hall, N., ed. (1992). *The NewScientist Guide to Chaos* (Penguin Books).
Henderson-Sellers, A. and McGuffie, K. (1987). *A Climate Modelling Primer* (Wiley).
Kaneko, K., ed. (1993). *Theory and Application of Coupled Map Lattices* (Wiley).
Kauffman S. (1995). *At Home in the Universe: The Search for the Laws of Self-Organization and Complexity* (Oxford University Press). [For further reading.]
Murray J.D. (1993). *Mathematical Biology* (Springer-Verlag).
Peixoto, J.P. and Oort, A.H. (1992). *Physics of Climate* (AIP Publishing).
Ottino, J.M. (1989). *The kinematics of Mixing: Stretching, Chaos and Transport* (Cambridge University Press).
Philander, S.G. (2000). *Is The Temperature Rising? The Uncertain Science of Global Warming* (Princeton University Press).
Rothman, D.H. and Zaleski, S. (1997). *Lattice-gas Cellular Automata. Simple Models of Complex Hydrodynamics* (Cambridge University Press, 1997).
Stanley, H.E. (1988). *Introduction to Phase Transitions and Critical Phenomena* (Oxford University Press).
Winfree, A.T. (1990). *The Geometry of Biological Time* (Springer). [Classical reference, for further reading.]
Wolfram, S., ed. (1986). *Theory and Application of Cellular Automata* (World Scientific).

Index

adiabatic elimination, 100, 129, 353
advection, 7, 86, 281
aliasing, 414
analytical mechanics, 45
aspect ratio, 103, 126, 353
   large, 109, 127, 227
   small, 104, 127
attractor, 14, 45
   fixed point, 14, 45, 49, 100, 350
   limit cycle, 15, 61, 138, 350
   limit torus, 151, 222, 351
   strange, 17, 154, 171, 175, 351
averaging
   ensemble, 321
   time, 178, 321
averaging method, 65, 81

baker map, 174
   dissipative, 177
Belousov–Zhabotinsky reaction, 25, 97
Benjamin–Feir instability, 247
bifurcation, 1, 56, 349
   Hopf, 15, 138, 242, 287, 350
   Neimark–Sacker, 152
   perfect/imperfect, 137, 166
   saddle-node, 137, 164
   super/sub-critical, 14, 100, 139, 230, 284
   transcritical, 165, 230
bifurcation cascade, 14, 135, 296
bifurcation diagram, 14, 63

Blasius boundary layer, 260
boundary conditions
   lateral, 126, 238
   no-slip/stress-free, 116, 119
boundary layer, 19, 259, 292
   laminar, 260, 327
   thickness, 260, 293
   transition, 293, 296
   turbulent, 23, 112, 327, 329
Boussinesq approximation, 115
Briggs–Bers criterion, 281, 311
Burgers equation, 36, 253
Busse balloon, 102
bypass transition, 295, 354

Cantor set, 158, 175, 221
catastrophe theory, 48, 137, 350, 371
cellular automata, 31
centre manifold, 131
chaos, 5, 97, 104, 154, 172, 347
chaos control, 205
   delayed feedback, 214
   OGY method, 215
chaotic advection, 194
chemical kinetics, 25
Chua circuit, 424
closure problem, 23, 324, 342, 355
codimension, 129
coherence length, 28, 34, 95, 238, 352
coherent structures (turbulence), 333
configuration space, 3, 41
confinement effects, 19, 87, 126, 353

435

conjugate variables, 41
control parameter, 9, 42, 86, 129, 349
Couette flow
    cylindrical, 122, 302
    plane
        laminar, 258
        transition, 21, 299
        turbulent, 335
coupled map lattice, 30
cross-roll instability, 240, 252

defect (in pattern), 108, 241
defect mediated turbulence, 29, 248
degree of freedom, 3, 41, 70
    effective, 12, 95, 126, 353
    in developed turbulence, 339
delay systems, 211
demodulation, 180
determinism, 4, 41
diffusion, 7
    molecular (Fick law), 7, 119, 191
    thermal (Fourier law), 6, 83
    viscous (Stokes law), 8, 83, 257
diffusivity, 6
    turbulent vs. molecular, 23, 326
dimension
    correlation, 189
    embedding, 182
    fractal, 176, 220
    topological, 176
dispersion relation, 87, 95, 228, 243, 263, 278
dissipative structure, 11, 230
drag coefficient, 288
driving/slave mode, 100, 129, 349
Duffing oscillator, 2, 64, 71
    forced, 142
dyadic map, 18, 171
dynamical system, 2, 347
    autonomous/forced, 41, 140
    conservative/dissipative, 45
    discrete-time, 3, 42
    gradient flow, 47, 241
    linear, 51, 378

Earth's climate, 356

Eckhaus instability, 244
eddy viscosity, 326, 355
elliptical instability, 285
energy method (stability), 56, 73
ensemble average, 321
ENSO phenomenon, 358
envelope equation, 237, 353
ergodicity, 322
Euler equation, 8

false neighbour method, 187
Fast Fourier Transform, 180, 414
first harmonic approximation, 62, 142
fixed point
    centre (elliptic point), 53, 145
    focus (spiral point), 53
    improper node, 54
    node, 52
    saddle, 48, 52, 145, 197
Fjørtoft criterion, 267, 308
Floquet theory, 284
fluid particle, 7, 22, 86, 193
forcing
    external, 42
    parametric, 41
    periodic, 141, 277, 360, 419
Fourier equation, 35, 86, 412
fractal structure, 18, 175, 221, 351
Fredholm alternative, 69, 389, 397
free energy, 32
    Ginzburg–Landau, 33, 241
    Landau, 33, 137, 167
frequency locking, 105, 146, 153, 169
friction, 43
frozen turbulence (Taylor), 318, 345

Galerkin method, 117
Ginzburg–Landau equation, 29, 243, 310, 354
global mode, 20
globally super/sub-critical transition, 296, 299, 354
gradient flow, 47, 241
greenhouse effect, 362
group velocity, 96, 243

Hénon map, 17, 175
Hamilton equations, 46
harmonic oscillator, 2, 39
    damped, 43
    forced, 142
Hilbert transform, 181
homo/heteroclinic
    points, 155, 157
    tangle, 157, 197, 205
    trajectory, 155
Hopf bifurcation, 138, 242
Howard theorem, 267

incompressible flow, 8, 257
inertial cascade, 22, 316, 332, 355
inertial range intermittency, 332
instability
    cellular, 29, 96, 114, 237, 352, 425
    convective/absolute, 20, 256, 271, 277, 281–283, 353
    homogeneous, 29, 96
    local/global, 20, 282
    oscillatory, 29, 55, 89, 138
    phase/amplitude, 244
    primary/secondary, 10, 101
    stationary, 29, 55, 89
    wave, 96, 256, 278, 352
instability threshold, 11, 92, 275, 349
intermittency, 107, 161, 421
    spatiotemporal, 249, 295

jet, 259, 286, 310
    turbulent, 327

Kármán log-law, 23, 330, 355
Kalman decomposition (control), 394
Kelvin–Helmholtz instability, 256, 270
kernel of an operator, 378
Kolmogorov scale, 319
Kolmogorov spectrum, 23, 318, 355
Korteweg–de Vries equation, 247
Kuramoto–Sivashinsky equation, 246, 253, 428

Lagrange equations, 46
Lagrangian chaos, 194

Lagrangian vs. Eulerian description, 7, 193
Landau expansion, 13, 48, 135, 350
Langevin equation, 42, 184
large eddy simulation (LES), 339
Lewis number, 120
lift-up mechanism, 307
limit set, 14, 125, 155, 171
linear control, 391
logistic map, 18, 160, 169, 209
    delayed, 170
Lorenz map, 421
Lorenz model, 17, 99, 420
Lotka–Volterra system, 27
Lyapunov exponent, 189
    conditional, 207
Lyapunov functional, 75

material derivative, 7, 28, 193
micro/meso/macroscopic scale, 6, 347
mixing
    in dynamical systems, 193
    in physical systems, 191
mixing layer, 19, 269, 286, 309
    turbulent, 327
mixing length, 326, 355
momentum space, 3, 41
multiple scale method, 70, 230
mutual information, 186

Navier–Stokes equation, 8, 257, 322
Newell–Whitehead–Segel equation, 238
noise
    extrinsic, 4, 42, 184, 191
    intrinsic, 107, 109
noise amplifier, 20, 271, 281
non-normal operator, 304, 388, 398
normal form
    Jordan, 53, 75, 379, 397
    Jordan–Arnold, 386
    nonlinear, 14, 132, 243, 349
normal mode analysis, 87, 263, 348
numerical scheme
    Adams–Bashford, 402, 411, 426
    Crank–Nicholson, 411, 418

Euler, 400, 418
  explicit/implicit, 400, 407, 418
  finite difference, 406, 425
  first/second order, 400–402
  leap-frog, 402
  pseudo-spectral, 414, 428
  quasi-linearised, 411
  Runge–Kutta, 404, 418
  spectral, 411
Nusselt number, 111

open flow, 19, 255, 277, 353
open/closed loop (control), 392
order parameter, 32, 97, 236
Orr–Sommerfeld equation, 265, 273

pattern, 11, 103, 127, 234, 354
  modulation, 237
  selection, 236, 242
Peclet number, 192
pendulum, 2, 57
  damped, 58
phase instability, 244
phase portrait, 42
phase space, 3, 40
phase transitions (thermodynamic), 32, 48, 97, 137, 350
phase turbulence, 29, 247, 428
phase velocity, 96
phase-winding solution, 251
Poincaré section & map, 4, 149, 351
Poincaré–Bendixson theorem, 139
Poincaré–Lindstedt method, 67, 81
Poiseuille channel flow
  laminar, 258
  transition, 275, 292
Poiseuille pipe flow
  laminar, 258
  transition, 21, 297
Prandtl number, 93, 101, 326
predator-prey systems, 27
predictability, 5, 17, 173, 356, 357
proper orthogonal decomposition, 188

Rössler attractor, 424
rank of a matrix, 378

Rayleigh criterion, 19, 266, 308
Rayleigh equation, 266, 271
Rayleigh number, 85, 92
Rayleigh–Bénard instability, 11, 83
reaction-diffusion system, 28, 121
repeller, 45, 224
resonance, 54, 134, 145, 379
  primary/secondary, 144
  spatial, 236, 293
  weak/strong, 152, 160
resonance tongue, 147
resonant term, 68, 132, 229
Reynolds averaging, 23, 322, 338, 355
Reynolds number, 9, 109, 192, 255, 262, 269, 287, 300, 315
  Taylor's R_λ, 320
Reynolds stress, 23, 323, 329, 336, 355
Reynolds–Orr equation, 274

scenario, 17, 126
  intermittency, 107
  intermittent, 107, 161, 421
  Landau, 16, 154, 351
  Ruelle–Takens, 16, 105, 351
  subharmonic, 104, 160, 351, 421
Schmidt number, 192
secular term, 54, 68, 69, 383, 397
self-sustained oscillator, 20, 60, 271
sensitivity to initial conditions, 17, 171, 351
separatrix, 58
singular value decomposition, 188
soliton, 78
spatiotemporal chaos, 29, 108, 227, 354
Squire theorem, 264, 308
stability, 4
  asymptotic, 55, 74
  conditional, 10
  global, 9, 292, 307
  global method, 56, 73, 266
  linear, 10, 51, 85, 262, 348
  numerical, 343, 403
  orbital, 56, 74
  structural, 5, 53, 56, 171, 370
stable/unstable manifold, 155, 197

state space (phase space), 3, 41
statistical ensemble, 321
stroboscopic analysis, 4, 42, 141, 420
Strouhal number, 288
structural stability (robustness), 5, 53, 56, 171, 370
subharmonic cascade, 105, 160, 351, 421
susceptibility, 1, 33
Swift–Hohenberg model, 113, 418, 425
symbolic dynamics, 159
synchronisation, 205
   identical/generalised, 207
   strong/weak, 208
synergetics, 129

tangent dynamics, 50, 189, 219
Taylor–Couette instability, 97, 122
temporal *vs.* spatial stability analysis, 264, 276
texture, 108, 127, 236
thermal convection, 10, 83, 353
thermodynamic branch, 9, 50, 295
thermodynamic equilibrium, 5, 347
   local, 6
thermohaline circulation, 366
time average, 178, 321
Tollmien–Schlichting wave, 256, 275
tracer (mixing), 191, 202
transfer function, 393
transient chaos, 204, 223
transient energy growth, 304, 388, 398
turbulence modelling, 336
turbulent convection, 111

turbulent scale, 22, 313
   dissipation, 22, 319, 355
   inertial, 22, 316
   production, 22, 314
   Taylor microscale, 320
turbulent spots, 22, 295, 354
   in boundary layer flow, 298
   in pipe flow (puffs & slugs), 298
   in plane Couette flow, 21, 301
   in Poiseuille channel flow, 292, 298
Turing pattern, 121

universality
   in developed turbulence, 318, 330
   in dynamical systems, 13, 51, 134
   in pattern dynamics, 29, 236, 243
   in phase transitions, 34

van der Pol oscillator, 59, 67, 72
   forced, 147
variation calculus, 241
vector field, 14, 41, 139
   divergence, 45, 73
   reconstruction, 189
viscosity
   dynamic, 8, 257
   kinematic, 8, 315, 322, 326
   turbulent (eddy), 326, 329, 355

wake, 20, 259, 287
wave dispersion, 96, 243
weather forecast, 356, 360

zigzag instability, 246